INTRODUCTION TO PID CONTROLLERS – THEORY, TUNING AND APPLICATION TO FRONTIER AREAS

Edited by **Rames C. Panda**

Introduction to PID Controllers – Theory, Tuning and Application to Frontier Areas
Edited by Rames C. Panda

Published by InTech
Janeza Trdine 9, 51000 Rijeka, Croatia

Publishing Process Manager Irena Voric

Technical Editor Teodora Smiljanic

Cover Designer InTech Design Team

Additional hard copies can be obtained from orders@intechopen.com

Introduction to PID Controllers – Theory, Tuning and Application to Frontier Areas,
Edited by Rames C. Panda
p. cm.
ISBN 978-953-307-927-1

For EU product safety concerns:
IN TECH d.o.o., Prolaz Marije Krucifikse Kozulić 3, 51000 Rijeka, Croatia,
info@intechopen.com or visit our website at intechopen.com.

Meet the editor

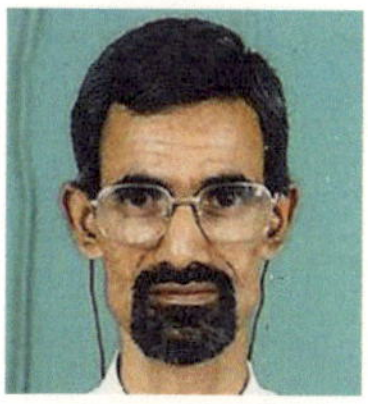

Dr Rames C Panda obtained his BTech degree from Jadavpur University, Kolkata, in 1987. He received his Masters before completing his PhD in Instrumentation and Process Control from the Indian Institute of Technology, Madras, in 1994. He joined CLRI as a scientist and was involved in making the manual unit operations of wet processing of leather to automatic. He provided solutions to many engineering problems related to instrumentation through consultancy. He has conducted much sponsored research, and has guided undergraduates and postgraduates towards their masters and doctoral degrees. He has also worked in many countries (Germany, Taiwan, and Australia) in various capacities and positions. He has written a number of research publications and book chapters on modeling and control topics, and he has received many awards.

Contents

Preface

With the development of process industries to meet market demand, new efficient technologies have come up innovative ways to enhance production capacity by giving special effort to process intensification. Maximizing production by allowing optimal inputs to flow to the process has saved energy, raw materials, and utility in process industries. In order to reduce global warming, various measures and preventive actions are initiated around the world.

This book provides the most recent information about the theory, history, state of the art, and best practices in implementing precise controllers in modern industrial processes, financial sectors, and bio processes. Presently monitoring and controlling are not only limited to industrial systems, but also have been extended to bio-medical, financial market, and areas related to forecasting and control. The main purpose of writing this book is to create an understanding of theory behind PID controller, how to tune, and where to apply the controller. The concept of applying PID controller to various fields and to design or tune the controller for specific purposes by fulfilling closed-loop performance specifications is the objective of this book.

The book covers the general topic of PID control configuration and tuning as well as new requirements that originate from allied application areas. This book contains eight preliminary sections each of which has one or more chapters. Thus, a total of 11 chapters has been accommodated in this book. It is proposed that section one will have chapter one and two. Chapter one will introduce PID controller design methods under specified performance, especially, time-domain and frequency domain specifications, optimization of PID parameters using error criteria. Chapter two presents the general discussion on the family of PID controllers with a path to proceed for optimal control. The architecture (series, parallel, and cascade) and structure of PID controllers for first and second order processes are explained here. Chapter three describes the PID-like controller tuning for second order unstable systems with different model structures, relying on assignment of dominant poles, as well as time domain specifications, providing sufficient stability margins. The controller parameters are obtained in terms of process parameters, or by using iterative techniques. Chapter four introduces concepts of tuning of PID controllers based on magnitude optimum for integrating unstable and Smith predictor systems in the continuous domain. The technique is combined with the concept of movement for cases on set point tracking and disturbance rejection. Model based tuning criteria of PID controllers for conventional controllers (P, PI, PID, PD etc) for multi input multi output systems are described in chapter five. Features and methods of auto tuning of PID controller, and the method of calculating performance of individual loops are also stated here. Tuning methods, using subspace identification techniques, different types of multiloop controllers with their design methods, and tuning of those controllers, are discussed in this chapter. Chapter six describes robust decentralized controller design for MIMO systems. Performances of individual loops and for the overall system are discussed here. Application of Nyquist type design for robust stability and nominal performance is discussed here. Chapter seven accounts for various intelligent controllers, namely using fuzzy logic based on the Mamdani structure. A method to make a pseudo-equivalence between the linear PID controllers and the fuzzy PID controllers is given here. Chapter eight presents discrete PID controller tuning using piecewise linearization methods using neural networks. PID controller is used using pole assignment. Design method of fractional order PID controllers for fractional order process is addressed in chapter nine. The difficulties in designing fractional order PID due to presence of fractional derivatives is explained here. Application of PID controllers in financial sectors is described in chapter 10. In this chapter, a portfolio management model with the aim to obtain good returns and decrease portfolio risk through stabilization of returns, by means of the PID control applied to pure returns, has been illustrated. The possible extensions in practical application of PID controllers in other fields, for example, polymerization and production of bromelin are described in Chapter 11. Each of the above chapters contains applications of the respective theory presented in it.

I sincerely thank the publisher and book-process-manager for their cooperation in bringing this book to a presentable manner.

Rames C. Panda
Scientist, Dept of Chemical Engineering, CLRI, Adyar, Chennai
India

Part 1

The Theory of PID Controllers and Their Design Methods

1

PID Controller Design for Specified Performance

Štefan Bucz and Alena Kozáková
Institute of Control and Industrial Informatics,
Faculty of Electrical Engineering and Information Technology,
Slovak University of Technology, Bratislava
Slovak Republic

1. Introduction

„How can proper controller adjustments be quickly determined on any control application?" The question posed by authors of the first published PID tuning method J.G.Ziegler and N.B.Nichols in 1942 is still topical and challenging for control engineering community. The reason is clear: just every fifth controller implemented is tuned properly but in fact:

- 30% of improper performance is due to inadequate selection of controller design method,
- 30% of improper performance is due to neglected nonlinearities in the control loop,
- 20% of improper closed-loop dynamics is due to poorly selected sampling period.

Although there are 408 various sources of PID controller tuning methods (O´Dwyer, 2006), 30% of controllers permanently operate in manual mode and 25% use factory-tuning without any up-date with respect to the given plant (Yu, 2006). Hence, there is natural need for effective PID controller design algorithms enabling not only to modify the controlled variable but also achieve specified performance (Kozáková et al., 2010), (Osuský et al., 2010). The chapter provides a survey of 51 existing practice-oriented methods of PID controller design for specified performance. Various options for design strategy and controller structure selection are presented along with PID controller design objectives and performance measures. Industrial controllers from ABB, Allen&Bradley, Yokogawa, Fischer-Rosemont commonly implement built-in model-free design techniques applicable for various types of plants; these methods are based on minimum information about the plant obtained by the well-known relay experiment. Model-based PID controller tuning techniques acquire plant parameters from a step-test; useful tuning formulae are provided for commonly used system models (FOPDT - first-order plus dead time, IPDT - integrator plus dead time, FOLIPDT - first-order lag and integrator plus dead time and SOPDT - second-order plus dead time). Optimization-based PID tuning approaches, tuning methods for unstable plants, and design techniques based on a tuning parameter to continuously modify closed-loop performance are investigated. Finally, a novel advanced design technique based on closed-loop step response shaping is presented and discussed on illustrative examples.

2. PID controller design for performance

Time response of the controlled variable y(t) is modifiable by tuning proportional gain K, and integrating and derivative time constants T_i and T_d, respectively; the objective is to achieve a zero steady-state control error e(t) irrespective if caused by changes in the reference w(t) or the disturbance d(t). This section presents practice-oriented PID controller design methods based on various perfomance criteria. Consider the control-loop in Fig. 1 with control action u(t) generated by a PID controller (switch SW in position "1").

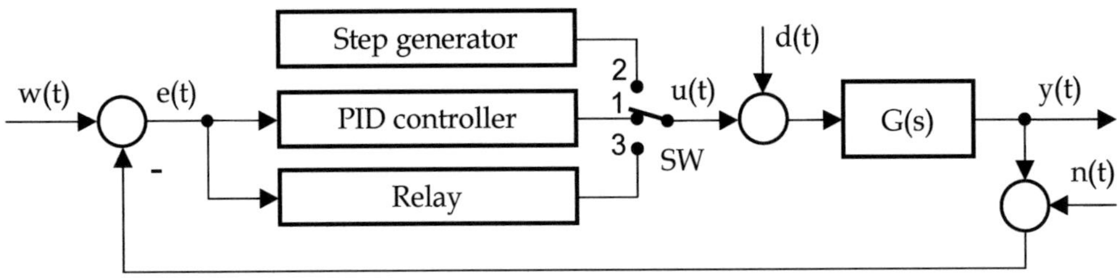

Fig. 1. Feedback control-loop with load disturbance d(t) and measurement noise n(t)

A controller design is a two-step procedure consisting of controller structure selection (P, PI, PD or PID) followed by tuning coefficients of the selected controller type.

2.1 Selection of PID controller structure

Appropriate structure of the controller $G_R(s)$ is usually selected with respect to zero steady-state error condition ($e(\infty)=0$), type, and parameters of the controlled plant.

2.1.1 Controller structure selection based on zero steady-state error condition

Consider the feedback control loop in Fig. 1 where G(s) is the plant transfer function. According to the Final Value Theorem, the steady-state error

$$e(\infty) = \lim_{s\to 0} sE(s) = \lim_{s\to 0} s\frac{1}{1+L(s)}W(s) = q!w_q \lim_{s\to 0}\frac{s^{\nu-q}}{s^{\nu}+K_L} \tag{1}$$

is zero if in the open-loop $L(s)=G(s)G_R(s)$, the integrator degree $\nu_L=\nu_S+\nu_R$ is greater than the degree q of the reference signal $w(t)=w_q t^q$, i.e.

$$\nu_L > q \tag{2}$$

where ν_S and ν_R are integrator degrees of the plant and controller, respectively, K_L is open-loop gain and w_q is a positive constant (Harsányi et al., 1998).

2.1.2 Principles of controller structure selection based on the plant type

Industrial process variables (e.g. position, speed, current, temperature, pressure, humidity, level etc.) are commonly controlled using PI controllers. In practice, the derivative part is usually switched off due to measurement noise. For pressure and level control in gas tanks, using P controller is sufficient (Bakošová & Fikar, 2008). However, adding derivative part improves closed-loop stability and steepens the step response rise (Balátě, 2004).

2.1.3 PID controller structure selection based on plant parametres

Consider the FOPDT (j=1) and FOLIPDT (j=3) plant models given as $G_{FOPDT}=K_1e^{-D_1s}/[T_1s+1]$ and $G_{FOLIPDT}=K_3e^{-D_3s}/\{s[T_3s+1]\}$ with following parameters

$$\tau_1=\frac{D_1}{T_1};\ \rho_1=K_1K_c;\ \tau_3=\frac{D_3}{T_3};\ \rho_3=\frac{\lim_{s\to 0} sG(s)}{\omega_c\left|G(j\omega_c)\right|}=\frac{T_cK_3K_c}{2\pi};\ \tau_3=\frac{\frac{2}{\pi}+arctg\sqrt{\rho_3^2-1}}{\sqrt{\rho_3^2-1}} \tag{3}$$

where K_c and ω_c are critical gain and frequency of the plant, respectively. Normed time delay τ_j and parameter ρ_j can be used to select appropriate PID control strategy. According to Tab. 1 (Xue et al., 2007), the derivative part is not used in presence of intense noise and a PID controller is not appropriate for plants with large time delays.

Ranges for τ and ρ	No precise control necessary	Precise control needed		
		High noise	Low saturation	Low measurement noise
$\tau_1>1;\ \rho_1<1{,}5$	I	I+B+C	PI+B+C	PI+B+C
$0{,}6<\tau_1<1;\ 1{,}5<\rho_1<2{,}25$	I or PI	I+A	PI+A	(PI or PID)+A+C
$0{,}15<\tau_1<0{,}6;\ 2{,}25<\rho_1<15$	PI	PI	PI or PID	PID
$\tau_1<0{,}15;\ \rho_1>15$ or $\tau_3>0{,}3;\ \rho_3<2$	P or PI	PI	PI or PID	PI or PID
$\tau_3<0{,}3;\ \rho_3>2$	PD+E	F	PD+E	PD+E

Table 1. Controller structure selection with respect to plant model parameters:
A: forward compensation suggested, B: forward compensation necessary, C: dead-time compensation suggested, D: dead-time compensation necessary, E: set-point weighing necessary, F: pole-placement

2.2 PID controller design objectives

Consider the following most frequently used PID controller types: ideal PID (4a), real interaction PID with derivative filtering (4b) and ideal PID in series with a first order filter (4c)

$$G_R(s)=K\left(1+\frac{1}{T_is}+T_ds\right);\ G_R(s)=K\left(1+\frac{1}{T_is}+\frac{T_ds}{1+\frac{T_d}{N}s}\right);\ G_R(s)=K\left(1+\frac{1}{T_is}+T_ds\right)\left(\frac{1}{T_fs+1}\right) \tag{4}$$

In practical cases $N\in\langle 8;16\rangle$ (Visoli, 2006). The PID controller design objectives are:

1. tracking of setpoint or reference variable w(t) by y(t),
2. rejection of disturbance d(t) and noise n(t) influence on the controlled variable y(t).

The first objective called also „servo-tuning" is frequent in motion systems (e.g. tracking required speed); techniques to guarantee the second objective are called „regulator-tuning".

2.3 Performance measures in the time domain

Performance measures indicating satisfactory quality of setpoint tracking (Fig. 2a) and disturbance rejection (Fig. 2b) are small maximum overshoot and small decay ratio, respectively, given as

$$\eta_{\max} = 100\frac{|y_{\max} - y(\infty)|}{y(\infty)} \ [\%]; \ \delta_{DR} = \frac{A_{i+1}}{A_i} \tag{5}$$

where y(∞) denotes steady state of y(t). The ratio of two successive amplitudes A_{i+1}/A_i is measure of y(t) decaying, where i=1...N, and N is half of the number of y(∞) crossings by y(t) (Fig. 2b). A time-domain performance measure is the settling time t_s, i.e. the time after which the output y(t) remains within $\pm\varepsilon$% of its final value (Fig. 2a); typically ε=[1%÷5%]y(∞), $\delta_{DR}\in$(1:4;1:2), $\eta_{max}\in$(0%;50%). Fig. 2c depicts underdamped (curve 1), overdamped (curve 2) and critically damped (curve 3) closed-loop step responses.

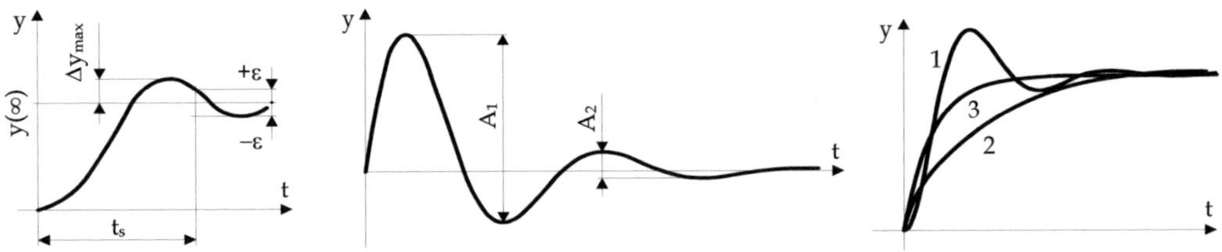

Fig. 2. Performance measures: δ_{DR}, t_s, η_{max} and e(∞); a) setpoint step response; b) load disturbance step response; c) over-, critically- and underdamped closed-loop step-responses

2.4 Model-free PID controller design techniques with guaranteed performance

Model-free tuning PID controller techniques are used if plant dynamics is not complicated (without oscillations, vibrations, large overshoots) or if plant modelling is time demanding, uneconomical or even unfeasible. To find PID controller coefficients, instead of a full model usually 2-4 characteristic plant parameters are used obtained from the relay test.

2.4.1 Tuning rules based on critical parameters of the plant

Consider the closed-loop in Fig. 1 with proportional controller. If the controller gain K is successively increased until the process variable oscillates with constant amplitudes, critical parameters can be specified: the period of oscillations T_c and the corresponding gain K_c. If the controller (4a) is considered, coefficients of P, PI and PID controllers are calculated according to Tab. 2, where $\omega_c=2\pi/T_c$ is critical frequency of the plant.

No.	Design method, year	Controller	K	T_i	T_d	Performance or response
1.	(Ziegler & Nichols, 1942)	P	0,5K_c	-	-	Quarter decay ratio
2.	(Ziegler & Nichols, 1942)	PI	0,45K_c	0,8T_c	-	Quarter decay ratio
3.	(Ziegler & Nichols, 1942)	PID	0,6K_c	0,5T_c	0,125T_c	Quarter decay ratio
4.	(Pettit & Carr, 1987)	PID	K_c	0,5T_c	0,125T_c	Underdamped
5.	(Pettit & Carr, 1987)	PID	0,67K_c	T_c	0,167T_c	Critically damped
6.	(Pettit & Carr, 1987)	PID	0,5K_c	1,5T_c	0,167T_c	Overdamped
7.	(Chau, 2002)	PID	0,33K_c	0,5T_c	0,333T_c	Small overshoot
8.	(Chau, 2002)	PID	0,2K_c	0,55T_c	0,333T_c	Without overshoot
9.	(Bucz, 2011)	PID	0,54K_c	0,79T_c	0,199T_c	Overshoot $\eta_{max}\leq$20%
10.	(Bucz, 2011)	PID	0,28K_c	1,44T_c	0,359T_c	Settling time $t_s\leq 13/\omega_c$

Table 2. Controller tuning based on critical parametres of the plant

Rules No. 1 - 3 represent the famous Ziegler-Nichols frequency-domain method with fast rejection of the disturbance d(t) for δ_{DR}=1:4 (Ziegler & Nichols, 1942). Related methods (No.

4 - 10) use various weighing of critical parameters thus allowing to vary closed-loop performance requirements. Methods (No. 1 - 10) are applicable for various plant types, easy-to-use and time efficient.

2.4.2 Specification of critical parameters of the plant using relay experiment

To quickly determine critical parameters K_c and T_c, industrial autotuners apply a relay test (Rotach, 1984) either with ideal relay (IR) or a relay with hysteresis (HR). In the loop in Fig. 1 when adjusting the setpoint w(t) in manual mode and switching SW into „3", a stable limit cycle around $y(\infty)$ arises. Due to switching between the levels -M, +M, G(s) is excited by a periodic rectangular signal u(t), (Fig. 3a). Then, ω_c and K_c can be calculated from

$$\omega_c = \frac{2\pi}{T_c}; \quad K_{c\_IR} = \frac{4M}{\pi A_c}; \quad K_{c\_HR} = \frac{4(M - 0{,}5\Delta_{DB})}{\pi A_c} \tag{6}$$

where the period and amplitude of oscillations T_c and A_c, respectively, can be obtained from a record of y(t) (Fig. 3b); Δ_{DB} is the width of the hysteresis. Relay amplitude M is usually adjusted at 3%÷10% of the control action limit. A relay with hysteresis is used if y(t) is corrupted by measurement noise n(t) (Yu, 2006); the critical gain is calculated using (6c).

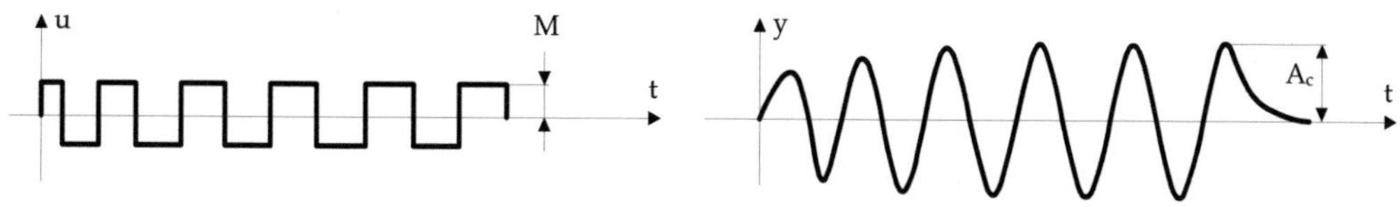

Fig. 3. A detailed view of u(t) and y(t) to determine critical parameters K_c and T_c

2.5 Model-based PID controller design with guaranteed performance

Steday-state and dynamic properties of real processes are described by simple FOPDT, IPDT, FOLIPDT or SOPDT models. Model parameters further used to calculate PID controller coefficients can be found e.g. from the plant step responses (Fig. 4 and 5).

2.5.1 Specification of FOPDT, IPDT and FOLIPDT plant model parameters

According to Fig. 1, the plant step response is obtained by switching SW into „2" and performing a step change in u(t). Plant model parameters are obtained by evaluating the particular step response (Fig. 4).

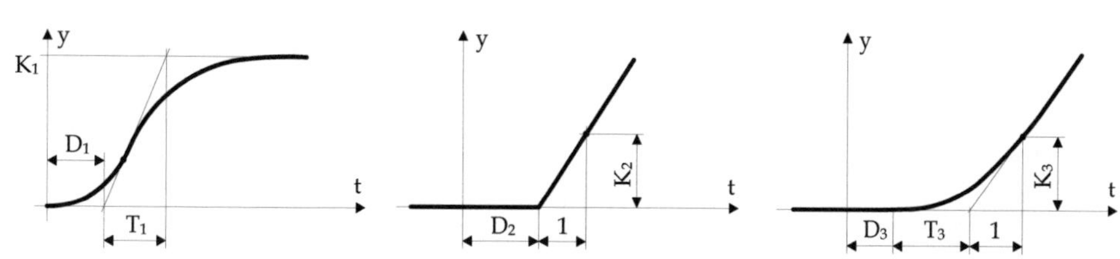

Fig. 4. Typical step responses of a) FOPDT; b) IPDT and c) FOLIPDT models

From the read-off parameters, transfer functions of individual models have been obtained

$$G_{FOPDT}(s) = \frac{K_1 e^{-D_1 s}}{T_1 s + 1}; \; G_{IPDT}(s) = \frac{K_2 e^{-D_2 s}}{s}; \; G_{FOLIPDT}(s) = \frac{K_3 e^{-D_3 s}}{s(T_3 s + 1)} \tag{7}$$

2.5.2 Tuning formulae for FOPDT models

FOPDT models (7a) are used for chemical processes, thermal systems, manufacturing processes etc. Corresponding P, PI and PID coefficients are calculated using formulae in Tab. 3.

No.	Design method, year, control purpose	Cont-roller	K	T_i	T_d	Performance
11.	(Ziegler & Nichols, 1942)	P	$1/\kappa_1$	-	-	Quarter decay ratio (δ_{DR}=1:4)
12.		PI	$0{,}9/\kappa_1$	$3D_1$	-	
13.		PID	$1{,}2/\kappa_1$	$2D_1$	$0{,}5D_1$	
14.		PI	$0{,}6/\kappa_1$	$4D_1$	-	η_{max}=0%,
15.	(Chien et al., 1952),	PID	$0{,}95/\kappa_1$	$2{,}38D_1$	$0{,}42D_1$	$D_1/T_1 \in (0{,}1;1)$
16.	Regulator tuning	PI	$0{,}7/\kappa_1$	$2{,}33D_1$	-	η_{max}=20%,
17.		PID	$1{,}2/\kappa_1$	$2D_1$	$0{,}42D_1$	$D_1/T_1 \in (0{,}1;1)$
18.		PI	$0{,}35/\kappa_1$	$1{,}17D_1$	-	η_{max}=0%,
19.	(Chien et al., 1952),	PID	$0{,}6/\kappa_1$	D_1	$0{,}5D_1$	$D_1/T_1 \in (0{,}1;1)$
20.	Servo tuning	PI	$0{,}6/\kappa_1$	D_1	-	η_{max}=20%,
21.		PID	$0{,}95/\kappa_1$	$1{,}36D_1$	$0{,}47D_1$	$D_1/T_1 \in (0{,}1;1)$
22.	(ControlSoft Inc.,	PID	$2/K_1$	T_1+D_1	$\max(D_1/3;T_1/6)$	Slow loop
23.	2005)	PID	$2/K_1$	T_1+D_1	$\min(D_1/3;T_1/6)$	Fast loop

Table 3. PID tuning rules based on FOPDT model; $\kappa_1=K_1D_1/T_1$ is the normed process gain

Formulae No. 11 – 13 represent the time-domain (or reaction curve) Ziegler-Nichols method (Ziegler & Nichols, 1942) and usually give higher open-loop gains than the frequency-domain version. Algorithms by Chien-Hrones-Reswick provide different settings for setpoint regulation and disturbance rejection for two representative maximum overshoot values.

2.5.3 Tuning formulae for IPDT and FOLIPDT models

While dynamics of slow industrial processes (polymer production, heat exchangers) can be described by IPDT model (7b), electromechanic subsystems of turning machines and servodrives are typical examples for using FOLIPDT model (7c).

No.	Design method, year, model	Cont-roller	K	T_i	T_d	Perfor-mance
24.	(Haalman, 1965), IPDT	P	$0{,}66/(K_2D_2)$	-	-	M_s=1,9
25.	(Ziegler & Nichols, 1942), IPDT	PI	$0{,}9/(K_2D_2)$	$3{,}33D_2$	-	δ_{DR}=1:4
26.	(Ford, 1953), IPDT	PID	$1{,}48/(K_2D_2)$	$2D_2$	$0{,}37D_2$	δ_{DR}=1:2,7
27.	(Coon, 1956), FOLIPDT	P	$x_3/[K_3(D_3+T_3)]$	-	-	δ_{DR}=1:4
28.	(Haalman, 1965), FOLIPDT	PD	$0{,}66/(K_3D_3)$	-	T_3	M_s=1,9

Table 4. Tuning rules based on IPDT and FOLIPDT model parameters

According to Haalman (rules No. 24 and 28), controller transfer function $G_R(s)=L(s)/G(s)$, where $L(s)=0{,}66e^{-Ds}/(Ds)$ is the ideal loop transfer function guaranteeing maximum closed-loop sensitivity $M_s=1{,}9$ to disturbance d(t), (see subsection 2.8.1). For various G(s), various controller structures are obtained. The gain K in rule No. 27 depends on the normed time delay $\upsilon_3=D_3/T_3$ of the FOLIPDT model; for corresponding couples hold: $(\upsilon_3;x_3)$={(0,02;5), (0,053;4); (0,11;3); (0,25;2,2); (0,43;1,7); (1;1,3); (4;1,1)}. Due to integrator contained in IPDT and FOLIPDT models, I-term in the controller structure is needed just to achieve zero steady-state error $e(\infty)$ under steady-state disturbance $d(\infty)$.

2.5.4 Tuning formulae for SOPDT plant models

Flexible systems in wood processing industry, automotive industry, robotis, shocks and vibrations damping are often modelled by SOSPTD models with transfer functions

$$G_{SOPDT}(s)=\frac{K_4e^{-D_4s}}{(T_4s+1)(T_5s+1)};\quad G_{SOPDT}(s)=\frac{K_6e^{-D_6s}}{T_6^2s^2+2\xi_6T_6s+1} \tag{8}$$

For SOPDT model (8b), the relative damping $\xi_6\in(0;1)$ indicates oscillatory step response.

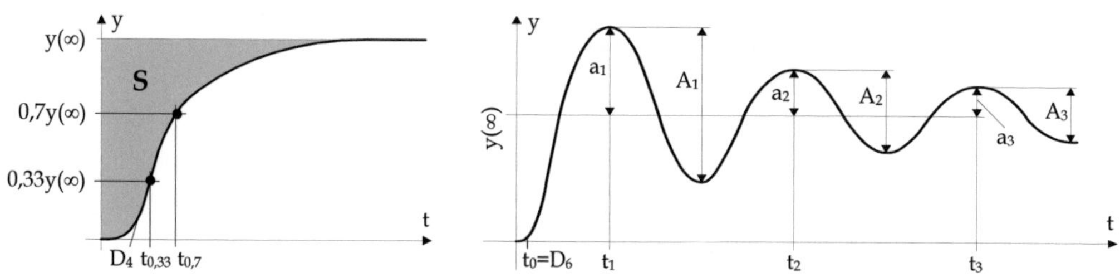

Fig. 5. Step response of SOPDT model: a) non-oscillatory, b) oscillatory

If $\xi_6>1$, SOPDT model (8a) is used; its parameters are found from the non-oscillatory step response in Fig. 5a using the following relations

$$T_{4,5}=\frac{1}{2}\left(C_2\pm\sqrt{C_2^2-4C_1^2}\right);\; D_4=\frac{t_{0,33}}{0{,}516}-\frac{t_{0,7}}{1{,}067};\; C_1=\frac{(t_{0,33}-t_{0,7})}{1{,}259};\; C_2=\frac{S}{y(\infty)}-D_4 \tag{9}$$

where $S=K_4(T_4+T_5+D_4)$ denotes the area above the step response of y(t), and $y(\infty)$ is its steady-state value. Parameters of the SOPDT model (8b) can be found from evaluation of 2-4 periods of step response oscillations (Fig. 5b) using following rules (Vítečková, 1998)

$$\xi_6=\frac{-\ln\frac{a_{i+1}}{a_i}}{\sqrt{\pi^2+\ln^2\frac{a_{i+1}}{a_i}}};\; T_6=\frac{\sqrt{1-\xi_6^2}}{\pi N}(t_{N+1}-t_1);\; D_6=\frac{1}{N}\left[\sum_{i=1}^{N}t_i-\frac{N+1}{2}(t_{N+1}-t_1)\right] \tag{10}$$

Quality of identification improves with increasing number of read-off amplitudes N. If N>2 several values ξ_6, T_6 and D_6 are obtained and their average is taken for further calculations. Tab. 5 summarizes useful tuning formulae for both oscillatory and non-oscillatory systems with SOPDT model properties.

No.	Method, year	Cont-roller	K	T_i	T_d	Performance for
29.	(Suyama, 1992)	PID	$\frac{T_4+T_5}{2K_4D_4}$	T_4+T_5	$\frac{T_4T_5}{T_4+T_5}$	Closed-loop step response overshoot $\eta_{max}=10\%$
30.	Vítečková, (1999),	PID	$x_4\frac{T_4+T_5}{K_4D_4}$	T_4+T_5	$\frac{T_4T_5}{T_4+T_5}$	Overdamped plants; $T_5>T_4$ $\eta_{max}=0\%$: $x_4=0{,}368$ $\eta_{max}=30\%$: $x_4=0{,}801$
31.	Vítečková et al., (2000)	PID	$\frac{x_6\xi_6T_6}{K_6D_6}$	$2\xi_6T_6$	$\frac{T_6}{2\xi_6}$	Underdamped plants ($0{,}5<\xi_6\leq1$) $\eta_{max}=0\%$: $x_6=0{,}736$ $\eta_{max}=30\%$: $x_6=1{,}602$
32.	(Wang & Shao, 1999)	PID	$\frac{x_6\xi_6T_6}{K_6D_6}$	$2\xi_6T_6$	$\frac{T_6}{2\xi_6}$	[$G_M=2$, $\phi_M=45°$]: $x_6=1{,}571$ [$G_M=5$, $\phi_M=72°$]: $x_6=0{,}628$
33.	(Chen et al., 1999)	PID	$\frac{x_6\xi_6T_6}{K_6D_6}$	$2\xi_6T_6$	$\frac{D_6}{2\xi_6}$	[$G_M;\phi_M;M_s$]=[3,14;61,4°;1]: $x_6=1{,}0$ [$G_M;\phi_M;M_s$]=[1,96;44,1°;1,5]: $x_6=1{,}6$

Table 5. Tuning rules based on SOPDT model parameters

2.6 PID controller design based on optimization techniques

Optimal PID controller tuning can be found by minimizing the performance index

$$I(K,T_i,T_d)=\int_0^{\infty}\left[t^n e(K,T_i,T_d,t)\right]^2 dt \tag{11}$$

Its particular cases are known as integral square error (ISE) for n=0; integral squared time weighed error (ISTE) for n=1, and integral squared time-squared weighed error (IST^{2}E) for n=2. Some tuning formulae for PID controller in form (4a) are shown in Tab. 6. Settling time t_s in rules No. 40 and 41 is affected by D_2.

No.	Method, year, model	K	T_i	T_d	Performance
34.	(Zhuang & Atherton,	$1{,}473\nu_1^{0{,}970}/K_1$	$0{,}897T_1\nu_1^{0{,}753}$	$0{,}550T_1\nu_1^{0{,}948}$	Minimum ISE
35.	1993), FOPDT model,	$1{,}468\nu_1^{0{,}970}/K_1$	$1{,}062T_1\nu_1^{0{,}725}$	$0{,}443T_1\nu_1^{0{,}939}$	Minimum ISTE
36.	$\nu_1\in\langle0{,}1;1\rangle$	$1{,}531\nu_1^{0{,}960}/K_1$	$1{,}030T_1\nu_1^{0{,}746}$	$0{,}413T_1\nu_1^{0{,}933}$	Minimum IST^{2}E
37.	(Zhuang & Atherton,	$1{,}524\nu_1^{0{,}735}/K_1$	$0{,}885T_1\nu_1^{0{,}641}$	$0{,}552T_1\nu_1^{0{,}851}$	Minimum ISE
38.	1993), FOPDT model,	$1{,}515\nu_1^{0{,}730}/K_1$	$1{,}045T_1\nu_1^{0{,}598}$	$0{,}444T_1\nu_1^{0{,}847}$	Minimum ISTE
39.	$\nu_1\in\langle1{,}1;2\rangle$	$1{,}592\nu_1^{0{,}705}/K_1$	$1{,}045T_1\nu_1^{0{,}597}$	$0{,}414T_1\nu_1^{0{,}850}$	Minimum IST^{2}E
40.	(Wang a Cluett, 1997),	$0{,}9588/[K_2D_2]$	$3{,}0425D_2$	$0{,}3912D_2$	$t_s=D_2$
41.	IPDT model	$0{,}3144/[K_2D_2]$	$11{,}1637D_2$	$0{,}1453D_2$	$t_s=5D_2$

Table 6. Tuning rules based on minimizing performance indices

2.7 PID controller setting for unstable FOPDT models

Minimization of performance indices can be applied also for unstable FOPDT models

$$G_{FOPDT\_US}(s)=\frac{K_1e^{-D_1s}}{T_1s-1} \tag{12}$$

leading to simple tuning rules for PID controller (4a) (No. 42 – 44 in Tab. 7). Tuning rules No. 45 and 46 for PID controller (4c) show that settling time t_s increases with growing normed time delay $\nu_1=D_1/T_1$ of the FOPDT model (12).

No.	Method, year	K	T_i	T_d	T_f	Performance
42.	(Visoli, 2001),	$1{,}37\nu_1/K_1$	$2{,}42T_1\nu_1^{1{,}18}$	$0{,}60T_1$	-	Minimum ISE
43.	Regulator	$1{,}37\nu_1/K_1$	$4{,}12T_1\nu_1^{0{,}90}$	$0{,}55T_1$	-	Minimum ISTE
44.	tuning	$1{,}70\nu_1/K_1$	$4{,}52T_1\nu_1^{1{,}13}$	$0{,}50T_1$	-	Minimum IST^2E
45.	(Chandrashekar	$10{,}3662/K_1$	$0{,}3874T_1$	$0{,}0435T_1$	$0{,}0134T_1$	$t_s=0{,}1T_1$: $\nu_1=0{,}1$
46.	et al., 2002)	$2{,}0217/K_1$	$4{,}65T_1$	$0{,}2366T_1$	$0{,}0696T_1$	$t_s=0{,}8T_1$: $\nu_1=0{,}5$

Table 7. Tuning rules for unstable FOPDT model

Using tuning methods shown in Tab. 2 – 7, achieved performance is a priori given by the chosen metod (e.g. a quarter decay ratio if using Ziegler-Nichols methods No. 11 – 13 in Tab. 3), or guaranteed performance however not specified by the designer (e.g. in Chen method No. 33 in Tab. 5, a gain margin $G_M=1{,}96$, a phase margin $\phi_M=44{,}1°$, and a maximum peak of the sensitivity to disturbance d(t) $M_s=1{,}5$).

2.8 PID controller design for specified performance

These methods provide tuning rules are based on a single tuning parameter that enables to systematically affect closed-loop performance by step response shaping.

2.8.1 Performance measures used as a PID tuning parameter

Most frequent parameters for tuning PID controllers are following performance measures (Åström & Hägglund, 1995):

- ϕ_M and G_M: phase and gain margins, respectively,
- M_s and M_t: maximum peaks of sensitivity $S(j\omega)$ and complementary sensitivity $T(j\omega)$ magnitudes, respectively,
- λ: required closed-loop time constant.

If a controller $G_R(j\omega)$ guarantees that $|S(j\omega)|$ or $|T(j\omega)|$ do not overrun prespecified values M_s or M_t, respectively, defined by

$$M_s = \sup_{\omega}|S(j\omega)| = \sup_{\omega}\left|\frac{1}{1+L(j\omega)}\right|; \quad M_t = \sup_{\omega}|T(j\omega)| = \sup_{\omega}\left|\frac{L(j\omega)}{1+L(j\omega)}\right| \tag{13}$$

over $\omega\in\langle 0,\infty)$, then the Nyquist plot $L(j\omega)$ of the open-loop $L(s)=G(s)G_R(s)$ avoids the respective circle M_S or M_T, each given by the their center and radius as follows

$$C_S = [-1, j0],\ R_S = \frac{1}{M_s}; \quad C_T = \left[-\frac{M_t^2}{M_t^2-1}, j0\right],\ R_T = -\frac{M_t}{\left|1-M_t^2\right|} \tag{14}$$

If $L(j\omega)$ avoids entering the circles corresponding to M_S or M_T, a safe distance from the point C_S is kept (Fig. 6a). Typical $|S(j\omega)|$ and $|T(j\omega)|$ plots for properly designed controller are plotted in Fig. 6b. The disturbance d(t) is sufficiently rejected if $M_s\in(1{,}2;2)$. The reference w(t) is properly tracked by the process output y(t) if $M_t\in(1{,}3;2{,}5)$. With further increasing of M_t the closed-loop tends to be oscillatory.

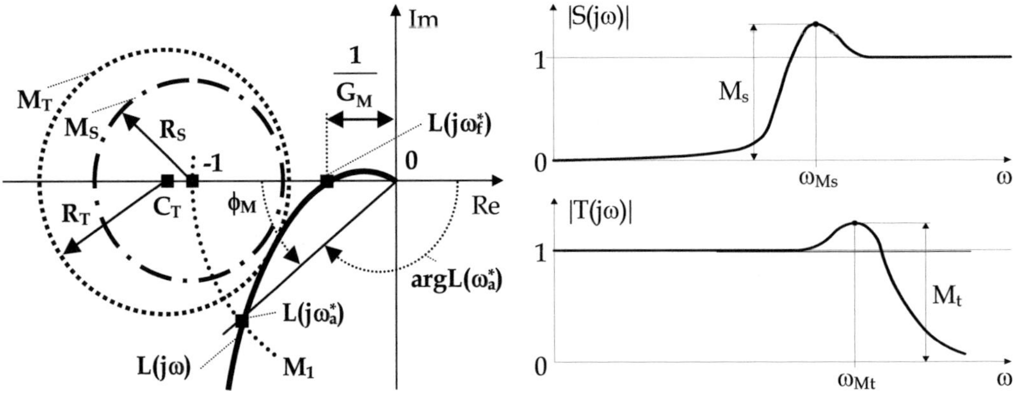

Fig. 6. a) Definition and geometrical interpretation of ϕ_M and G_M in the complex plane; b) Sensitivity and complementary sensitivity magnitudes $|S(j\omega)|$, $|T(j\omega)|$ and performance measures M_s, M_t

From Fig. 6a results, that increasing open-loop phase margin ϕ_M causes moving the gain crossover $L(j\omega_a^*)$ lying on the unit circle M_1 away from the critical point (-1,j0). Increasing open-loop gain margin G_M causes moving the phase crossover $L(j\omega_f^*)$ away from (-1,j0). Therefore, parameters ϕ_M or G_M given by

$$\phi_M = 180° + \arg L(\omega_a^*)\,; \quad G_M = \frac{1}{\left|L(j\omega_f^*)\right|} \tag{15}$$

are frequently used performance measures, their typical values are $\phi_M \in (20°;90°)$, $G_M \in (2;5)$. Relations between them are given by following inequalities

$$\phi_M \geq 2\arcsin\left(\frac{1}{2M_s}\right); \quad \phi_M \geq 2\arcsin\left(\frac{1}{2M_t}\right); \quad G_M \geq \frac{M_s}{M_s-1}; \quad G_M \geq 1+\frac{1}{M_t} \tag{16}$$

The point at which the Nyquist plot $L(j\omega)$ touches the M_T circle defines the closed-loop resonance frequency ω_{Mt}.

2.8.2 Tuning formulae with performance specification

Table 8 shows open formulae for PID controller design. The coefficients tuning is carried out with respect to closed-loop performance specification. Rules No. 47 – 49 consider tuning of ideal PID controller (4a). To apply the Rotach method, knowledge of the plant magnitude $|G(j\omega)|$ is supposed as well as of the roll-off of $\arg G(\omega)$ at $\omega=\omega_{Mt}$, where the maximum peak M_t of the complementary sensitivity is required. Method No. 50 is based on so-called λ-tuning, with the resulting closed-loop expressed as a 1st order system with time constant λ; this rule considers a real PID controller (4b) with filtering constant in the derivative part $T_f=T_d/N=0{,}5\lambda D_1/(1+D_1)$ where λ is to be chosen to meet following conditions: $\lambda>0{,}25D_1$; $\lambda>0{,}25T_1$ (Morari & Zafiriou, 1989). The λ-tuning technique is used also in the rule No. 51 to design interaction PI controller.

No.	Design method, year, model	K	T_i	T_d
47.	(Hang & Åström, 1988), Non-model	$K_c \sin\phi_M$	$\frac{T_c(1-\cos\phi_M)}{\pi \sin\phi_M}$	$\frac{T_c(1-\cos\phi_M)}{4\pi \sin\phi_M}$
48.	(Rotach, 1994), Non-model	$\frac{M_t\lvert G(j\omega_{Mt})\rvert}{\sqrt{M_t^2-1}}$	$\frac{-2}{\omega_{Mt}^2\left(\frac{d\left[\arg G(\omega_{Mt})\right]}{d\omega_{Mt}}\right)}$	$-\frac{1}{2}\frac{d\left[\arg G(\omega_{Mt})\right]}{d\omega_{Mt}}$
49.	(Wojsznis et al., 1999), FOPDT	$\frac{K_c \cos\phi_M}{G_M}$	$\frac{T_c}{\pi}\left(tg\phi_M+\sqrt{1+tg^2\phi_M}\right)$	$\frac{T_c}{4\pi}\left(tg\phi_M+\sqrt{1+tg^2\phi_M}\right)$
50.	(Morari & Zafiriou, 1989), FOPDT	$\frac{T_1+0,5D_1}{K_1(\lambda+D_1)}$	$T_1+\frac{1}{2}D_1$	$\frac{T_1 D_1}{2T_1+D_1}$
51.	(Chen & Seborg, 2002), FOPDT	$\frac{T_1D_1+2T_1\lambda-\lambda^2}{K_1(\lambda+D_1)^2}$	$\frac{T_1D_1+2T_1\lambda-\lambda^2}{T_1+D_1}$	-

Table 8. PID design formulae for specified performance based on tuning parameters ϕ_M, G_M, M_t and λ

2.8.3 Performance evaluation

Phase margin ϕ_M is the most wide-spread performance measure in PID controller design. Maximum overshoot η_{max} and settling time t_s of the closed-loop step response are related with ϕ_M according to Reinisch relations

$$\eta_{\max}=\begin{cases}-0,91\phi_M+64,55 \;\; for\; \phi_M\in\langle 38°;71°\rangle \\ -1,53\phi_M+88,46 \;\; for\; \phi_M\in\langle 12°;38°\rangle\end{cases};\;\; \eta_{\max}=100e^{-2\pi b^2 M_t};\;\; t_s\in\left(\frac{\pi}{\omega_a^*},\frac{4\pi}{\omega_a^*}\right) \tag{17}$$

valid for 2nd order closed-loop with relative damping $\xi\in(0,25;0,65)$ where ω_a^* is the gain crossover frequency (Hudzovič, 1982). Relations

$$\eta_{\max}\le 100\frac{1.18M_t-\lvert T(0)\rvert}{\lvert T(0)\rvert}\;[\%];\;\; t_s\approx\frac{3}{\omega_a^*}\;\; for\; M_t\in(1,3;1,5) \tag{18}$$

(Hudzovič, 1982); (Grabbe et al., 1959-61) are general for any order of the closed-loop T(s); if the controller has the integral part then $|T(0)|=|T(\omega=0)|=1$.

The engineering practice is persistently demanding for PID controller design methods simultaneously guaranteeing several performance criteria, especially maximum overshoot η_{max} and settling time t_s. However, we ask the question: how to suitably transform the above-mentioned engineering requirements into frequency domain specifications applicable for PID controller coefficients tuning? The response can be found in Section 3 where a novel original PID controller design method is presented.

3. Advanced PID controller design method based on sine-wave identification

The presented method is applicable for linear stable SISO systems even with unknown mathematical model. The control objective is to provide required maximum overshoot η_{max} and settling time t_s of the process variable y(t). The method enables the designer to prescribe η_{max} and t_s within following ranges (Bucz et al., 2010b, 2010c), (Bucz, 2011)

- $\eta_{max}\in\langle 0\%; 90\%\rangle$ and $t_s\in\langle 6{,}5/\omega_c; 45/\omega_c\rangle$ for systems without integrator,
- $\eta_{max}\in\langle 9{,}5\%; 90\%\rangle$ and $t_s\in\langle 11{,}5/\omega_c; 45/\omega_c\rangle$ for systems with integrator,

where ω_c is the plant critical frequency. The PID controller design provides guaranteed phase margin ϕ_M. The tuning rule parameter is a suitably chosen point of the plant frequency response obtained by a sine-wave signal with excitation frequency ω_n. The designed controller then moves this point into the gain crossover with the required phase margin ϕ_M. With respect to engineering requirements, the pair $(\omega_n;\phi_M)$ is specified on the closed-loop step response in terms of η_{max} and t_s according to parabolic dependencies in Fig. 11 and Fig. 14-16. A multipurpose loop for the proposed sine-wave method is in Fig. 7.

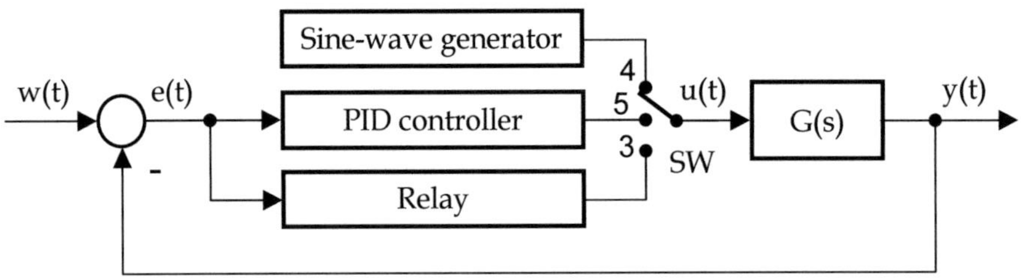

Fig. 7. Multipurpose loop for identification and control using the sine-wave method

3.1 Plant identification by a sinusoidal excitation input

By switching SW into "4", the loop in Fig. 7 opens; a stable plant with unknown model G(s) is excited by a persistent sinusoid $u(t)=U_n\sin(\omega_n t)$ (Fig. 8a) where U_n denotes the amplitude and ω_n excitation frequency. The plant output $y(t)=Y_n\sin(\omega_n t+\varphi)$ is also a persistent sinusoid with the same frequency ω_n, amplitude Y_n and phase shift φ with respect to the input excitation sinusoid (Fig. 8b). From the stored records of y(t) and u(t) it is possible to read-off the amplitude Y_n and phase shift φ_n and thus to identify a particular point of the plant frequency response $G(j\omega)$ under excitation frequency ω_n with coordinates $G\equiv G(j\omega_n)$

$$G(j\omega_n)=\left|G(j\omega_n)\right|e^{j\arg G(\omega_n)}=\frac{Y_n(\omega_n)}{U_n(\omega_n)}e^{j\arg G(\omega_n)} \tag{19}$$

where $\varphi=\arg G(\omega_n)$. The point $G(j\omega_n)$ can be plotted in the complex plane (Fig. 8c).

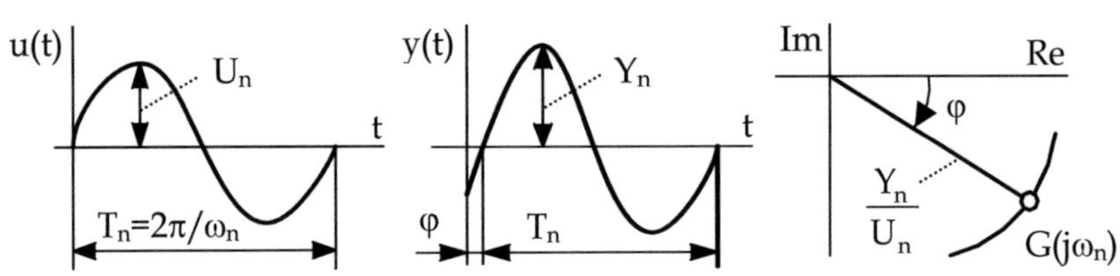

Fig. 8. Time responses of a) u(t); b) y(t), and c) location of $G(j\omega_n)$ in the complex plane

The output sinusoid amplitude Y_n can be affected by the amplitude U_n of the excitation sinusoid generated by the sine wave generator; it is recommended to use U_n=3÷7%u_{max}. Identified plant parameters are represented by the triple $\{\omega_n, Y_n(\omega_n)/U_n(\omega_n), \varphi(\omega_n)\}$. In the SW position „4", identification is performed in the open-loop. Hence, this method is applicable only for stable plants. The excitation frequency ω_n is to be adjusted prior to identification and taken from the empirical interval (29) (Bucz et al., 2010a, 2010b, 2011).

3.2 Sine-wave method tuning rules

In the control loop in Fig. 7, let us switch SW in „5"and put the PID controller into manual mode. The closed-loop characteristic equation $1+L(j\omega)=1+G(j\omega)G_R(j\omega)=0$ at the gain crossover frequency ω_a^* can be broken down into the amplitude and phase conditions as follows

$$\left|G(j\omega_a^*)\right|\left|G_R(j\omega_a^*)\right| = 1; \quad \arg G(\omega_a^*) + \arg G_R(\omega_a^*) = -180^\circ + \phi_M \tag{20}$$

where ϕ_M is the required phase margin, $L(j\omega_n)$ is the open-loop transfer function. Denote $\Theta=\arg G_R(\omega_a^*)$. We are searching for K, T_i and T_d of the ideal PID controller (4a). Comparing frequency transfer functions of the PID controller in parallel and polar forms

$$G_R(j\omega) = K + jK\left[T_d\omega - \frac{1}{T_i\omega}\right]; \quad G_R(j\omega) = \left|G_R(j\omega)\right|e^{j\Theta} = \left|G_R(j\omega)\right|\cos\Theta + j\left|G_R(j\omega)\right|\sin\Theta \tag{21}$$

coefficients of PID controller can be obtained from the complex equation

$$K + jK\left[T_d\omega_a^* - \frac{1}{T_i\omega_a^*}\right] = \frac{\cos\Theta}{\left|G(j\omega_a^*)\right|} + j\frac{\sin\Theta}{\left|G(j\omega_a^*)\right|}, \tag{22}$$

at $\omega=\omega_a^*$ using the substitution $|G_R(j\omega_a^*)|=1/|G(j\omega_a^*)|$ resulting from the amplitude condition (20a). The complex equation (22) is solved as a set of two real equations

$$K = \frac{\cos\Theta}{\left|G(j\omega_a^*)\right|}; \quad K\left[T_d\omega_a^* - \frac{1}{T_i\omega_a^*}\right] = \frac{\sin\Theta}{\left|G(j\omega_a^*)\right|} \tag{23}$$

where (23a) is a general rule for calculation of the controller gain K. Using (23a) and the ratio of integration and derivative times $\beta=T_i/T_d$ in (23b), a quadratic equation in T_d is obtained after some manipulations

$$T_d^2\left(\omega_a^*\right)^2 - T_d\omega_a^* tg\Theta - \frac{1}{\beta} = 0 \tag{24}$$

A positive solution of (24) yields the rule for calculating the derivative time T_d

$$T_d = \frac{tg\Theta}{2\omega_a^*} + \frac{1}{\omega_a^*}\sqrt{\frac{tg^2\Theta}{4} + \frac{1}{\beta}}; \quad \Theta = -180^\circ + \phi_M - \arg G(\omega_a^*) \tag{25}$$

where $\Theta=\arg G_R(\omega_a^*)$ is found from the phase condition (20b). Thus, using the PID controller with coefficients $\{K;T_i=\beta T_d;T_d\}$, the identified point $G(j\omega_n)$ of the plant frequency response with coordinates (19) can be moved on the unit circle M_1 into the gain crossover $L_A \equiv L(j\omega_a^*)$; the required phase margin ϕ_M is guaranteed if the following identity holds between the excitation and amplitude crossover frequencies ω_n and ω_a^*, respectively

$$\omega_a^* = \omega_n \tag{26}$$

Thus

$$\left|G(j\omega_a^*)\right| = \left|G(j\omega_n)\right|; \quad \arg G(\omega_a^*) = \arg G(\omega_n) = \varphi; \quad \Theta = -180° + \phi_M - \varphi \tag{27}$$

and coordinates of the gain crossover L_A are

$$L_A \equiv L(j\omega_a^* \equiv j\omega_n) = \left[\left|L(j\omega_n)\right|, \arg L(\omega_n)\right] = \left[\left|1\right|, -180° + \phi_M\right] \tag{28}$$

Substituting (27a) and (27b) into (23a) and (23b), respectively, and (26) into (25a), tuning rules in Table 9 are obtained (Bucz et al., 2010a, 2010b, 2010c, 2011), (Bucz, 2011). Resulting PID controller coefficients guarantee required phase margin ϕ_M for $\beta=4$.

No.	Design method, year	Controller	K	T_i	T_d	Range of Θ; $\Theta=-180°+\phi_M-\varphi$
52.	Sine-wave method, 2010	PI	$\dfrac{\cos\Theta}{\left|G(j\omega_n)\right|}$	$\dfrac{-1}{\omega_n tg\Theta}$	–	$\left(-\dfrac{\pi}{2};0\right)$
53.	Sine-wave method, 2010	PD	$\dfrac{\cos\Theta}{\left|G(j\omega_n)\right|}$	–	$\dfrac{1}{\omega_n}tg\Theta$	$\left(0;\dfrac{\pi}{2}\right)$
54.	Sine-wave method, 2010	PID	$\dfrac{\cos\Theta}{\left|G(j\omega_n)\right|}$	βT_d	$\dfrac{tg\Theta}{2\omega_n}+\dfrac{1}{\omega_n}\sqrt{\dfrac{tg^2\Theta}{4}+\dfrac{1}{\beta}}$	$\left(-\dfrac{\pi}{2};\dfrac{\pi}{2}\right)$

Table 9. PI, PD and PID controller tuning rules according to the sine-wave method

Note that PI controller tuning rules were derived for $T_d=0$, and PD tuning rules for $T_i \to \infty$ in (21a). The excitation frequency is taken from the interval (Bucz et al., 2011), (Bucz, 2011)

$$\omega_n \in \langle 0,2\omega_c; 0,95\omega_c \rangle \tag{29}$$

obtained empirically by testing the sine-wave method on benchmark examples (Åström & Hägglund, 2000). Shifting the point $G(j\omega_n)=|G(j\omega_n)|e^{j\varphi}$ into the gain crossover $L_A(j\omega_n)$ on the unit circle M_1 is depicted in Fig. 9a.

3.3 Controller structure selection using the „triangle ruler“ rule

The argument Θ appearing in tuning rules in Tab. 9 indicates, what angle is to be contributed to the identified phase φ by the controller at ω_n to obtain the resulting open-loop phase $(-180°+\phi_M)$ needed to provide the required phase margin ϕ_M. The working range of PID controller argument is the union of PI and PD controllers phase ranges symmetric with respect to 0°

$$\Theta_{PID} \in \Theta_{PI} \cup \Theta_{PD} = (-90°,0°) \cup (0°,+90°) = (-90°,+90°) \tag{30}$$

The working range (30) can be interpreted by means of an imaginary transparent triangular ruler turned as in Fig. 9b; its segments to the left and right of the axis of symmetry represent the PD and PI working ranges, respectively. Put this ruler on Fig. 9a, the middle of the hypotenuse on the complex plane origin and turn it so that its axis of symmetry merges with the ray (0,G). Thus, the ruler determines in the complex plane the cross-hatched area representing the full working range of the PID controller argument. The controller type is chosen depending on the situation of the ray ($0,L_A$) forming the angle ϕ_M with the negative real halfaxis: situation of the ray ($0,L_A$) in the left-hand-sector suggests PD controller, and in the right-hand sector the PI controller. The case when the phase margin ϕ_M is achievable using both PI or PID controller is shown in Fig. 9b (Bucz et al., 2010b, 2011), (Bucz, 2011).

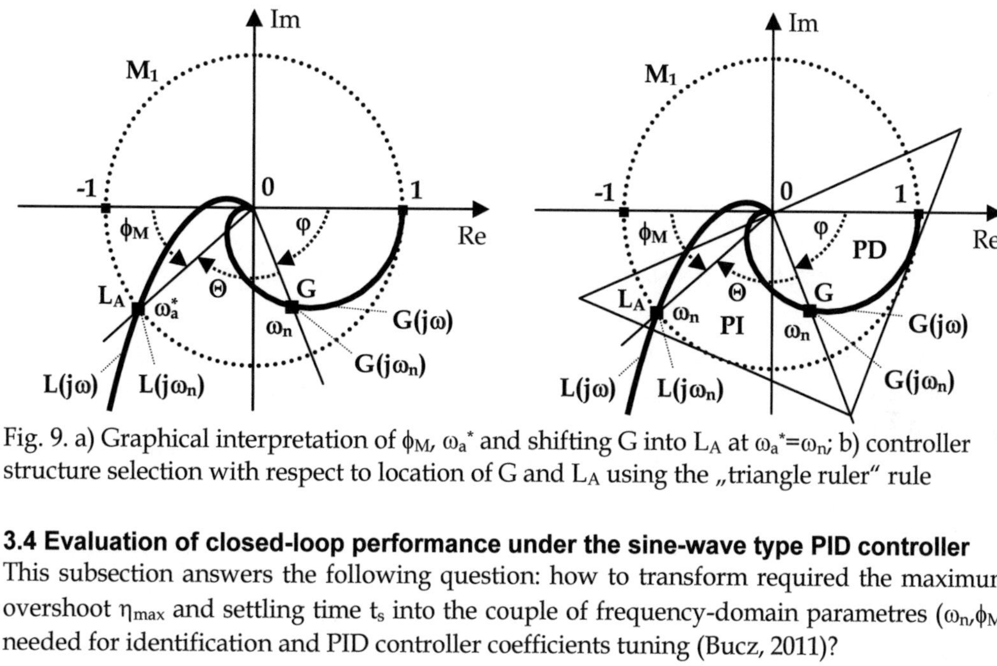

Fig. 9. a) Graphical interpretation of ϕ_M, ω_a^* and shifting G into L_A at $\omega_a^*=\omega_n$; b) controller structure selection with respect to location of G and L_A using the „triangle ruler" rule

3.4 Evaluation of closed-loop performance under the sine-wave type PID controller

This subsection answers the following question: how to transform required the maximum overshoot η_{max} and settling time t_s into the couple of frequency-domain parametres (ω_n,ϕ_M) needed for identification and PID controller coefficients tuning (Bucz, 2011)?

3.4.1 Systems without integrator

Looking for appropriate transformation $\Re$: $(\eta_{max},t_s)\rightarrow(\omega_n,\phi_M)$ we have considered typical phase margins ϕ_M given by the set

$$\{\phi_{Mj}\} = \{20°,30°,40°,50°,60°,70°,80°,90°\},\ j=1...8 \tag{31}$$

split into 5 equal sections $\Delta\omega_n=0,15\omega_c$; let us generate the set of excitation frequencies

$$\{\omega_{nk}\} = \{0,2\omega_c;0,35\omega_c;0,5\omega_c;0,65\omega_c;0,8\omega_c;0,95\omega_c\},\ k=1...6 \tag{32}$$

Elements of (32) divided by the plant critical frequency ω_c determine the set of so-called excitation levels

$$\{\sigma_k = \omega_{nk}/\omega_c\} \Rightarrow \{\sigma_k\} = \{0,2;0,35;0,5;0,65;0,8;0,95\}\ ,\ k=1...6 \tag{33}$$

Fig. 10 shows closed-loop step responses under PID controllers designed for the plant

$$G_1(s) = \frac{1}{(s+1)(0,5s+1)(0,25s+1)(0,125s+1)} \tag{34}$$

for three different phase margins ϕ_M=40°,60°,80° each on three excitation levels $\sigma_1=\omega_{n1}/\omega_c=0,2$; $\sigma_3=\omega_{n3}/\omega_c=0,5$ and $\sigma_5=\omega_{n5}/\omega_c=0,8$. Qualitative effect of ω_{nk} and ϕ_{Mj} on closed-loop step response is demonstrated.

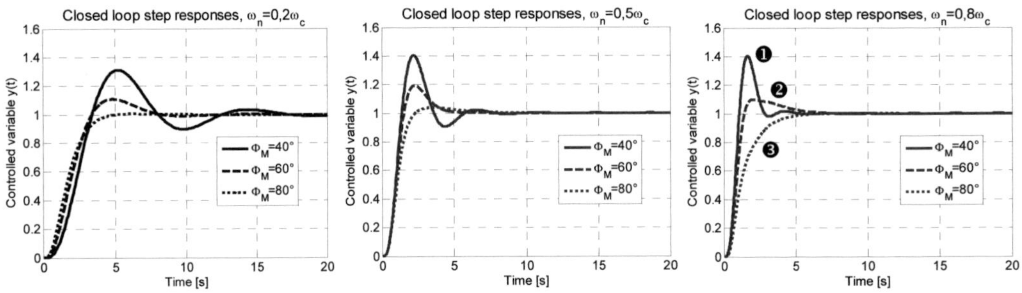

Fig. 10. Closed-loop step responses of $G_1(s)$ under PID controllers designed for various ϕ_M and ω_n

Achieving t_s and η_{max} was tested by designing PID controller for a vast set of benchmark examples (Åström & Hägglund, 2000) at excitation frequencies and phase margins expressed by a Cartesian product $\phi_{Mj}\times\omega_{nk}$ of (31) and (32) for j=1...8, k=1...6. Acquired dependencies $\eta_{max}=f(\phi_M,\omega_n)$ and $t_s=(\phi_M,\omega_n)$ are plotted in Fig. 11 (Bucz et al., 2010b, 2011).

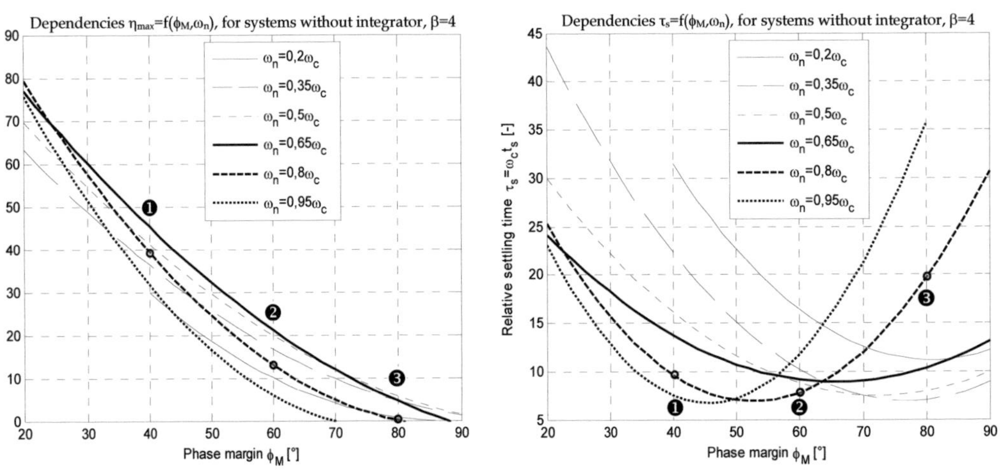

Fig. 11. Dependencies: a) $\eta_{max}=f(\phi_M,\omega_n)$; b) $\tau_s=\omega_c t_s=f(\phi_M,\omega_n)$ for $\omega_{nk}\times\phi_{Mj}$, j=1...8, k=1...6 (relative settling time τ_s is t_s weighed by the critical frequency ω_c of the plant)

Considering (26) resulting from the assumptions of the engineering method, the settling time can be expressed by the relation

$$t_s = \frac{\gamma\pi}{\omega_n} \tag{35}$$

similar to (17c) (Hudzovič, 1989), γ is the curve factor of the step response. In (17c) valid for a 2nd order closed-loop, γ is from the interval (1;4) and depends on the relative damping (Hudzovič, 1989). In case of the proposed sine-wave method, γ varies in a considerably broader interval (0,5;16) found empirically, and strongly depends on ϕ_M, i.e. $\gamma=f(\phi_M)$ at the given excitation frequency ω_n. To examine closed-loop settling times of plants with various dynamics, it is advantageous to define the relative settling time (Bucz et al., 2011)

$$\tau_s = t_s\omega_c \tag{36}$$

Substituting $\omega_n=\sigma\omega_c$ into (35), the following relation for the relative settling time is obtained

$$t_s\omega_c = \frac{\pi}{\sigma}\gamma \Rightarrow \tau_s = \frac{\pi}{\sigma}\gamma \tag{37}$$

where t_s is related to the critical frequency ω_c. By substituting ω_c in (37) its left-hand side is constant for the given plant, independent of ω_n. Fig. 11b depicts (37b) empirically evaluated for different excitation frequencies ω_{nk}; it is evident that at every excitation level σ_k with increasing phase margin ϕ_M the relative settling time τ_s first decreases and after achieving its minimum τ_{s_min} it increases again. Empirical dependencies in Fig. 11 were approximated by quadratic regression curves and called B-parabolas. B-parabolas are a useful design tool to carry out the transformation $\Re:(\eta_{max},t_s)\rightarrow(\omega_n,\phi_M)$ that enables choosing appropriate values of phase margin and excitation frequencies ϕ_M and ω_n, respectively, to provide performance specified in terms of maximum overshoot η_{max} and settling time t_s (Bucz et al., 2011). Note that pairs of B-parabolas at the same level (Fig. 11a, Fig. 11b) are always to be used.

Procedure 1. Specification of ϕ_M and ω_n from η_{max} and t_s from B-parabolas prior to designing the controller

1. Set the PID controller into manual mode. Find the plant critical frequency ω_c using the multipurpose loop in Fig. 7 (SW in position „3").
2. From the required settling time t_s calculate the relative settling time $\tau_s=\omega_c t_s$.
3. On the vertical axis of the plot in Fig. 11b find the value of τ_s calculated in Step 2.
4. Choose the excitation level σ (e.g. $\sigma_5=\omega_{n5}/\omega_c=0{,}8$).
5. For τ_s, find the corresponding phase margin ϕ_M on the parabola $\tau_s=f(\phi_M,\omega_n)$ with the chosen excitation level found in Step 4.
6. Find ϕ_M from Step 5 on the horizontal axis of the plot in Fig. 11a.
7. For ϕ_M, find the corresponding maximum overshoot η_{max} on the parabola $\eta_{max}=f(\phi_M,\omega_n)$ with the chosen excitation level found in Step 4.
8. If the found η_{max} is inappropriate, repeat Steps 4 to 7 for other parabolas $\tau_s=f(\phi_M,\omega_n)$ and $\eta_{max}=f(\phi_M,\omega_n)$ corresponding to other levels $\sigma_k=\omega_{nk}/\omega_c$ (related with the choice $\sigma_5=\omega_{n5}/\omega_c=0{,}8$ for $\sigma_k=\{0{,}2;0{,}35;0{,}50;0{,}65;0{,}95\}$, k=1...4,6). Repeat until both the required performance measures η_{max} and t_s are satisfied.

9. Calculate the excitation frequency ω_n according to the relation $\omega_n=\sigma\omega_c$ using the critical frequency ω_c (from Step 1) and the chosen excitation level σ (from Step 4).

Discussion

When choosing $\phi_M=40°$ on the B-parabola corresponding to the excitation level $\sigma_5=\omega_{n5}/\omega_c=0{,}8$ (further denoted as $B_{0,8}$ parabola), maximum overshoot $\eta_{max}=40\%$ and relative settling time $\tau_s\approx10$ are expected. Point ❶ corresponding to these parameters is located on the left (falling) portion of $B_{0,8}$ yielding oscillatory step response (see response ❶ in Fig. 10c). If the phase margin increases up to $\phi_M=60°$, the relative settling time decreases up to the point ❷ on the right (rising) portion of the $B_{0,8}$ parabola; the corresponding step response ❷ in Fig. 10c is weakly-aperiodic. For the phase margin $\phi_M=80°$ the $B_{0,8}$ parabola indicates a zero maximum overshoot, the relative settling time $\tau_s=20$ corresponds to the position ❸ on the $B_{0,8}$ parabola with aperiodic step response ❸ (Fig. 10c). If the maximum overshoot $\eta_{max}=20\%$ is acceptable then $\phi_M=53°$ yields the least possible relative settling time $\tau_s=6{,}5$ on the given level $\sigma_5=0{,}8$ ("at the bottom" of $B_{0,8}$) (Bucz et al., 2011), (Bucz, 2011).

Procedure 2. PID controller design using the sine-wave engineering method

1. From the required values (η_{max},t_s) specify the couple (ω_n;ϕ_M) using Procedure 1.
2. Identify the plant using the sinusoidal excitation signal with frequency ω_n specified in Procedure 1. The switch SW is in position „4".
3. Specify $\varphi=\arg G(\omega_n)$, and$|G(j\omega_n)|$. Calculate the controller argument Θ by substituting φ and ϕ_M into (27c); if Θ is within the range shown in the last column of Tab. 9, go to Step 4, if not, change (ω_n;ϕ_M) and repeat Steps 1-3.
4. Substitute the identified values $\varphi=\arg G(\omega_n)$, $|G(j\omega_n)|$ and specified ϕ_M into the tuning rules in Tab. 9 to calculate PID controller parameters.
5. Adjust the resulting PID controller values, switch into automatic mode and complete the controller by switching SW into position „5".

Example 1

Using the sine-wave method, ideal PID controller (4a) is to be designed for the operating amplifier modelled by the transfer function $G_A(s)$

$$G_A(s)=\frac{1}{(T_A s+1)^3}=\frac{1}{(0{,}01s+1)^3} \tag{38}$$

The controller has to be designed for two values of the maximum overshoot of the closed-loop step response $\eta_{max1}=30\%$ (Design No. 1) and $\eta_{max2}=5\%$ (Design No. 2) and maximum relative settling time $\tau_s=12$ in both cases.

Solution

1. Critical frequency of the plant identified by the Rotach test is $\omega_c=173{,}216$[rad/s] (the process is "fast"). The prescribed settling time is $t_s=\tau_s/\omega_c=12/173{,}216$[s]=69,3[ms].
2. For the Design No. 1 (η_{max1};τ_s)=(30%;12), a suitable choice is (ϕ_{M1};ω_{n1})=(50°;0,5ω_c) resulting from the $B_{0,5}$ parabola in Fig. 11. The performance in Design No. 2 (η_{max2};τ_s)=(5%;12) can be achieved for (ϕ_{M2};ω_{n2})=(70°;0,8ω_c) chosen from the $B_{0,8}$ parabola in Fig. 11.
3. Identified points for the Designs No. 1 and No. 2 are $G_A(j0{,}5\omega_c)=0{,}43e^{-j120^\circ}$ and $G_A(j0{,}8\omega_c)=0{,}19e^{-j165^\circ}$, respectively. According to Fig. 12a, both points are located in the

Quadrant II of the complex plane, on the Nyquist plot $G_A(j\omega)$ (solid line) which verifies the identification.

4. Using the PID controller designed for $(\phi_{M1};\omega_{n1})=(50°;0{,}5\omega_c)$, the point $G_A(j0{,}5\omega_c)$ is moved into the gain crossover $L_{A1}(j0{,}5\omega_c)=1e^{-j130°}$ on the unit circle M_1, which verifies achieving the phase margin $\phi_{M1}=180°-130°=50°$ (dashed line in Fig. 12a). The point $G_A(j0{,}8\omega_c)$ has been moved into $L_{A2}(j0{,}8\omega_c)=1e^{-j110°}$ by the PID controller designed for $(\phi_{M2};\omega_{n2})=(80°;0{,}8\omega_c)$ yielding the phase margin $\phi_{M2}=180°-110°=70°$ (dotted line in Fig. 12a).
5. Achieved performance according to the closed-loop step response in Fig. 12b (dashed line) is $\eta_{max1}^*=29{,}7\%$, $t_{s1}^*=58{,}4$[ms]. Performance in terms of $\eta_{max2}^*=4{,}89\%$, $t_{s2}^*=60{,}5$[ms] identified from the closed-loop step response in Fig. 12b (dotted line) fulfils the performance requirements.

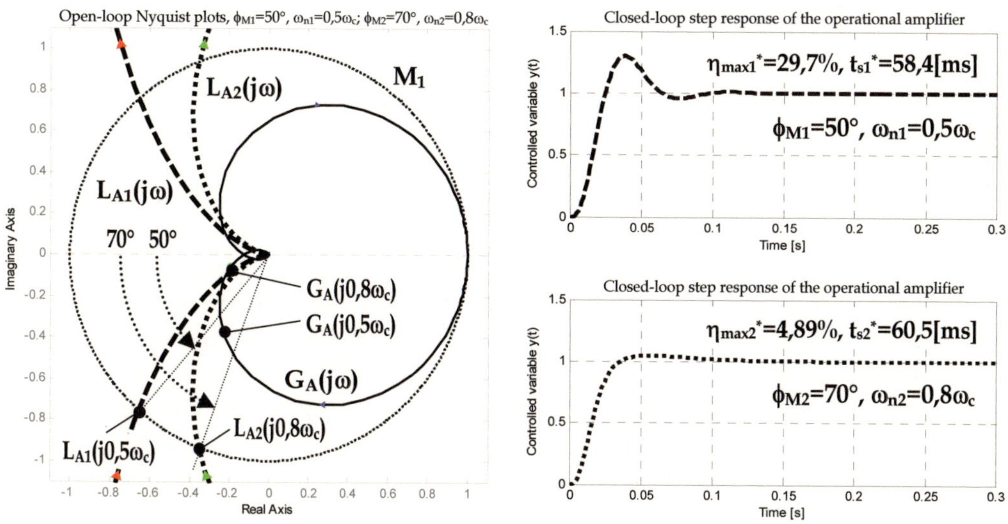

Fig. 12. a) Open-loop Nyquist plots; b) closed-loop step responses of the operational amplifier, required performance $\eta_{max1}=30\%$, $\eta_{max2}=5\%$ and $\tau_s=12$

3.4.2 Systems with time delay

The sine-wave method is applicable also for plants with time delay considered as difficult-to-control systems. It is a well-known fact, that the time delay D turns the phase at each frequency $\omega_n \in \langle 0,\infty)$ by $\omega_n D$ with respect to the delay-free system. For time delayed plants, phase condition of the sine-wave method (20b) is extended by additional phase $\varphi_D=-\omega_n D$

$$\left(\varphi' + \varphi_D\right) + \Theta = -180° + \phi_M \tag{39}$$

where φ' is the phase of the delay-free system and

$$\varphi = \varphi' + \varphi_D \tag{40}$$

is the identified phase of the plant including the time delay. The added phase $\varphi_D=-\omega_n D$ can be associated with the required phase margin ϕ_M

$$\varphi' + \Theta = -180° + (\phi_M + \omega_n D) \tag{41}$$

The only modification in using the PID tuning rules in Tab. 9 is that increased required phase margin is to be specified (Bucz, 2011)

$$\phi_M' = \phi_M + \omega_n D \tag{42}$$

and the controller working angle Θ is computed using the relation

$$\Theta = -180° - \varphi' + (\phi_M + \omega_n D) \tag{43}$$

The phase delay $\omega_n D$ increases with increasing frequency of the sinusoidal signal ω_n.
To lessen the impact of time delay on closed-loop dynamics, it is recommended to use the smallest possible added phase $\varphi_D=-\omega_n D$.

Discussion

Time delay D can easily be specified during critical frequency identification as the time $D=T_y-T_u$, that elapses since the start of the test at time T_u until time T_y, when the system output starts responding to the excitation signal u(t). A small added phase $\varphi_D=-\omega_n D$ due to time delay can be secured by choosing the smallest possible ω_n attenuating effect of D in (43) and subsequently in the PID controller design.
Therefore, when designing PID controller for time delayed systems according to Procedure 1, in Step 4 it is recommended to choose the lowest possible excitation level on the performance B-parabolas (most frequently ω_n/ω_c=0,2 resp. 0,35) and corresponding couples of B-parabolas in Fig. 11. Procedure 2 is used for plant identification and PID controller design. ϕ_M is specified from the given couple (η_{max};t_s) using the chosen couple of B-parabolas, however its increased value ϕ_M' given by (42) is to be supplied in the design algorithm thus minimizing effect of the time delay on closed-loop dynamics.

Example 2

Using the sine-wave method, ideal PID controllers (4a) are to be designed for the distillation column modelled by the transfer function $G_B(s)$

$$G_B(s) = \frac{K_B e^{-D_B s}}{T_B s + 1} = \frac{1{,}11 e^{-6{,}5s}}{3{,}25s + 1} \tag{44}$$

Control objectives are the same as in Example 1.

Solution

1. Critical frequency of the plant is ω_c=0,3521[rad/s]. Based on comparison of critical frequencies, $G_B(s)$ is 500-times slower than $G_A(s)$. Required settling time is $t_s=\tau_s/\omega_c$= =12/0,3521[s]=34,08[s].
2. Because D_B/T_B=2>1, the plant is a so-called „dead-time dominant system". Due to a large the time delay, it is necessary to choose the lowest possible excitation frequency ω_n to minimize the added phase $\omega_n D_B$ in (43). Hence, for the required performance (η_{max2};τ_s)=(5%;12) (Design No. 2) we choose the $B_{0,2}$ parabolas in Fig. 11 at the lowest possible level ω_n/ω_c=0,2 to find (ϕ_{M2};ω_{n2})=(70°;0,2ω_c). The added phase is

$\omega_{n2}D_B(180°/\pi)=0,2\omega_cD_B(180°/\pi)=0,2.0,3521.6,5.180°/\pi=26,2°$, hence the phase supplied to the PID design algorithm is $\phi'_{M2}=\phi_{M2}+\omega_{n2}D_B(180°/\pi)=70°+26,2°=96,2°$ (instead of $\phi_{M2}=70°$ for a delay-free system). The required performance $(\eta_{max1};\tau_s)=(30\%;12)$ (Design No. 1) can be achieved by choosing $(\phi_{M1};\omega_{n1})=(55°;0,35\omega_c)$ from the $B_{0,35}$ parabolas in Fig. 11 (i.e. $\omega_n/\omega_c=0,35$). The phase margin $\phi'_{M1}=55°+45,9°$ supplied into the design algorithm was increased by $\omega_{n1}D_B(180°/\pi)=0,35\omega_cD_B(180°/\pi)=0,35.0,3521.6,5.180°/\pi= =45,9°$ compared with $\phi_{M1}=55°$ in case of delay-free system.

3. Identified points $G_B(j0,35\omega_c)=1,03e^{-j23°}$ and $G_B(j0,2\omega_c)=1,09e^{-j13°}$ in Fig. 13a are located in the Quadrant I of the complex plane at the beginning of the frequency response $G_B(j\omega)$ (solid line). The point $G_B(j0,2\omega_c)$ (Design No. 2) was shifted by the PID controller to the open-loop gain crossover $L_{B2}(j0,2\omega_c)=1e^{-j110°}$ (dotted line in Fig. 13a). Note that L_{B2} has the same location in the complex plane as L_{A2} in Fig. 12a, however at a considerably lower frequency $\omega_{n2B}=0,2.0,3521=0,07$[rad/s] compared to $\omega_{n2A}=0,8.173,216= =138,6$[rad/s] ($t_{s2\_B}^*=28,69$[s] is almost 500 times larger than $t_{s2\_A}^*=0,0584$[s] which demonstrates the key role of the excitation frequency ω_n in achieving required closed-loop dynamics). The identified point $G_B(j0,35\omega_c)$ (Design No. 1) was moved into the gain crossover $L_{B1}(j0,35\omega_c)=1e^{-j125°}$ (dashed line in Fig. 13a).
4. Achieved performances ($\eta_{max1}^*=18,6\%$, $t_{s1}^*=24,78$[s], dashed line), ($\eta_{max2}^*=0,15\%$, $t_{s2}^*=28,69$[s], dotted line) in terms of closed-loop step responses in Fig. 13b comply with the required performance specification.

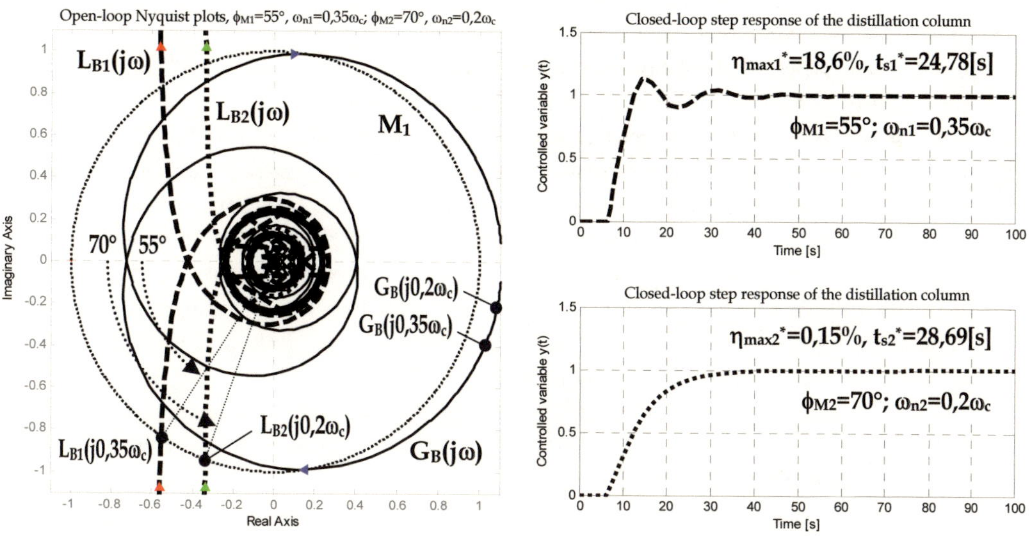

Fig. 13. a) Open-loop Nyquist plots; b) closed-loop step responses of the distillation column, required performance $\eta_{max1}=30\%$, $\eta_{max2}=5\%$ and $\tau_s=12$

3.4.3 Systems with 1st order integrator

By testing the sine-wave method on benchmark systems with 1st order integrator, the B-parabolas in Fig. 14 - 16 were obtained (for Cartesian product $\phi_{Mj}\times\omega_{nk}$ of sets (31) and (32), j=1...8, k=1...6 and three various ratios T_i/T_d: β=4, 8 and 12).

Discussion

Inspection of Fig. 14a, 15a and 16a reveals, that increasing β results in decreasing of the maximum overshoot η_{max}, narrowing of the B-parabolas of relative settling times $\tau_s=f(\phi_M,\omega_n)$ for each identification level ω_n/ω_c, and consequently settling time increasing. Consider e.g. the $B_{0,95}$ parabolas in Fig. 14b, Fig. 15b and Fig. 16b: if $\phi_M=70°$ and $\beta=4$, relative settling time is $\tau_s=30$, for $\beta=8$ it grows to $\tau_s=40$, and for $\beta=12$ even to $\tau_s=45$. If a 10% maximum overshoot is acceptable, then the standard interaction PID controller can be used with no need to use a setpoint filter; however a larger settling time is to be expected.
Procedure 1 is used to specify the performance in terms of (ϕ_M,ω_n) from (η_{max},t_s) using pertinent B-parabolas in Fig. 14 - 16. Procedure 2 is used for plant identification and PID controller design.

Example 3

Using the sine-wave method, design ideal PID controller for the flow valve modelled by the transfer function $G_C(s)$ (system with integrator and time delay)

$$G_C(s)=\frac{K_C e^{-D_C s}}{s(T_C s+1)}=\frac{1{,}3e^{-2{,}1s}}{s(7{,}51s+1)} \tag{45}$$

Control objective is to provide the maximum overshoots of the closed-loop step response $\eta_{max1}=30\%$, $\eta_{max2}=20\%$ and a maximum relative settling time $\tau_s=20$.

Solution

1. Critical frequency of the plant identified by the Rotach test is $\omega_c=0{,}2407$[rad/s]. Then, the required settling time is $t_s=\tau_s/\omega_c=20/0{,}2407$[s]$=83{,}09$[s].
2. For $G_C(s)$ the time delay/time constant ratio is $D_C/T_C=2{,}1/7{,}51=0{,}28<1$, hence, the influence of the time constant prevails - $G_C(s)$ is a so-called „lag-dominant system" with integrator, therefore B-parabolas are to be chosen carefully. From one side, due to time delay it would be desirable to choose B-parabolas from Fig. 14, Fig. 15 or Fig. 16 with the lowest identification level $\omega_n/\omega_c=0{,}2$. However, the minima of $B_{0,2}$ parabolas in Fig. 14b (for $\beta=4$), Fig. 15b (for $\beta=8$) and Fig. 16b (for $\beta=12$) indicate the smallest achievable relative settling time $\tau_s=36{,}5$ (for $\beta=4$), $\tau_s=33$ (for $\beta=8$) and $\tau_s=34$ (for $\beta=12$), which do not satisfy the required value $\tau_s=20$.
3. Identified points $G_C(j0{,}35\omega_c)=12{,}7e^{-j122°}$ and $G_C(j0{,}5\omega_c)=8{,}10e^{-j129°}$ are located on the plant frequency response $G_C(j\omega)$ (solid line) in Fig. 17a, verifying correctness of the sine-wave type identification.
4. The first performance specification $(\eta_{max1};\tau_s)=(30\%;20)$ can be provided using the $B_{0,35}$ parabolas for $\beta=12$ (Fig. 16b) at the level $\omega_n/\omega_c=0{,}35$ and for parameters $(\phi_{M1};\omega_{n1})=$ $=(53°;0{,}35\omega_c)$ (Design No. 1), supplying the augmented open-loop phase margin $\phi'_{M1}=\phi_{M1}+(180°/\pi)\omega_{n1}D_C=53°+10{,}1°=63{,}1°$ into the controller design algorithm. The second performance specification $(\eta_{max2};\tau_s)=(20\%,20)$ is achievable using the $B_{0,5}$ parabolas in Fig. 16 for $\beta=12$ and $\omega_n/\omega_c=0{,}5$ and parametres $(\phi_{M2};\omega_{n2})=(62°;0{,}5\omega_c)$ (Design No. 2). To reject the influence of D_C, instead of $\phi_{M2}=62°$ the augmented open-loop phase margin $\phi'_{M2}=\phi_{M2}+(180°/\pi)\omega_{n2}D_C=62°+14{,}5°=76{,}5°$ was supplied into the PID controller design algorithm.

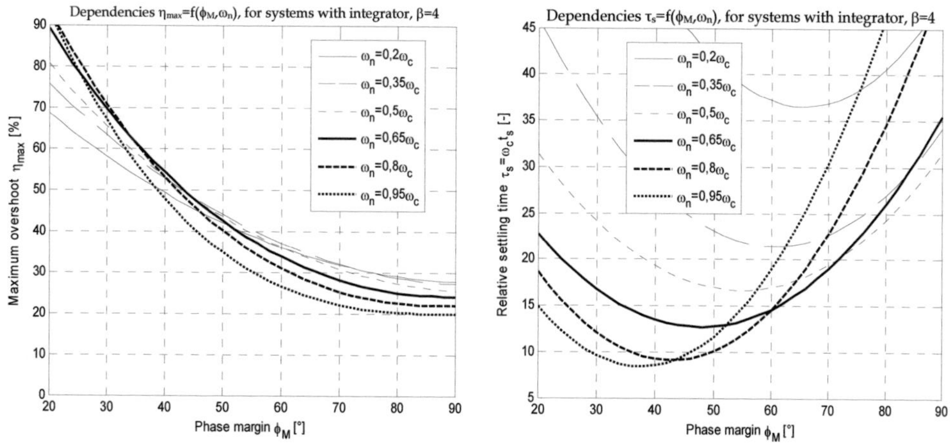

Fig. 14. B-parabolas: a) η_{max}=f(ϕ_M,ω_n); b) τ_s=$\omega_c t_s$=f(ϕ_M,ω_n) for systems with integrator, β=4

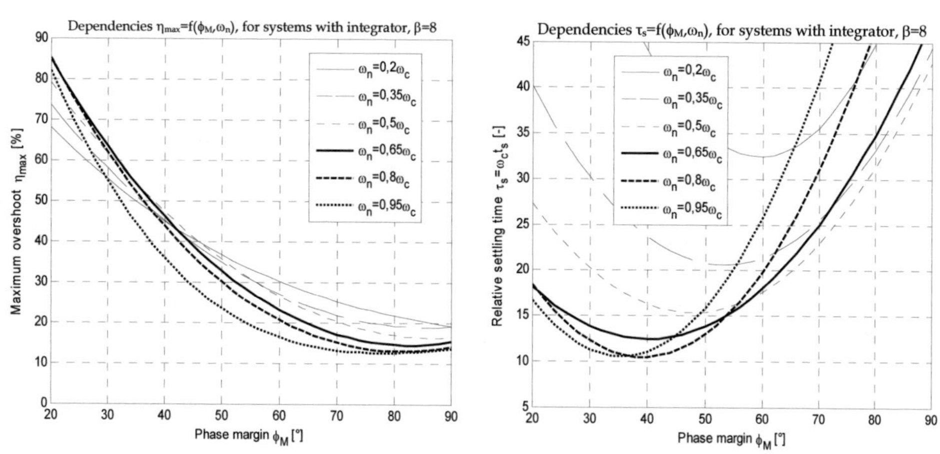

Fig. 15. B-parabolas: a) η_{max}=f(ϕ_M,ω_n); b) τ_s=$\omega_c t_s$=f(ϕ_M,ω_n) for systems with integrator, β=8

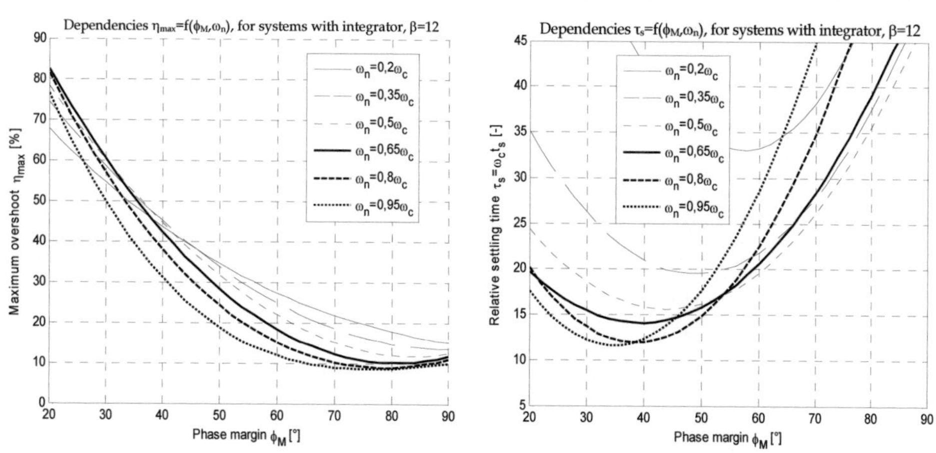

Fig. 16. B-parabolas: a) η_{max}=f(ϕ_M,ω_n); b) τ_s=$\omega_c t_s$=f(ϕ_M,ω_n) for systems with integrator, β=12

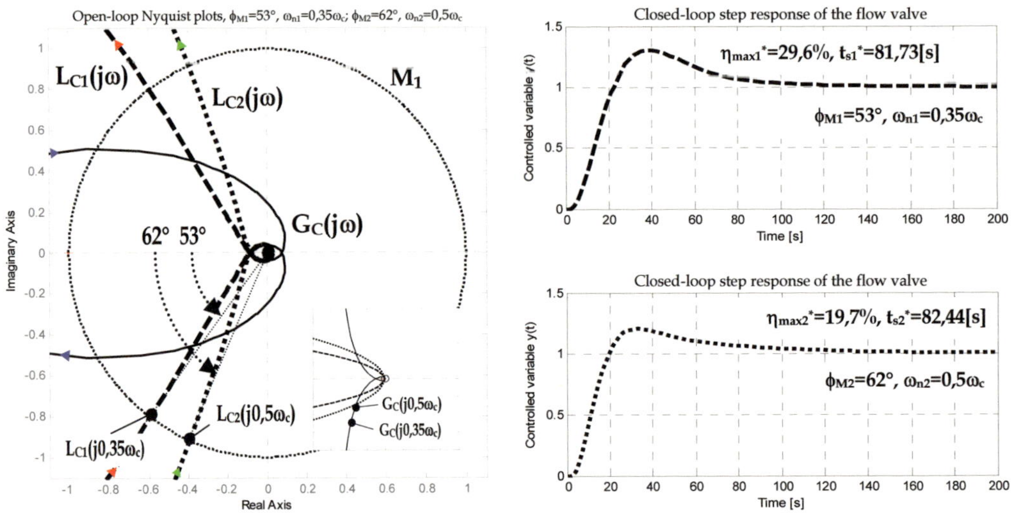

Fig. 17. a) Open-loop Nyquist plots; b) closed-loop step responses of the flow valve, required performance η_{max1}=30%, η_{max2}=20% and τ_s=20

5. Using the PID controller, the first identified point $G_C(j0{,}35\omega_c)$ (Design No. 1) was moved into the gain crossover $L_{C1}(j0{,}35\omega_c)=1e^{-j127^\circ}$ located on the unit circle M_1; this verifies achieving the phase margin ϕ_{M1}=180°-127°=53° (dashed line in Fig. 17a). Achieved performance in terms of the closed-loop step response in Fig. 17b is η_{max1}^*=29,6%, t_{s1}^*=81,73[s] (dashed line). The second identified point $G_C(j0{,}5\omega_c)$ (Design No. 2) was moved into $L_{C2}(j0{,}5\omega_c)=1e^{-j118^\circ}$ achieving the phase margin ϕ_{M2}=180°-118°=62° (dotted line in Fig. 17a). Achieved performance in terms of the closed-loop step response parameters η_{max2}^*=19,7%, t_{s2}^*=82,44[s] (dotted line in Fig. 17b) meets the required specification. Frequency characteristics $L_{C1}(j\omega)$, $L_{C2}(j\omega)$ begin near the negative real half-axis of the complex plane, because both open-loops contain a 2nd order integrator.

Discussion

All data necessary to design two PID controllers of all three plants $G_A(s)$, $G_B(s)$ and $G_C(s)$ along with specified and achieved performance measure values are summarized in Tab. 10 where η_{max} and t_s in the last two columns marked with „*" indicate closed-loop performance complying with the required one.

Model	η_{max};τ_s	ω_c[rad/s]	t_s[s]	B-par.	ϕ_M	ω_n/ω_c	$G(j\omega_n)$	$G_R(j\omega_n)$	η_{max}^*	t_s^*[s]
$G_A(s)$	30%;12	173,22	0,0693	Fig. 11	50°	0,5	$0{,}43e^{-j120^\circ}$	$2{,}31e^{-j10^\circ}$	29,7%	0,0584
$G_A(s)$	5%;12	173,22	0,0693	Fig. 11	70°	0,8	$0{,}19e^{-j165^\circ}$	$5{,}20e^{j55^\circ}$	4,89%	0,0605
$G_B(s)$	30%;12	0,3521	34,08	Fig. 11	55°+45,9°	0,35	$1{,}03e^{-j23^\circ}$	$0{,}97e^{-j56^\circ}$	18,6%	24,78
$G_B(s)$	5%;12	0,3521	34,08	Fig. 11	70°+26,2°	0,2	$1{,}09e^{-j13^\circ}$	$0{,}92e^{-j71^\circ}$	0,15%	28,69
$G_C(s)$	30%;20	0,2407	83,09	Fig. 16	53°+10,1°	0,35	$12{,}7e^{-j122^\circ}$	$0{,}08e^{j5{,}8^\circ}$	29,6%	81,73
$G_C(s)$	20%;20	0,2407	83,09	Fig. 16	62°+14,5°	0,5	$8{,}10e^{-j129^\circ}$	$0{,}12e^{-j28^\circ}$	19,7%	82,44

Table 10. Summary of required and achieved performance measure values, identification parametres and PID controller tunings for $G_A(s)$, $G_B(s)$ and $G_C(s)$

4. Conclusion

The proposed new engineering method based on the sine-wave identification of the plant provides successful PID controller tuning. The main contribution has been construction of empirical charts to transform engineering time-domain performance specifications (maximum overshoot and settling time) into frequency domain performance measures (phase margin). The method is applicable for shaping the closed-loop response of the process variable using various combinations of excitation signal frequencies and required phase margins. Using B-parabolas, it is possible to achieve optimal time responses of processes with various types of dynamics and improve their performance. When applying digital PID controller, it is recommended to set the sampling period T_s from the interval

$$T_s \in \left\langle \frac{0,2}{\omega_c}, \frac{0,6}{\omega_c} \right\rangle \tag{46}$$

where ω_c is the critical frequency of the controlled plant (Wittenmark, 2001).
By applying appropriate PID controller design methods including the above presented 51+3 tuning rules for prescribed performance, it is possible to achieve cost-effective control of industrial processes. The presented advanced sine-wave design method offers one possible way to turn the unfavourable statistical ratio between properly tuned and all implemented PID controllers in industrial control loops.

5. Acknowledgment

This research work has been supported by the Scientific Grant Agency of the Ministry of Education of the Slovak Republic, Grant No. 1/1241/12.

6. References

Åström, K.J. & Hägglund, T. (1995). *PID Controllers: Theory, Design and Tuning* (2nd Edition), Instrument Society of America, Research Triangle Park, ISBN 1-55617-516-7

Åström, K.J. & Hägglund, T. (2000). Benchmark Systems for PID Control. *IFAC Workshop on Digital Control PID'00,* pp. 181-182, Terrassa, Spain, April, 2000

Bakošová, M. & Fikar, M. (2008). *Riadenie procesov (Process Control),* Slovak University of Technology in Bratislava, ISBN 978-80-227-2841-6, Slovak Republic (in Slovak)

Balátě, J. (2004). *Automatické řízení (Automatic Control)* (2nd Edition), BEN - technická literatúra, ISBN 80-7300-148-9, Praha, Czech Republic (in Czech)

Bucz, Š.; Marič, L.; Harsányi, L. & Veselý, V. (2010). A Simple Robust PID Controller Design Method Based on Sine Wave Identification of the Uncertain Plant. *Journal of Electrical Engineering,* Bratislava, Vol. 61, No. 3, (2010), pp. 164-170, ISSN 1335-3632

Bucz, Š.; Marič, L.; Harsányi, L. & Veselý, V. (2010). A Simple Tuning Method of PID Controllers with Prespecified Performance Requirements. *9th International Conference Control of Power Systems 2010,* High Tatras, Slovak Republic, May 18-20, 2010

Bucz, Š.; Marič, L.; Harsányi, L. & Veselý, V. (2010). Design-oriented Identification Based on Sine Wave Signal and its Advantages for Tuning of the Robust PID Controllers. *International Conference Cybernetics and Informatics,* Vyšná Boca, Slovak Republic, 2010

Bucz, Š.; Marič, L.; Harsányi, L. & Veselý, V. (2011). Easy Tuning of Robust PID Controllers Based on the Design-oriented Sine Wave Type Identification. *ICIC Express Letters,* Vol. 5, No. 3 (March 2011), pp. 563-572, ISSN 1881-803X, Kumamoto, Japan

Bucz, Š. (2011). *Engineering Methods of Robust PID Controller Tuning for Specified Performance.* Doctoral Thesis, Slovak University of Technology in Bratislava, Slovak Republic (in Slovak)

Coon, G.A. (1956). How to Find Controller Settings from Process Characteristics, In: *Control Engineering,* Vol. 3, No. 5, (May 1956), pp. 66-76

Chandrashekar, R.; Sree, R.P. & Chidambaram, M. (2002). Design of PI/PID Controllers for Unstable Systems with Time Delay by Synthesis Method, *Indian Chemical Engineer Section A,* Vol. 44, No. 2, pp. 82-88

Chau, P.C. (2002). Process Control - a First Course with MATLAB (1st edition), *Cambridge University Press,* ISBN 978-0521002554, New York

Chen, D. & Seborg, D.E. (2002). PI/PID Controller Design Based on Direct Synthesis and Disturbance Rejection, *Industrial Engineering Chemistry Research,* 41, pp. 4807-4822

Chien, K.L.; Hrones, J.A. & Reswick, J.B. (1952). On the Automatic Control of Generalised Passive Systems. *Transactions of the ASME,* Vol. 74, February, pp. 175-185

Ford, R.L. (1953). The Determination of the Optimum Process-controller Settings and their Confirmation by Means of an Electronic Simulator, *Proceedings of the IEE, Part 2,* Vol. 101, No. 80, pp. 141-155 and pp. 173-177, 1953

Grabbe, E.M.; Ramo, S. & Wooldrige, D.E. (1959-61). *Handbook of Automation Computation and Control,* Vol. 1,2,3, New York

Haalman, A. (1965). Adjusting Controllers for a Deadtime Process, *Control Engineering,* July

Hang, C.C. & Åström, K.J. (1988). Practical Aspects of PID Auto-tuners Based on Relay feedback, *Proceedings of the IFAC Adaptive Control of Chemical Processes Conference,* pp. 153-158, Copenhagen, Denmark, 1998

Harsányi, L.; Murgaš, J.; Rosinová, D. & Kozáková, A. (1998). *Teória automatického riadenia (Control Theory),* Slovak University of Technology in Bratislava, ISBN 80-227-1098-9, Slovak Republic (in Slovak)

Hudzovič, P. (1982). *Teória automatického riadenia I. Lineárne spojité systémy (Control theory: Linear Continuous-time Systems),* Slovak University of Technology in Bratislava, Slovak Republic (in Slovak)

Kozáková, A.; Veselý, V. & Osuský, J. (2010). Decentralized Digital PID Design for Performance. *In: 12th IFAC Symposium on Large Scale Systems: Theory and Applications,* Lille, France, 12.-14.7.2010, Ecole Centrale de Lille, ISBN 978-2-915-913-26-2

Morari, M., Zafiriou, E. (1989). *Robust Processs Control.* Prentice-Hall Inc., Englewood Cliffs, ISBN 0137821530, 07632 New Jersey, USA

O'Dwyer, A. (2006). *Handbook of PI and PID Controllers Tuning Rule* (2nd Edition), Imperial College Press, ISBN 1860946224, London

Osuský, J.; Veselý, V. & Kozáková, A. (2010). Robust Decentralized Controller Design with Performance Specification, *ICIC Express Letters,* Vol. 4, No. 1, (2010), pp. 71-76, ISSN 1881-803X, Kumamoto, Japan

Pettit, J.W. & Carr, D.M. (1987). Self-tuning Controller, *US Patent* No. 4669040

Rotach, V. (1984). Avtomatizacija nastrojki system upravlenija. *Energoatomizdat,* Moskva, Russia (in Russian)

Rotach, V. (1994). Calculation of the Robust Settings of Automatic Controllers, *Thermal Engineering (Russia),* Vol. 41, No. 10, pp. 764-769, Moskva, Russia

Suyama, K. (1992). A Simple Design Method for Sampled-data PID Control Systems with Adequate Step Responses, *Proceedings of the International Conference on Industrial Electronics, Control, Instrumentation and Automation,* pp. 1117-1122, 1992

Veselý, V. (2003). Easy Tuning of PID Controller. *Journal of Electrical Engineering,* Vol. 54, No. 5-6, (2003), pp. 136-139, ISSN 1335-3632, Bratislava, Slovak Republic

Visioli, A. (2001). Tuning of PID Controllers with Fuzzy Logic, *IEE Proceedings-Control Theory and Applications,* Vol. 148, No. 1, pp. 180-184, 2001

Visoli, A. (2006). *Practical PID Control, Advances in Industrial Control,* Springer-Verlag London Limited, ISBN 1-84628-585-2

Vítečková, M. (1998). *Seřízení regulátorů metodou požadovaného modelu (PID Controllers Tuning by Desired Model Method),* Textbook, VŠB - Technical University of Ostrava, ISBN 80-7078-628-0, Czech Republic (in Czech)

Vítečková, M. (1999). Seřízení číslicových i analogových regulátorů pro regulované soustavy s dopravním zpožděním (Tuning Discrete and Continuous Controllers for Processes with Time Delay). *Automatizace,* Vol. 42, No. 2, (1999), pp. 106-111, Czech Republic (in Czech)

Vítečková, M.; Víteček, A. & Smutný, L. (2000). Controller Tuning for Controlled Plants with Time Delay, *Preprints of Proceedings of PID'00: IFAC Workshop on Digital Control,* pp. 83-288, Terrassa, Spain, April 2000

Wang, L. & Cluett, W.R. (1997). Tuning PID Controllers for Integrating Processes, *IEE Proceedings - Control Theory and Applications,* Vol. 144, No. 5, pp. 385-392, 1997

Wang, Y.-G. & Shao, H.-H. (1999). PID Autotuner Based on Gain- and Phase-margin Specification, *Industrial Engineering Chemistry Research,* 38, pp. 3007-3012

Wittenmark, B. (2001). A Sample-induced Delays in Synchronous Multirate Systems, *European Control Conference,* Porto, Portugal, pp. 3276–3281, 2001

Wojsznis, W.K.; Blevins, T.L. & Thiele, D. (1999). Neural Network Assisted Control Loop Tuner, *Proceedings of the IEEE International Conference on Control Applications,* Vol. 1, pp. 427-431, USA, 1999

Xue, D.; Chen, Y. & Atherton, D.P. (2007). Linear Feedback Control. Analysis and Design with MATLAB, SIAM Press, ISBN 978-0-898716-38-2

Yu, Ch.-Ch. (2006). *Autotuning of PID Controllers. A Relay Feedback Approach* (2nd Edition), Springer-Verlag London Limited, ISBN 1-84628-036-2

Ziegler, J.G. & Nichols, N.B. (1942). Optimum Settings for Automatic Controllers, *ASME Transactions,* Vol. 64 (1942), pp. 759-768

Zhuang, M. & Atherton, D.P. (1993). Automatic Tuning of Optimum PID Controllers, *IEE Proceedings, Part D: Control Theory and Applications,* Vol. 140, No. 3, pp. 216-224, ISSN 0143-7054, 1993

2

Family of the PID Controllers

Ilan Rusnak
RAFAEL, Advanced Defense Systems, Haifa
Israel

1. Introduction

The PID controllers (P, PD, PI, PID) are very widely used, very well and successfully applied controllers to many applications, for many years, almost from the beginning of controls applications (D'Azzo & Houpis, 1988)(Franklin et al., 1994). (The facts of their successful application, good performance, easiness of tuning are speaking for themselves and are sufficient rational for their use, although their structure is justified by heuristics: "These ... controls - called proportional-integral-derivative (PID) control - constitute the heuristic approach to controller design that has found wide acceptance in the process industries." (Franklin et al., 1994, pp. 168)).

In this chapter we state a problem whose solution leads to the PID controller architecture and structure, thus avoiding heuristics, giving a systematic approach for explanation of the excellent performance of the PID controllers and gives insight why there are cases the PID controllers do not work well. Namely, by the use of Linear Quadratic Tracking (LQT) theory (Kwakernaak & Sivan, 1972)(Anderson & Moore, 1989) control-tracking problems are formulated and those cases when their solution gives the PID controllers are shown.

Further, problem of controlling-tracking high order polynomial inputs and rejecting high order polynomial disturbances is formulated. By applying the LQT theory extended family of PID controllers - the family of generalized PID controllers denoted PI^mD^{n-1} is derived. This provides tool for application of optimal controllers for those systems that the conventional PID controllers are not satisfactory, for generalization and derivation of further results. The notation of generalized PID controllers, PI^mD^{n-1}, is consistent with the notation of controllers for fractional order systems (Podlubny, 1999).

The present work is strongly motivated by problems-question tackled by the author during a continuous work on high performance servo and motion control applications. Some of the theoretical results that have had motivated and led to the present work have been documented in (Rusnak, 1998, 1999, 2000a, 2000b). Some of the presented architectures appear and are recommended for use in (Leonhard, 1996, pp. 80, 347) without rigorous rationale and were partial trigger for the presented approach.

By Architecture we mean, loosely, the connections between the outputs/sensors and the inputs/actuators; Structure deals with the specific realization of the controllers' blocks; and Configuration is a specific combination of architecture and structure. These issues fall within the control and feedback organization theory that have been reviewed and presented in a concise form in (Rusnak, 2002, 2005) and in a widened form in (Rusnak, 2006, 2008). It is beyond the scope of this chapter. It is used here as a basis at a system theoretic level to

enable formulation of the control-feedback loops organization problem that leads to the family of generalized PID controllers. This article does not deal with the numerical values of the controllers' filters coefficients/gains; rather it concentrates in organization of the control loop and structure of the filters. This is the way the optimal LQT theory is used.
The LQT theory requires a reference trajectory generator. The reference trajectory is generated by a system that reflects the nominal behavior of the plant. The differences are the initial conditions, the input to the reference trajectory generator and the deviation of the actual plant from the nominal one. The zero steady state error is imposed by integral action of a required order on the state tracking error.
The generalized controllers derived by the presented methodology have been applied to high performance motion control in (Nanomotion, 2009a, 2009b) and to high performance missile autopilot in (Rusnak and Weiss, 2011).
The novelty of the results in this approach is that it shows for what problems a controller from the family of PID controllers is the optimal controller and for which it is not.
The importance of this result is:

1. From theoretical point of view it is important to know that widely used control architecture can be derived from an optimal control/tracking problem.
2. The solution shows for what kind of systems the PID controller is optimal and for which systems it is not, thus showing why a PID controller does not perform well for all systems. This will enable to forecast what control designs not to apply a PID controller.
3. For those systems that the PID is not the controller architecture derived from the optimal control approach shows what is the optimal controller architecture and structure, thus achieving generalization.
4. The present approach advises how to design PID controller on finite time interval, i.e. when the gains are time varying.
5. The generalization can be used in deriving generalized PID controllers for high order SISO systems, for SIMO and MIMO systems (Rusnak, 1999, 2000a), for time–varying and non-linear systems; thus enabling a systematic generalization of the PID controller paradigm.
6. The design procedures of PID controllers are assuming noise free environment. The presented approach advises how to generalize the PID controller configuration in presence of noises by the use of the Linear Quadratic Gaussian Tracking-LQGT theory (Rusnak, 2000b).
7. The conventional PID paradigm introduces integral action in order to drive the steady state tracking errors in presence of constant reference trajectory or disturbance. The present approach enables to systematically generalize the controller to drive the steady state tracking errors to zero for high order polynomial inputs and disturbances.
8. Choosing the optimal generalized PID controller reduces the quantity of controller parameters-gains that are required for tuning, Thus saving time during the design process.
9. The LQT motivated architecture enables separate treatment of the transient, by the trajectory generator, and the steady state performance by introducing integrators into the controllers (Rusnak and Weiss, 2011).

The results on the architecture and structure of the PID controllers' family for 1st and 2nd order are rederived in the article. Specifically, it is shown that the classical one block PID controller is optimal for Linear Quadratic Tracking problem of a 2nd order minimum phase plant. For plants with non-minimum phase zero the family of PID controllers is only suboptimal. Multi output single input architectures are proposed that are optimal.

Throughout this chapter the same notation for time domain and Laplace domain is used, and the explicit Laplace variable (s) is stated to avoid confusion wherever necessary.

2. The optimal tracking problem

The optimal tracking problem is introduced in (Kwakernaak & Sivan, 1972) (Anderson & Moore, 1989). The nth order system is

$$\begin{aligned} \dot{x} &= Ax + Bu; \ x(t_o) = x_o, \\ y &= Cx \end{aligned} \tag{1}$$

where x is the state; u is the input and y is the measured output, x_o is a zero mean random vector.
The νth order reference trajectory generator is

$$\begin{aligned} \dot{x}_r &= A_r x_r + B_r w_r; \ x_r(t_o) = x_{ro}, \\ y_r &= C_r x_r \end{aligned} \tag{2}$$

where x_r is the state; w_r is the input and y_r is the reference output; w_r is a zero mean stochastic process, x_{ro} is zero mean random vector. Further it is assumed that n=ν. The case n≠ν is beyond the scope of this chapter.
The integral action is introduced into the control in order to "force" zero tracking errors for polynomial inputs, and to attenuate disturbances (Kwakernaak & Sivan, 1972)(Anderson & Moore, 1989). This is done by introducing the auxiliary variables, integrals of the tracking error. This way the generalized PID controller, denoted PI^mD^{n-1}, is created. That is, the state is augmented by

$$\begin{aligned} e_x &= x_r - x \\ \dot{\eta}_1 &= C_{e1}\left[x_r - x\right] = C_{e1}e_x \\ \dot{\eta}_2 &= C_{e2}\eta_1; \\ &\vdots \\ \dot{\eta}_m &= C_{em}\eta_{m-1} \end{aligned} \tag{3}$$

$$\eta = \begin{bmatrix} \eta_1 \\ \eta_2 \\ \vdots \\ \eta_m \end{bmatrix}; \quad \eta(t_o) = \eta_o, \tag{4}$$

where (m) is the number of integrators that are introduced on the tracking error.
The control objective is

$$J = \frac{1}{2}E\left\{ \begin{aligned} &\left[y_r(t_f) - y(t_f)\right]^T G_1 \left[y_r(t_f) - y(t_f)\right] + \eta(t_f)^T G_2 \eta(t_f) \\ &+ \int_{t_o}^{t_f} \left[\left[y_r(t) - y(t)\right]^T Q_1 \left[y_r(t) - y(t)\right] + \eta(t)^T Q_2 \eta(t) + u(t)^T R u(t) \right] \end{aligned} \right\} \tag{5}$$

The optimal tracking problem (Kwakernaak & Sivan, 1972) is to find an admissible input ***u(t)*** such that the tracking objective (5) is minimized subject to the dynamic constraints (1-4).
All vectors and matrices are of the proper dimension.

3. Solution of the optimal tracking problem

In order to solve the Optimal Tracking Problem we augment the state variables (Kwakernaak & Sivan, 1972) and further assume that $A=A_r$, $B=B_r$ and $C=C_r$. This assumption states that the nominal values of the plant's parameters are known. The case $A\neq A_r$, $B\neq B_r$ and $C\neq C_r$ is beyond the scope of this chapter.
We have the error system

$$\dot{e}_x = Ae_x + Bw_r - Bu;\ e_x(t_o) = x_{ro} - x_o, \tag{6}$$

$$X = \begin{bmatrix} e_x \\ \eta_1 \\ \eta_2 \\ \\ \eta_m \end{bmatrix};\ \bar{A} = \left[\begin{array}{c|cccc} A & 0 & 0 & \cdots & 0 \\ \hline C_{e1} & 0 & 0 & & 0 \\ 0 & C_{e2} & 0 & & 0 \\ \vdots & & \ddots & & \vdots \\ 0 & 0 & \cdots & C_{em} & 0 \end{array}\right];\ \bar{B} = \begin{bmatrix} -B \\ 0 \\ 0 \\ \\ 0 \end{bmatrix};\ \bar{C} = \begin{bmatrix} C & 0 & 0 & & 0 \end{bmatrix} \tag{7}$$

then the problem is minimization of (5) subject to (1-4) is the problem of minimization of

$$J = \tfrac{1}{2}E\{X(t_f)^T GX(t_f) + \int_{t_o}^{t_1} [X(t)^T QX(t) + u(t)^T Ru(t)]dt\} \tag{8}$$

subject to

$$\dot{X} = \bar{A}X + \bar{B}u - \bar{B}\bar{w}_r,\quad \bar{x}(t_o) = \bar{x}_o, \tag{9}$$

where

$$Q = \bar{C}^T Q_1 \bar{C} + \begin{bmatrix} \underline{0} \\ \underline{1} \end{bmatrix} Q_2 [\underline{0}\ \ \underline{1}],\ \ G = \bar{C}^T G_1 \bar{C} + \begin{bmatrix} \underline{0} \\ \underline{1} \end{bmatrix} G_2 [\underline{0}\ \ \underline{1}]. \tag{10}$$

The solution is (Kwakernaak & Sivan, 1972) (Bryson & Ho, 1969)

$$\begin{aligned} &u = -R^{-1}\bar{B}PX \\ &-\dot{P} = P\bar{A} + \bar{A}^T P + Q - P\bar{B}R^{-1}\bar{B}P,\quad P(t_f) = G. \end{aligned} \tag{11}$$

If we appropriately partition ***P***, then

$$u = R^{-1}B^T \begin{bmatrix} P_{11} & P_{12} \end{bmatrix} \begin{bmatrix} e_x \\ \eta \end{bmatrix} = \begin{bmatrix} K_1 & K_2 \end{bmatrix} \begin{bmatrix} e_x \\ \eta \end{bmatrix} \tag{12}$$

Notice that the solution is independent of the reference trajectory generator input, $\bar{w}_r$.

4. Architectures

As stated in the introduction Architecture deals with the connections between the outputs/sensors and the inputs/actuators; Structure deals with the specific realization of the controllers' blocks; and Configuration is a specific combination of architecture and structure. These issues fall within the of control and feedback organization theory (Rusnak, 2006, 2008), and are beyond the scope of this chapter.

In this chapter we deal with three specific architectures. These are:

1. Parallel controller architecture;
2. Cascade controller architecture;
3. One block controller architecture.

4.1 Parallel controller architecture

This control architecture is directly derived from the Solution of the Optimal Tracking Problem as derived in (Asseo, 1970) and in (12). The parallel controller can be written directly from (12) in Laplace domain as

$$\begin{aligned} u(s) &= \sum_{i=1}^{n} C_i(s)\left(x_{ri}(s) - x_i(s)\right) = \sum_{i=1}^{n} C_i(s)e_i(s) \\ &= C_1(s)\left(x_{r1}(s) - x_1(s)\right) + \dots\dots + C_n(s)\left(x_{rn}(s) - x_n(s)\right) \end{aligned} \tag{13}$$

For 2nd order system the parallel controller architecture takes the form.

$$u(s) = C_1(s)\left(x_{r1}(s) - x_1(s)\right) + C_2(s)\left(x_{2r}(s) - x_2(s)\right) \tag{14}$$

Figure 1 presents the block diagram of the parallel controller architecture for a 2nd order system.

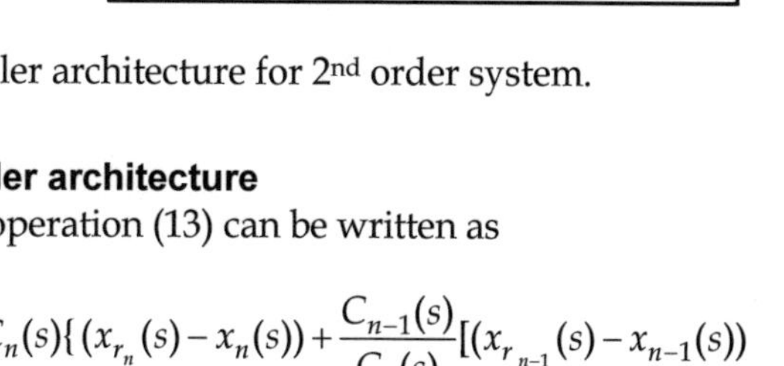

Fig. 1. Parallel controller architecture for 2nd order system.

4.2 Cascade controller architecture

By elementary block operation (13) can be written as

$$\begin{aligned} u(s) = C_n(s)\{ (x_{r_n}(s) - x_n(s)) &+ \frac{C_{n-1}(s)}{C_n(s)}[(x_{r_{n-1}}(s) - x_{n-1}(s)) \\ + \frac{C_{n-2}(s)}{C_{n-1}(s)}[(x_{r_{n-2}}(s) - x_{n-2}(s)) &+ \frac{C_{n-3}(s)}{C_{n-2}(s)}[(x_{r_{n-3}}(s) - x_{n-3}(s)) + \dots \\ &+ \frac{C_1(s)}{C_2(s)}(x_{r_1}(s) - x_1(s))]\dots]\} \end{aligned} \tag{15}$$

This is the cascade controller architecture. For 2nd order system the cascade controller architecture takes the form.

$$u(s) = C_2\left\{(x_{r2} - x_2) + \frac{C_1}{C_2}(x_{r1} - x_1)\right\} = C_v\{(x_{r2} - x_2) + C_p(x_{r1} - x_1)\} \tag{16}$$

Figure 2 presents the block diagram of the cascade controller architecture for a 2nd order system. The rationale for the notation of C_p (position) and C_v (velocity) will be presented in the sequel.

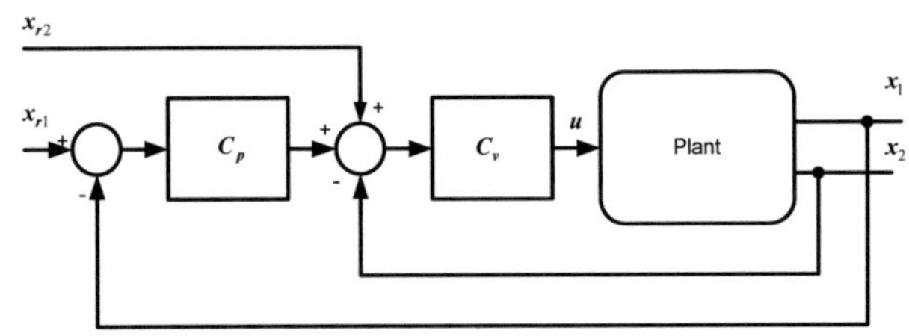

Fig. 2. Cascade controller architecture for 2nd order system.

4.3 One block controller architecture

By elementary operation on (13), and exploiting the relations between the state space variables, the one block controller architecture can be written as

$$u(s) = C(s)(x_{1r}(s) - x_1(s)) \tag{17}$$

Figure 3 presents the block diagram of the one block controller architecture.

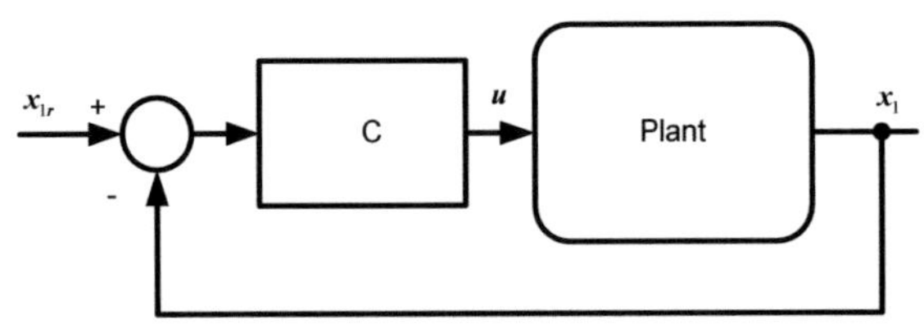

Fig. 3. One block controller architecture.

4.4 Discussion

Although from input-output transfer function point-of-view, there is no formal difference between the different architectures, there is difference with respect to the response to initial conditions, effects of saturation and nonlinearities, robustness, and more.

5. Controllers for first order system

As a first order system is considered, this leads to the one block controller architecture only.

5.1 P controller

Here we have

$$u = k_1 e_x = k_1 (x_r - x) \tag{18}$$

This is the proportional - P controller.

$$C(s) = k_1 \tag{19}$$

5.2 PI controller

Here we have

$$\begin{aligned} &\dot{\eta}_1 = e_x \\ &u = R^{-1} B^T \begin{bmatrix} P_{11} & P_{12} \end{bmatrix} \begin{bmatrix} e_x \\ \eta \end{bmatrix} = \begin{bmatrix} k_1 & k_2 \end{bmatrix} \begin{bmatrix} e_x \\ \eta_1 \end{bmatrix} = k_1 e_x + k_2 \int e_x dt \end{aligned} \tag{20}$$

This is the proportional + Integral - PI controller.

$$C(s) = k_1 + \frac{k_2}{s} = \frac{k_1 s + k_2}{s} \tag{21}$$

5.3 PI^2 controller

Here we have

$$\begin{aligned} &\dot{\eta}_1 = e_x \\ &\dot{\eta}_2 = \eta_1, \; or \; \ddot{\eta}_2 = e_x \\ &u = R^{-1} B^T \begin{bmatrix} P_{11} & P_{12} \end{bmatrix} \begin{bmatrix} e_x \\ \eta \end{bmatrix} = \begin{bmatrix} k_1 & k_{21} & k_{22} \end{bmatrix} \begin{bmatrix} e_x \\ \eta_1 \\ \eta_2 \end{bmatrix} = k_1 e_x + k_{21} \int e_x dt + k_{22} \iint e_x dt \end{aligned} \tag{22}$$

This is the proportional + double integrator - PI^2 controller.

$$C(s) = k_1 + \frac{k_{21}}{s} + \frac{k_{22}}{s^2} = \frac{k_1 s^2 + k_{21} s + k_{22}}{s^2} \tag{23}$$

5.4 PI^m controller

Here we have

$$\begin{aligned} &\dot{\eta}_1 = e_x \\ &\dot{\eta}_2 = \eta_1, \; or \; \ddot{\eta}_2 = e_x, \\ &\vdots \\ &\dot{\eta}_m = \eta_{m-1}, \; or \; \eta_m^{(m)} = e_x \end{aligned} \tag{24}$$

$$u = R^{-1}B^T \begin{bmatrix} P_{11} & P_{12} \end{bmatrix} \begin{bmatrix} e_x \\ \eta \end{bmatrix} = \begin{bmatrix} k_1 & k_{21} & k_{22} & \cdots & k_{2m} \end{bmatrix} \begin{bmatrix} e_x \\ \eta_1 \\ \eta_2 \\ \vdots \\ \eta_m \end{bmatrix} \tag{25}$$

$$= k_1 e_x + k_{21} \int e_x d\tau + \ldots + k_{2m} \int \cdots \int e_x d\tau_1 \cdots d\tau_m$$

This is the proportional + (m) integrators - PI^m controller.

$$C(s) = k_1 + \frac{k_{21}}{s} + \ldots + \frac{k_{2m}}{s^m} = \frac{k_1 s^m + k_{21} s^{m-1} + \ldots + k_{2m}}{s^m} \tag{26}$$

Table 1 summarizes the one block generalized PID controller structure for first order system.

controller	
P	k_1
PI	$\frac{k_1 s + k_2}{s}$
PI^2	$\frac{k_1 s^2 + k_{21} s + k_{22}}{s^2}$
PI^m	$\frac{k_1 s^m + k_{21} s^{m-1} + \ldots + k_{2m}}{s^m}$

Table 1. One block generalized PID controllers for 1st order system.

6. Controllers for second order system

Second order plant and the trajectory generator are assumed and are represented in the companion form

$$A = A_r = \begin{bmatrix} 0 & 1 \\ -a_2 & -a_1 \end{bmatrix}, B = \begin{bmatrix} b_1 \\ b_2 \end{bmatrix}, \; C = C_r = \begin{bmatrix} 1 & 0 \end{bmatrix}, \tag{27}$$

and we have

$$H(s) = H_r(s) = \frac{x_1}{u} = \frac{y}{u} = \frac{b_1 s + (b_2 + a_1 b_1)}{s^2 + a_1 s + a_2}, \tag{28}$$

$$\frac{x_2}{u} = \frac{b_2 s - a_2 b_1}{s^2 + a_1 s + a_2} \tag{29}$$

$$\frac{x_2}{x_1} = \frac{b_2 s - a_2 b}{b_1 s + (b_2 + a_1 b_1)} \tag{30}$$

The plant's and trajectory's state generator are denoted

$$\begin{bmatrix} x_1 \\ x_2 \end{bmatrix} = \begin{bmatrix} y \\ x_2 \end{bmatrix}; \begin{bmatrix} x_{r1} \\ x_{r2} \end{bmatrix} = \begin{bmatrix} y_r \\ x_{r2} \end{bmatrix} \tag{31}$$

The reason for selecting the state space representation (27) is that plant without zero, i.e. $b_1 = 0$, is a case that is often met in motion control with electrical and PZT motors (Rusnak, 2000a). For plant without zero $x_2 = \dot{y}$, so that

$$\begin{bmatrix} x_1 \\ x_2 \end{bmatrix} = \begin{bmatrix} y \\ \dot{y} \end{bmatrix}; \begin{bmatrix} x_{r1} \\ x_{r2} \end{bmatrix} = \begin{bmatrix} y_r \\ \dot{y}_r \end{bmatrix} \tag{32}$$

and one can deal with position feedback, feedback on y, and velocity feedback, feedback on $\dot{y}$. For this reason in this chapter we will call, with slight abuse of nomenclature, the feedback loop on y the position loop and the feedback loop on x_2,($\dot{y}$), the velocity loop.

6.1 PD controller

Feedback without integral action is implemented. The tracking errors are

$$\begin{aligned} e_{x1} &= \mathrm{y_r\text{-}y} = \mathrm{x_{1r}\text{-}x_1} = e \\ e_{x2} &= \mathrm{x_{2r}\text{-}x_2} \end{aligned} \tag{33}$$

The controller is

$$u = \begin{bmatrix} k_1 & k_2 \end{bmatrix} \begin{bmatrix} e_{x1} \\ e_{x2} \end{bmatrix} = k_1 e + k_2 \left(\mathrm{x_{2r}\text{-}x_2} \right) \tag{34}$$

6.1.1 Parallel controller

$$u = k_1 \left(\mathrm{y_r\text{-}y} \right) + k_2 \left(\mathrm{x_{2r}\text{-}x_2} \right) \tag{35}$$

6.1.2 Cascade controller

$$u = k_2 \left[\left(\mathrm{x_{2r}\text{-}x_2} \right) + \frac{k_1}{k_2} \left(\mathrm{y_r\text{-}y} \right) \right] \tag{36}$$

6.1.3 One block controller

To get the one block controller we substitute (30) and get (in Laplace domain)

$$\begin{aligned} u(s) &= k_1 e + k_2 \left(\mathrm{x_{2r}\text{-}x_2} \right) = k_1 e + \frac{k_2}{s} \frac{x_2}{x_1} \left(\mathrm{x_{1r}\text{-}x_1} \right) \\ \mathrm{C}(s) &= \frac{u(s)}{e(s)} = k_1 + k_2 \frac{b_2 s - a_2 b}{b_1 s + (b_2 + a_1 b_1)} = k_P + \frac{k_D s}{s\tau_D + 1} \end{aligned} \tag{37}$$

This is the PD controller.

6.1.4 Discussion

1. We used the assumption that $x_2(s)/x_1(s)=x_{2r}(s)/x_{1r}(s)$ and ignored the response to initial conditions.
2. For 2nd order plant with a stable zero the optimal controller is a proper PD controller, i.e. no direct derivative is required.
3. The pole/filter of the derivative in (37) cancels out the zero of the plant (28). This is optimal/correct for deterministic (noiseless) systems. For systems with noisy measurements this cancelation is no more optimal (Rusnak, 2000b).
4. The cancellation of the plant's zero by the optimal controller creates an uncontrollable system. This may work (although is not good practice) for stable zero. However, when the plant has non-minimum phase (unstable) zero the optimal PD controller induces uncontrollable unstable mode, which means that the Optimal PD controller cannot/should not be implemented in the one block controller architecture.
5. As for a plant with unstable zero the optimal one block PID controller cannot be realized, then measurement of the two states, or an observer is required if one wishes to build the optimal controller.
6. If stable controller is required it is possible to implement the optimal PD one block architecture controller only for minimum phase plants!
7. For 2nd order system without zero the deterministic optimal controller is not proper, i.e. requires pure derivative.

6.2 PID controller

Zero steady state tracking error on the output is required. The tracking errors are

$$\begin{aligned} e_{x1} &= \mathrm{y_r}\text{-}\mathrm{y_r} = \mathrm{x_{1r}}\text{-}\mathrm{x_1} = e \\ e_{x2} &= \mathrm{x_{2r}}\text{-}\mathrm{x_2} \\ \dot{\eta}_1 &= e_{x1} \end{aligned} \tag{38}$$

The controller is

$$\mathrm{u}=\begin{bmatrix} k_1 & k_2 & k_3 \end{bmatrix}\begin{bmatrix} e_{x1} \\ e_{x2} \\ \eta_1 \end{bmatrix} = k_1 e + k_2\left(\mathrm{x_{2r}}\text{-}\mathrm{x_2}\right) + k_3\int e dt \tag{39}$$

and the controller in Laplace domain

$$u(s) = k_1 e + k_2\left(\mathrm{x_{2r}}\text{-}\mathrm{x_2}\right) + \frac{k_3}{s}e \tag{40}$$

6.2.1 Parallel controller

$$u = \left(k_1 + \frac{k_3}{s}\right)\left(\mathrm{y_r}\text{-}\mathrm{y}\right) + k_2\left(\mathrm{x_{2r}}\text{-}\mathrm{x_2}\right) \tag{41}$$

6.2.2 Cascade controller

$$u(s)=k_2\left[\left(x_{2r}-x_2\right)+\frac{k_1s+k_3}{k_2s}\left(y_r-y\right)\right] \tag{42}$$

6.2.3 One block controller

To get the one block output controller derive we substitute (30) and get (in Laplace domain)

$$\begin{aligned} u(s)&=k_1e+k_2\left(x_{2r}-x_2\right)+\frac{k_3}{s}e=k_1e+\frac{k_2}{s}\frac{x_2}{x_1}\left(x_{1r}-x_1\right)+\frac{k_3}{s}e \\ C(s)&=\frac{u(s)}{e(s)}=k_1+k_2\frac{b_2s-a_2b}{b_1s+(b_2+a_1b_1)}+\frac{k_3}{s}=k_P+\frac{k_I}{s}+\frac{k_Ds}{s\tau_D+1} \end{aligned} \tag{43}$$

This is the PID controller.

6.2.4 Discussion

1. Remarks in section 6.1.4 apply here mutatis mutandis.
2. For 2nd order plant with a stable zero, the optimal controller with one integrator is a stable proper PID controller, i.e. no direct derivative is required.

6.3 PID controller in PIV configuration

Zero steady state tracking error on the output and the second state (velocity) is required. The tracking errors are

$$\dot{\eta}_1=\begin{bmatrix} x_{1r}-x_1 \\ x_{2r}-x_2 \end{bmatrix}=\begin{bmatrix} e_{x1} \\ e_{x2} \end{bmatrix}, \qquad \eta_1=\begin{bmatrix} \int e_{x1}dt \\ \int e_{x2}dt \end{bmatrix}, \tag{44}$$

The controller is

$$u=\begin{bmatrix} k_1 & k_2 & k_3 & k_4 \end{bmatrix}\begin{bmatrix} e_{x1} \\ e_{x2} \\ \int e_{x1}dt \\ \int e_{x2}dt \end{bmatrix}=k_1e_{x1}+k_2e_{x2}+k_3\int e_{x1}dt+k_4\int e_{x2}dt \tag{45}$$

and in Laplace domain

$$u(s)=k_1e_{x1}+k_2e_{x2}+\frac{k_3}{s}e_{x1}+\frac{k_4}{s}e_{x2} \tag{46}$$

6.3.1 Parallel controller

$$u(s)=\left(k_1+\frac{k_3}{s}\right)\left(y_r-y\right)+\left(k_2+\frac{k_4}{s}\right)\left(x_{2r}-x_2\right) \tag{47}$$

6.3.2 Cascade controller - the PIV configuration

$$u(s)=\left(k_2+\frac{k_4}{s}\right)\left[\left(x_{2r}\text{-}x_2\right)+\frac{k_1 s+k_3}{k_2 s+k_4}\left(y_r\text{-}y\right)\right] \tag{48}$$

This is called the PIV configuration (Proportional feedback in position loop and proportional+integral feedback in the velocity loop) (configuration=combination of architecture and structure) as there is almost proportional feedback (Lead-Lag) on the position x_1 and then in the velocity loop on x_2 there is proportional and one integral feedback.

6.3.3 One block output controller

To get the one block controller we substitute (30) and get (in Laplace domain)

$$\begin{aligned} u(s)&=\left(k_1+\frac{k_3}{s}\right)e_{x1}+\left(k_2+\frac{k_4}{s}\right)e_{x2}\\ C(s)&=\frac{u(s)}{e(s)}=\left(k_1+\frac{k_3}{s}\right)+\left(k_2+\frac{k_4}{s}\right)\frac{x_2}{x_1}=k_P+\frac{k_I}{s}+\frac{k_D}{s\tau_D+1}\end{aligned} \tag{49}$$

This is the PID controller.

6.3.4 Discussion

1. Remarks in section 6.1.4 apply here mutatis mutandis.
2. Two different tracking problems (38, 44) lead to the same one block controller.
3. In the parallel architecture there is a PI controller in each of the errors (47).
4. Although formally the cascade architecture controller requires the tuning of six parameters in (48), the deterministic optimal PIV controller needs the tuning of four parameters only, as can be deduced from (46).

6.4 PI²D controller

Zero steady state tracking error on the output for ramp input or disturbance is required. The tracking errors are

$$\begin{aligned} \dot{\eta}_1 &= e_{x1}=y_r-y=x_{1r}-x_1=e\\ e_{x2} &= x_{2r}-x_2\\ \dot{\eta}_2 &= \eta_1 \end{aligned} \tag{50}$$

The controller is

$$\begin{aligned} u&=\begin{bmatrix}k_1 & k_2 & k_3 & k_4\end{bmatrix}\begin{bmatrix}e_{x1}\\ e_{x2}\\ \eta_1\\ \eta_2\end{bmatrix}=\begin{bmatrix}k_1 & k_2 & k_3 & k_4\end{bmatrix}\begin{bmatrix}e_{x1}\\ e_{x2}\\ \int e_{x1}dt\\ \iint e_{x1}dt\end{bmatrix}\\ &=k_1e_{x1}+k_2e_{x2}+k_3\int e_{x1}dt+k_4\iint e_{x1}dt \end{aligned} \tag{51}$$

and in Laplace domain

$$u(s) = k_1 e + k_2 (x_{2r} - x_2) + \frac{k_3}{s} e + \frac{k_4}{s^2} e \tag{52}$$

6.4.1 Parallel controller

$$u = \left(k_1 + \frac{k_3}{s} + \frac{k_4}{s^2} \right)(y_r - y) + k_2 (x_{2r} - x_2) \tag{53}$$

6.4.2 Cascade controller

$$u(s) = k_2 \left[(x_{2r} - x_2) + \frac{\left(k_1 s^2 + k_3 s + k_4\right)}{k_2 s^2} (y_r - y) \right] \tag{54}$$

Two integrators in the position loop and proportional feedback in the velocity loop.

6.4.3 One block output controller

To get the one block controller we substitute (30) and get (in Laplace domain)

$$\begin{aligned} u(s) &= \left(k_1 + \frac{k_3}{s} + \frac{k_4}{s^2} \right)(x_{1r} - x_1) + k_2 (x_{2r} - x_2) = \left(k_1 + \frac{k_3}{s} + \frac{k_4}{s^2} \right)(x_{1r} - x_1) + k_2 \frac{x_2}{x_1}(x_{1r} - x_1) \\ C(s) &= \frac{u(s)}{e(s)} = k_1 + k_2 \frac{b_2 s - a_2 b}{b_1 s + (b_2 + a_1 b_1)} + \frac{k_3}{s} + \frac{k_4}{s^2} = k_P + \frac{k_I}{s} + \frac{k_{I2}}{s^2} + \frac{k_D s}{s\tau_D + 1} \end{aligned} \tag{55}$$

This is the PI^2D controller.

6.4.4 Discussion

1. Remarks in section 6.1.4 apply here mutatis mutandis.

6.5 PI^2D controller in IPIV configuration

Here we want to force zero steady state tracking error on the second state, as well, however in different configuration, i.e.

$$\dot{\eta}_1 = \begin{bmatrix} x_{1r} - x_1 \\ x_{2r} - x_2 \end{bmatrix} = \begin{bmatrix} e_{x1} \\ e_{x2} \end{bmatrix}, \quad \eta_1 = \begin{bmatrix} \int e \\ \int (x_{2r} - x_2) \end{bmatrix}, \quad \dot{\eta}_2 = [1 \quad 0]\eta_1, \ or \ \ddot{\eta}_2 = \dot{\eta}_1 = e_{x1} = e \tag{56}$$

The controller is

$$\begin{aligned} u &= [k_1 \quad k_2 \quad k_3 \quad k_4] \begin{bmatrix} e_{x1} \\ e_{x2} \\ \eta_1 \\ \eta_2 \end{bmatrix} = [k_1 \quad k_2 \quad k_3 \quad k_4] \begin{bmatrix} e_{x1} \\ e_{x2} \\ \int e_{x1} dt \\ \iint e_{x1} dt \end{bmatrix} \\ &= k_1 e_{x1} + k_2 e_{x2} + k_3 \int e_{x1} dt + k_4 \iint e_{x1} dt \end{aligned} \tag{57}$$

and in Laplace domain

$$u(s) = k_1 e + k_2 \left(x_{2r} - x_2 \right) + \frac{k_3}{s} e + \frac{k_4}{s} \left(x_{2r} - x_2 \right) + \frac{k_5}{s^2} e \tag{58}$$

6.5.1 Parallel implementation

$$u = \left(k_1 + \frac{k_3}{s} + \frac{k_5}{s^2} \right) \left(y_r - y \right) + \left(k_2 + \frac{k_4}{s} \right) \left(x_{2r} - x_2 \right) \tag{59}$$

6.5.2 Cascade controller – the IPIV configuration

$$u(s) = \frac{k_2 s + k_4}{s} \left[\left(x_{2r} - x_2 \right) + \frac{k_1 s^2 + k_3 s + k_5}{s\left(k_2 s + k_4 \right)} \left(y_r - y \right) \right] \tag{60}$$

This is called the IPIV configuration (Proportional +integral feedback in position loop and proportional +integral feedback in the velocity loop) as there is almost proportional feedback on the position loop, y, and then in the velocity loop on, x_2, there is proportional and one integral feedback.

6.5.3 One block output controller

To get the one block output controller we substitute (30) and get (in Laplace domain)

$$\begin{aligned} u(s) &= k_1 e + k_2 \left(x_{2r} - x_2 \right) + \frac{k_3}{s} e + \frac{k_5}{s^2} e + \frac{k_4}{s} \left(x_{2r} - x_2 \right) \\ C(s) &= \frac{u(s)}{e(s)} = k_1 + k_2 \frac{x_2}{x_1} + \frac{k_3}{s} + \frac{k_5}{s^2} + \frac{k_4}{s} \frac{x_2}{x_1} = k_P + \frac{k_{I1}}{s} + \frac{k_{I2}}{s^2} + \frac{k_D s}{s\tau_D + 1} \end{aligned} \tag{61}$$

This is the PI^2D controller.

6.5.4 Discussion

1. Remarks in section 6.1.4 apply here mutatis mutandis.

6.6 PI^2D controller in PI^2V configuration

Here we want to force zero steady state tracking error on the rate of the output as well, and

$$\dot{\eta}_1 = \begin{bmatrix} x_1 - x_{1r} \\ x_2 - x_{2r} \end{bmatrix} = \begin{bmatrix} e_{x1} \\ e_{x2} \end{bmatrix}, \qquad \eta_1 = \begin{bmatrix} \int e_{x1} dt \\ \int e_{x2} dt \end{bmatrix},$$

$$\dot{\eta}_2 = \eta_1 \tag{62}$$

The controller is

$$u = \begin{bmatrix} k_1 & k_2 & k_3 & k_4 & k_5 & k_6 \end{bmatrix} \begin{bmatrix} e_{x1} \\ e_{x2} \\ \int e_{x1} dt \\ \int e_{x2} dt \\ \iint e_{x1} dt \\ \iint e_{x2} dt \end{bmatrix} \tag{63}$$

and in Laplace domain

$$u(s) = k_1 e + k_2 (x_{2r}\text{-}x_2) + \frac{k_3}{s} e + \frac{k_4}{s}(x_{2r}\text{-}x_2) + \frac{k_5}{s^2} e + \frac{k_6}{s^2}(x_{2r}\text{-}x_2) \tag{64}$$

6.6.1 Parallel controller

$$u = \left(k_1 + \frac{k_3}{s} + \frac{k_5}{s^2} \right)(y_r\text{-}y) + \left(k_2 + \frac{k_4}{s} + \frac{k_6}{s^2} \right)(x_{2r}\text{-}x_2) \tag{65}$$

6.6.2 Cascade controller – the PI^{2}V configuration

$$u = \frac{k_2 s^2 + k_4 s + k_6}{s^2} \left[(x_{2r}\text{-}x_2) + \frac{k_1 s^2 + k_3 s + k_5}{k_2 s^2 + k_4 s + k_6} (y_r\text{-}y) \right] \tag{66}$$

This is called the PI^{2}V configuration (Proportional feedback in position loop and proportional +double integral feedback in the velocity loop) as there is almost proportional feedback (Lead-Lag) in the position loop, on y, and then in the velocity loop, on x_2, there is proportional and two integrals feedback.

6.6.3 One block output controller

$$\begin{aligned} u(s) &= \left(k_1 + \frac{k_3}{s} + \frac{k_5}{s^2} \right)(y_r\text{-}y) + \left(k_2 + \frac{k_4}{s} + \frac{k_6}{s^2} \right)(x_{2r}\text{-}x_2) \\ C(s) &= \frac{u(s)}{e(s)} = \left(k_1 + \frac{k_3}{s} + \frac{k_5}{s^2} \right) + \left(k_2 + \frac{k_4}{s} + \frac{k_6}{s^2} \right) \frac{b_2 s - a_2 b}{b_1 s + (b_2 + a_1 b_1)} \\ &= k_P + \frac{k_{I1}}{s} + \frac{k_{I2}}{s^2} + \frac{k_D s}{s\tau_D + 1} \end{aligned} \tag{67}$$

This is the PI^{2}D controller.

6.6.4 Discussion

1. Remarks in section 6.1.4 apply here mutatis mutandis.

6.7 Summary

This section presented the family of the generalized PID controllers for 2nd order systems. The following tables summarize the structure of the controllers in the different architectures.

Table 2 presents the family of generalized PID controllers for 2nd order systems in the parallel architecture that are able to drive the tracking error to zero for up to constant acceleration input and disturbance. Formally, if all possible parallel configurations are enumerated then there are three more parallel structures as detailed in Table 3. However these additional structures are equivalent to the respective structures in Table 2 as detailed in the rightmost column. Therefore these configurations are not considered in the following.

Generalized PID controller - Parallel architecture (Figure 1)			
	C_{x1}	C_{x2}	§
PD	k_1	k_2	6.1
PID	k_1+k_3/s	k_2	6.2
PID- PIV	k_1+k_3/s	k_2+k_4/s	6.3
PI²D	$k_1+k_3/s+k_4/s^2$	k_2	6.4
PI²D-IPIV	$k_1+k_3/s+k_5/s^2$	k_2+k_4/s	6.5
PI²D- PI²V	$k_1+k_3/s+k_5/s^2$	$k_2+k_4/s+k_6/s^2$	6.6

Table 2. The structure of the parallel architecture controllers for 2nd order plant.

Generalized PID controller - Parallel architecture (Figure 1)			
	C_{x1}	C_{x2}	§
PID	k_1	k_2+k_4/s	6.2
PI²D	k_1	$k_2+k_4/s+k_6/s^2$	6.4
PI²D-IPIV	k_1+k_3/s	$k_2+k_4/s+k_6/s^2$	6.5

Table 3. The structure of the parallel architecture controllers for 2nd order plant.

Tables 4 and 5 present the family of generalized PID controllers for 2nd order systems in the cascade architecture and in the one block controller architecture, respectively, that are able to drive the tracking error to zero for up to constant acceleration input and disturbance.

Generalized PID controller - Cascade architecture (Figure 2)			
	C_p (position-outer loop)	C_v (velocity-inner loop)	§
PD	k_1	k_1/k_2	6.1
PID	$(k_1s+k_3)/k_2/s$	k_2	6.2
PIV	$(k_1s+k_3)/(k_2s+k_4)$	$(k_2s+k_4)/s$	6.3
PI²D	$(k_1s^2+k_3s+k_4)/k_2/s^2$	k_2	6.4
IPIV	$(k_1s^2+k_3s+k_5)/s(k_2s+k_4)$	$(k_2s+k_4)/s$	6.5
PI²V	$(k_1s^2+k_3s+k_5)/(k_2s^2+k_4s+k_6)$	$(k_2s^2+k_2s+k_6)/s^2$	6.6

Table 4. The structure of the cascade architecture controllers for 2nd order plant.

One block PD, PID and generalized PID controller (Figure 3)				
Controller type		plant	integral action(m)	§
PD	$k_P+k_D s$	no zero	0	6.1
PD	$k_P+k_D s/(s\ \tau_D+1)$	zero	0	6.1
PID	$k_P + k_I/s+k_D s$	no zero	1	6.2,3
PID	$k_P+ k_I/s+k_D s/(s\ \tau_D+1)$	zero	1	6.2,3
PI²D	$k_P +k_{I1}/s+k_{I2}/s^2+k_D s$	no zero	2	6.4,5,6
PI²D	$k_P+k_I/s+k_{I1}/s+k_{I2}/s^2+k_D s/(s\ \tau_D+1)$	zero	2	6.4,5,6

Table 5. The structure of one block generalized PID controller for 2nd order plant with and without minimum phase zero.

7. Reference trajectory generator

The reference trajectory generator encapsulates the required closed loop behavior as stated by the system specification-requirements. There can be two cases: the trajectory is either unknown or known in advance. The former case gives the well known pre-filter that creates the feed-forward as well. In the second case, for example, minimum time trajectories for limited acceleration or jerk, minimum acceleration or jerk energy trajectories, or any other profile can be required. Both cases are presented in (Leonhard, 1996, pp. 80, 347, 363-364, 367) and in many other publications.

8. Discussion

In this chapter the generalized PID controllers for 1st and 2nd order system that are able to drive the tracking error to zero for up to second order polynomials inputs and disturbances have been derived. This presented in detail a methodology to derive additional members of the family of generalized PID controllers for high order system (Rusnak, 1999) and high order polynomial inputs and disturbances. These are the PI^mD^{n-1} controllers.

Following the theory and the author's experience the full state feedback, especially the cascade architecture, Figure 2, is preferable over the one block controller, Figure 3. This may come at the expense of higher cost. However in modern digital control loop that are using absolute or incremental encoders the position and velocity information is derived at the same cost.

The motion control engineers prefer the cascade controller because of implementation and tuning easiness. The most important feature is that in the cascade architecture the feedback loop can be tuned sequentially. That is, start with the velocity-inner loop, that is usually high bandwidth, and then to proceed to the position-outer loop. The same apply to higher order generalized PID controllers.

9. Conclusions

By the use of LQR theory we formulated a control-tracking problem and showed those cases when their solution gives members of the PI^mD^{m-1} family of controllers. This way heuristics are avoided and a systematic approach to explanation for the excellent performance of the PID controllers is given. The well known one block PID controller architecture is optimal for Linear Quadratic Tracking problem of 2nd order systems with no zero or stable zero.

10. References

Anderson, B.D.O. and Moore, J.B. (1989). *Optimal Control: Linear Quadratic Methods*, Prentice-Hall. ISBN 0-13-638651-3

Asseo, S.J. (1970). Application of Optimal Control to Perfect Model Following, *Journal of Aircraft*, Vol. 7, No. 4, July-August, pp.308-313.

Bryson, A.E. and Ho, Y.C. (1969). *Applied Optimal Control*, Ginn and Company. ISBN 0891162283

D'Azzo, J.J. and Houpis, C.H. (1988). Linear Control Systems Analysis And Design: Conventional and Modern, McGraw-Hill Book Company. ISBN 978-0824740382

Franklin, G.F., Powell, J.D. and Emani-Naeini, A. (1994). *Feedback Control of Dynamic Systems*, 3rd Edition, Addison-Wesley Publishing Company, Inc. ISBN 0-201-52747-2

Kwakernaak, H. and Sivan, R. (1972). *Linear Optimal Control*, John Wiley & Sons, Inc. ISBN 0-471-51110-2

Leonhard, W. (1996). *Control of Electric Drives*, 2nd edition, Springer-Verlag. ISBN 3-540-59380-2

Nanomotion. (2009a). A Johnson Electric Company, *FlexDC, User Manual*, July, Available from http://www.nanomotion.com

Nanomotion. (2009b). A Johnson Electric Company, *FlexDC, Software User Manual*, July, Available from http://www.nanomotion.com

Podlubny, I. (1999). Fractional order systems and $PI^{\lambda}D^{\mu}$ controllers, *IEEE Transactions on Automatic Control*, Vol.44, No.1, pp.208–214.]

Rusnak, I. (1998). The Optimality of PID Controllers, *The 9th Mediterranean Electrotechnical Conference*, MELECON '98, May 18-20, Tel-Aviv, Israel.

Rusnak, I. (1999). Generalized PID Controller, *The 7th IEEE Mediterranean Conference on Control & Automation*, MED 99, 28-29 June, Haifa, Israel.

Rusnak, I. (2000a). The Generalized PID Controller and its Application to Control of Ultrasonic and Electric Motors, *IFAC Workshop on Digital Control, Past, Present and Future of PID Control*, PID' 2000, Tarrasa, Spain, 5-7 April.

Rusnak, I. (2000b). Generalized PID Controller for Stochastic Systems, *The 21st Convention of IEEE in Israel*, April 11-12, Tel-Aviv, Israel. (Invited paper).

Rusnak, I. (2002). On Feedback Loops Organization, *The 22nd Convention of IEEE in Israel*, December, Tel-Aviv, Israel. (Abstract)

Rusnak, I. (2005). Organization of the Feedback for Gantry System, *The 30th Israel Conference on Mechanical Engineering*, Tel-Aviv, Israel, May 29-30. (Extended abstract)

Rusnak, I. (2006). On Control and Feedback Organization, *The 24nd IEEE Convention of IEEE in Israel*, November, Eilat, Israel.

Rusnak, I. (2008). Control Organization - Survey and Application, *The 9th Biennial ASME Conference on Engineering Systems Design and Analysis*, ESDA 2008, July 7-9, Haifa, Israel.

Rusnak, I. and Weiss, H. (2011). New Control Architecture for High Performance Autopilot, *The 51th Israel Annual Conference on Aerospace Science*, February 23-24, Israel.

Part 2

Tuning Criteria

3

PID-Like Controller Tuning for Second-Order Unstable Dead-Time Processes

G.D. Pasgianos[1], K.G. Arvanitis[1] and A.K. Boglou[2]
[1]Department of Agricultural Engineering, Agricultural Univeristy of Athens,
[2]Kavala Institute of Technology, School of Applied Technology, Kavala,
Greece

1. Introduction

Several processes encountered in various fields of engineering exhibit an inherently unstable behaviour coupled with time delays. To approximate the open loop dynamics of such systems for the purpose of designing controllers, many of these processes can be satisfactorily described by unstable transfer function models. The most widely used models of this type is the unstable first order plus dead-time (UFOPDT) and the unstable second order plus dead-time (USOPDT) transfer function models, which take into account dead times that might appear in the model, due to measurement delay or due to the approximation of higher order dynamics of the process, by a simple transfer function model.

Research on tuning methods of two or three-term controllers for unstable dead-time processes has been very active in the last 20 years. The most widely used feedback schemes for the control of such processes are the Proportional-Integral-Differential (PID) controller with set-point filter (Jung et al, 1999; Lee et al, 2000), the Pseudo-Derivative Feedback (PDF) or I-PD controller (Paraskevopoulos et al, 2004), and the Proportional plus Proportional–Integral–Derivative (P-PID) controller (Jacob & Chidambaram, 1996; Park et al, 1998). These control schemes are identical in practice, provided that the parameters of the controllers and of the pre-filters needed in some cases are selected appropriately. Controller tuning for unstable dead-time processes has been performed according to several methods, the most popular of them being various modifications of the Ziegler-Nichols method (De Paor & O' Malley, 1989; Venkatashankar & Chidambaram, 1994; Ho & Xu, 1998), several variations of the direct synthesis tuning method (Jung et al, 1999; Prashanti & Chidambaram, 2000; Paraskevopoulos et al, 2004; Padma Sree & Chidambaram, 2004), the ultimate cycle method (Poulin & Pomerleau, 1997), the pole placement method (Clement & Chidambaram, 1997), the method based on the minimization of various integral criteria, the Internal Model Control (IMC) tuning method (Rotstein & Lewin, 1991; Lee et al, 2000; Yang et al, 2002; Tan et al, 2003), the optimization method (Jhunjhunwala & Chidambaram, 2001; Visioli, 2001), the two degrees of freedom method (Huang & Chen, 1997; Liu et al, 2005; Shamsuzzoha et al, 2007), etc. (see the work (O'Dwyer, 2009), and the references cited therein). Moreover, due to the wide practical acceptance of the gain and phase margins (GPM) in characterizing system robustness, some tuning methods for unstable dead-time models, based on the satisfaction of GPM specifications, have also been reported (Ho & Xu, 1998; Fung et al, 1998;

Wang & Cai, 2002; Lee & Teng, 2002; Paraskevopoulos et al, 2006). The vast majority of the tuning methods mentioned above refer to the design of controllers for UFOPDT models and less attention has been devoted to USOPDT models (Lee et al, 2000; Rao & Chidambaram, 2006). Usually these models are further simplified to second order ones without delay, or to UFOPDT models, in order to design controllers for this type of processes. However, this simplification is not possible when the time delay of the system and/or the stable dynamics (stable time constant) are significant.

The aim of this work is to present a variety of innovative tuning rules for designing PID-like controllers for USOPDT processes. These tuning rules are obtained by imposing various specifications on the closed-loop system, such as the appropriate assignment of its dominant poles, the satisfaction of several time response criteria (like the fastest settling time and the minimization of the integral of squared error), as well as the simultaneous satisfaction of stability margins specifications. In particular, the development of the proposed tuning methods relying on the assignment of dominant poles as well as on time response criteria is performed on the basis of the fact that (under appropriate selection of the derivative term), the delayed open loop response of a 3rd order system, with poles equal to the three dominant poles of the closed loop system, is identical to the closed loop step response of an USOPDT system. Simple numerical algorithms are, then, used to obtain the solution of the tuning problem. To reduce the computational effort and to obtain the controller settings in terms of the process parameters (a fact that permits on-line tuning), the obtained solution is further approximated by analytical functions of these parameters. Moreover, in the case of the method that relies on the satisfaction of stability margin specifications, the controller parameters are obtained using iterative algorithms, whose solutions, in a particular case, are further approximated quite accurately by analytic functions of the process parameters. The obtained approximate solutions have been obtained using appropriate curve-fitting optimization techniques. Furthermore, the admissible values of the stability robustness specifications for a particular process are also given in analytic forms. Finally, the tuning rules proposed in this work, are applied to the control of an experimental magnetic levitation system that exhibits highly nonlinear unstable behaviour. The experimental results obtained clearly illustrate the practical efficiency of the proposed tuning methods.

2. PID-like controller structures for USOPDT processes

The three main feedback configurations applied in the extant literature in order to control unstable processes with time delay are depicted in Fig. 1 (see Jacob & Chidambaram, 1996; Park et al, 1998, Paraskevopoulos et al, 2004). As it can easily verified, the loop transfer functions obtained by these control schemes are identical, provided that the following relations hold

$$\bar{K}'_C = \bar{K}_C(\bar{\tau}_I + \bar{\tau}_D) / \bar{\tau}_I = (1+\bar{k}_c)\bar{k}_{c,i} = \bar{K}_P$$

$$\bar{\tau}'_I = \bar{\tau}_I + \bar{\tau}_D = \bar{\tau}_i(1+\bar{k}_c) / \bar{k}_c = \bar{K}_P / \bar{K}_I \tag{1}$$

$$\bar{\tau}'_D = \bar{\tau}_D\bar{\tau}_I / (\bar{\tau}_I + \bar{\tau}_D) = \bar{\tau}_d\bar{k}_c / (1+\bar{k}_c) = \bar{K}_D / \bar{K}_P$$

$$G_{F,P-PID}(s) = G_{F,PID}\frac{(\overline{\tau}_I s+1)(\overline{\tau}_D s+1)}{\overline{\tau}_i\overline{\tau}_d s^2+\overline{\tau}_i s+1}$$
$$G_{F,PDF}(s) = G_{F,PID}(\overline{\tau}_I s+1)(\overline{\tau}_D s+1) \tag{2}$$

where $\overline{K}'_C$, $\overline{\tau}'_I$ and $\overline{\tau}'_D$ are the three controller parameters of the conventional PID controller in its parallel form. In the case of the series PID controller, the pre-filter $G_{F,PID}$ is used in order to cancel out all or some of the zeros introduced by the controller and to smoothen the set-point step response of the closed loop system. The pre-filters $G_{F,P-PID}$ and $G_{F,PDF}$ are the equivalent pre-filters of the corresponding control schemes. Note that, the pre-filter $G_{F,PDF}$ can be used only when the reference input is a known and differentiable signal. Therefore, is seldom used in real practice. From Fig. 1, one can easily recognize that in the case of regulatory control the three control schemes are identical when the controller parameters are chosen as suggested by (1), even if there are no pre-filters used. Moreover, one can also see that the stability properties of the closed loop system are not affected, in any case, by the respective pre-filter used, which is applied here, only to filter the set point and to prevent excessive overshoot in closed-loop responses to set-point changes, which are common in the case of unstable time-delay systems (Jacob & Chidambaram, 1996). Thus, the loop transfer functions obtained for the above three alternative control schemes are identical.

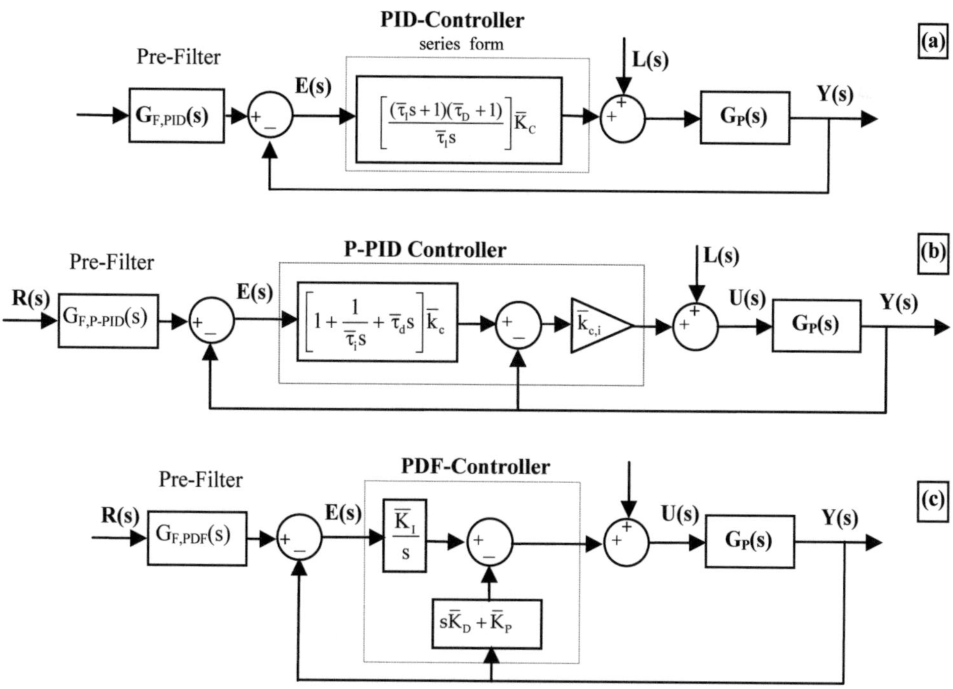

Fig. 1. Equivalent three-term controller schemes with appropriate pre-filters: (a) The series PID controller, (b) The PDF (or I-PD) controller, and (c) The P-PID controller.

Original Parameters	Normalized Parameters	Original Parameters	Normalized Parameters
$\overline{\tau}_U$	τ_U=1	ω	$w=\omega\overline{\tau}_U$
$\overline{\tau}_S$	$\tau_S=\overline{\tau}_S/\overline{\tau}_U$	s	$\hat{s}=s\overline{\tau}_U$
$\overline{d}$	$d=\overline{d}/\overline{\tau}_U$	$\overline{K}$	K=1
$\overline{\tau}_I$	$\tau_I=\overline{\tau}_I/\overline{\tau}_U$	$\overline{K}_C$	$K_C=\overline{K}\overline{K}_C$
$\overline{\tau}_D$	$\tau_D=\overline{\tau}_D/\overline{\tau}_U$		

Table 1. Normalized vs. original system parameters.

In the sequel, our focus of interest is the design of PID-like controllers when applied to control USOPDT process, with the following transfer function model

$$G_P(s)=\frac{\overline{K}\exp\left(-\overline{d}s\right)}{\left(\overline{\tau}_S s+1\right)\left(\overline{\tau}_U s-1\right)} \tag{3}$$

where $\overline{K}$, $\overline{d}$, $\overline{\tau}_S$ and $\overline{\tau}_U$ are the process gain, the time delay and the stable and unstable time constants, respectively. In order to simplify the analysis and in order to facilitate comparisons, all system and controller parameters are normalized with respect to $\overline{\tau}_U$ and $\overline{K}$. Thus, the original process and controller parameters are replaced with the dimensionless parameters shown in Table 1.

Observe now that, the loop transfer function of an USOPDT system in connection with a PID-like controller, is given by

$$G_L(\hat{s})=\frac{K_C\left(\tau_I\hat{s}+1\right)\left(\tau_D\hat{s}+1\right)\exp\left(-d\hat{s}\right)}{\tau_I\hat{s}\left(\tau_S\hat{s}+1\right)\left(\hat{s}-1\right)} \tag{4}$$

while, using the pre-filter $G_{F,\,PID}(\hat{s})=(\tau_I\hat{s}+1)^{-1}$, the closed-loop transfer function becomes

$$G_{CL}(\hat{s})=\frac{K_C(\tau_D\hat{s}+1)\exp(-d\hat{s})}{\tau_I\hat{s}(\tau_S\hat{s}+1)(\hat{s}-1)+K_C(\tau_I\hat{s}+1)(\tau_D\hat{s}+1)\exp(-d\hat{s})} \tag{5}$$

Relations (2) and (5) are next elaborated for the derivation of the tuning methods proposed in this work.

3. Frequency domain analysis of closed-loop USOPDT processes

The argument and the magnitude of the loop transfer function (4) are given by

$$\varphi_L(w)=-3\pi/2-dw-\mathrm{atan}(w)-\mathrm{atan}(\tau_S w)+\mathrm{atan}(\tau_I w)+\mathrm{atan}(\tau_D w) \tag{6}$$

$$A_L(w)=\left|G_L(jw)\right|=K_C\frac{\sqrt{1+\left(\tau_I w\right)^2}\sqrt{1+\left(\tau_D w\right)^2}}{\tau_I w^2\sqrt{1+w^2}\sqrt{1+\left(\tau_S w\right)^2}} \tag{7}$$

It is not difficult to recognize that the Nyquist plot of the $G_L(\hat{s})$ has tow crossover points with the real axis, which determine the critical (or crossover) frequencies w_{min} and w_{max}, and the critical gains $K_{C,min}=1/A_L(w_{min})$ and $K_{C,max}=1/A_L(w_{max})$. These crossover frequencies are obtained as the solutions of the equation $\varphi_L(w_C)=-\pi$, or equivalently, of the equation

$$-\pi/2-dw_C+atan(w_C)+atan(\tau_I w_C)+atan(\tau_D w_C)-atan(\tau_S w_C)=0 \tag{8}$$

when the values of the *atan* function are assigned in the range $(-\pi/2, \pi/2)$. Having computed w_{min} and w_{max}, one can determine the acceptable values for the controller gain K_C, for which the closed-loop system is stable. In particular $K_{C,min}<K_C<K_{C,max}$, where, with subscript "M" used for either "min" or "max"

$$K_{C,M}=\frac{\tau_I w_M\sqrt{1+w_M^2}\sqrt{1+\left(\tau_S w_M\right)^2}}{\sqrt{1+\left(\tau_I w_M\right)^2}\sqrt{1+\left(\tau_D w_M\right)^2}} \tag{9}$$

We next define the increasing gain margin GM_{inc}, the decreasing gain margin GM_{dec} and the gain margin product of the closed-loop system as follows

$$GM_{inc}=K_{C,max}/K_C \ , \ GM_{dec}=K_C/K_{C,min} \tag{10}$$

$$GM_{prod}= GM_{inc}GM_{dec}=K_{C,max}/K_{C,min} \tag{11}$$

Obviously for the closed loop system to be stable GM_{inc} and GM_{dec} should be grater than one. Note that, the largest the values of GM_{prod}, the more robust the system becomes with respect to the gain uncertainty, if the controller gain K_C is appropriately selected. Furthermore, the phase margin of the closed loop system is defined by $PM=\varphi_L(w_G)+\pi$, where w_G is the frequency at which $A_L(w_G)=1$. From (7), one can easily conclude that w_G is given by the maximum real root of the equation

$$\tau_I^2\tau_S^2\omega_G^6+\left(\tau_I^2\tau_S^2+\tau_I^2-K_C^2\tau_I^2\tau_D^2\right)\omega_G^4+\left[\tau_I^2-K_C^2\left(\tau_I^2+\tau_D^2\right)\right]\omega_G^2-K_C^2=0 \tag{12}$$

In order to obtain the maximum phase margin for given d, τ_S, τ_I and τ_D, the controller gain K_C should be selected as

$$K_C=\frac{\tau_I w_p\sqrt{1+w_p^2}\sqrt{1+\left(\tau_S w_p\right)^2}}{\sqrt{1+\left(\tau_I w_p\right)^2}\sqrt{1+\left(\tau_D w_p\right)^2}} \tag{13}$$

where w_p is the frequency at which the argument of the loop transfer function is maximized. From (6), one can easily conclude that w_p is given by the solution of $d\varphi_L/d\omega\,/_{w=wp}=0$, or equivalently of the equation

$$-d+\frac{1}{1+w_p^2}+\frac{\tau_I}{1+\tau_I^2 w_p^2}+\frac{\tau_D}{1+\tau_D^2 w_p^2}-\frac{\tau_S}{1+\tau_S^2 w_p^2}=0 \tag{14}$$

that results in a fourth order linear equation with respect to w_p^2, with only one acceptable positive real root. Substituting w_p in (6), the respective maximum argument $\varphi_L(w_p)$ is calculated.

When the maximum phase margin is zero, then the closed-loop system (with the appropriate selection of K_C) is marginally stable. The solution of $max(PM(d,\tau_I,\tau_D,\tau_S))=0$, yields the acceptable values of the controller parameters τ_I and τ_D, which render the close-loop system stable. Obviously these values depend on the rest of the system parameters. From (8) and for $\tau_I \rightarrow \infty$, one can easily verify that $w_{min}=0$ and $\varphi_L(0)=-\pi$. If, at $w_{min}=0$, the derivative of φ_L is positive, then, it is obvious that the system has a maximum phase margin grater than zero and can be stabilized with the appropriate K_C. With this observation, using (14), one can easily verify that, for $\tau_D > \tau_{D,min} \equiv 1-d-\tau_S$, the closed-loop system can be stabilized. Note here that, when $\tau_S \leq 1$, $\tau_{D,min}$ is also the smallest τ_D that renders the closed-loop system stable, while when $\tau_S > 1$, the system can be stabilized with smaller values of τ_D. Moreover, although the function $\varphi_L(\tau_D)$ is strictly increasing, the function $GM_{prod}(\tau_D)$ is not strictly monotonous. In fact, there exists a very large value of τ_D for which $GM_{prod}(\tau_D)=1$ and the system is no longer stabilizable. In the case where $\tau_I \rightarrow \infty$, then $K_{C,min}=1$. Solving the equation $K_{C,min}(\tau_D)=1$, one can determine the maximum value of τ_D, say $\tau_{D,max}$, for which the system can be stabilized. Unfortunately, the solution of $K_{C,min}(\tau_D)=1$ involves nonlinear equations that can only be solved using iterative algorithms. A simple and quite accurate approximate solution for $\tau_{D,max}$ has been obtained through fitting, using the optimization toolbox of MATLAB® and is given by

$$\hat{\tau}_{D,\max} \approx 0.85 + \tau_S \left(-0.46 + 1.5 / d\right) \tag{15}$$

The maximum normalized error of this approximation is 6%, when $0.1 < \tau_D < 10$ and $0.01 < d < 0.9$. In general, it is plausible to obtain a stable closed-loop system by selecting $\tau_{D,min} < \tau_D < \tau_{D,max}$. In real practice, when τ_D is close to $\tau_{D,min}$ or $\tau_{D,max}$, the stability region of the closed-loop system is very small. After extensive search, it has been found that a more suitable range for the selection of τ_D is the following

$$\tau_S \leq \tau_D \leq \tau_S + d/2 \tag{16}$$

When τ_D is selected in the range defined by (16), very large PM_{max} and GM_{prod} can be obtained. Moreover, with this selection the functions $max(PM(\tau_I))$ and $GM_{prod}(\tau_I)$ are strictly increasing with respect to τ_I. This is a very useful property for the design of PID-like controllers for USOPDT processes. It is worth noticing, at this point, that in order to tune PID-like controllers for USOPDT processes one can distinguish three cases depending on the values of d and τ_S. In the case where $\tau_S < 0.1$ the PID-type controllers can be tuned using tuning rules for UFOPDT systems, assuming that the new normalized dead time is equal to $d+\tau_S$. On the other hand, if $\tau_S > 10$, then it is possible to tune the PID-type controller assuming that the system is a second order one with no time delay. In this particular case, the inverse of the eigen-frequency of the closed loop system (without delay) must be at least five times larger than the time delay of the USOPDT system. Finally, in the case where $0.1 < \tau_S < 10$, the above approximate solutions do not provide accurate results, and it is recommended to use the more accurate tuning rules presented in the following Sections.

4. Controller tuning by assigning the closed-loop system dominant poles

A first method of tuning PID-like controllers for USOPDT processes is based on the appropriate placement of the dominants poles of the closed-loop system. This method is designated here as the DPC method, since it relies on the satisfaction of dominant poles criteria. In order to systematically present the DPC method, we start by selecting the derivative time constant τ_D equal to the lowest value in the range defined by (16). That is, $\tau_D = \tau_S$. With this selection, relations (4) and (5) take the forms

$$G_L(\hat{s}) = \frac{K_C(\tau_I\hat{s}+1)\exp(-d\hat{s})}{\tau_I\hat{s}(\hat{s}-1)} \tag{17}$$

$$G_{CL}(\hat{s}) = \frac{K_C\exp(-d\hat{s})}{\tau_I\hat{s}(\hat{s}-1)+K_C(\tau_I\hat{s}+1)\exp(-d\hat{s})} \tag{18}$$

Clearly, in this case, the closed-loop transfer function has no zeroes. Note also that, if initially $\tau_S >> 1$, then, the controller parameter τ_D takes very large values, a fact that is not desirable, for reasons of noise amplification. Unfortunately, as suggested by (16), in this case, large values of τ_D are inevitable and an appropriate filtered derivative should be considered.

Let us now select the controller gain K_C as the geometric middle point of the two ultimate gains, $K_{C,min}$ and $K_{C,max}$, of the closed loop system, that is

$$K_C = \sqrt{K_{C,\min}K_{C,\max}} \tag{19}$$

Note that this selection of K_C provides the same robustness against both increasing and decreasing parametric uncertainty of the system gain. This is particularly useful for systems with large values of d (i.e. $d>0.3$) where the region of stability is reduced significantly (Paraskevopoulos et al, 2006).

On the basis of (17), the two ultimate gains are, in this case, given by

$$K_{C,\min} = \frac{\tau_I w_{\min}\sqrt{1+w_{\min}^2}}{\sqrt{1+(\tau_I w_{\min})^2}} \quad , \quad K_{C,\max} = \frac{\tau_I w_{\max}\sqrt{1+w_{\max}^2}}{\sqrt{1+(\tau_I w_{\max})^2}} \tag{20}$$

In (20), w_{min} and w_{max} are the two critical frequencies given by the two solutions of the equation (8), when $\tau_D=\tau_S$ and when the values of the atan function are assigned in the range $(-\pi/2,\pi/2)$. For given d, the solution of (8), for $\tau_D=\tau_S$, exists only if τ_I is larger than a critical value $\tau_{I,min}(d)$ (Paraskevopoulos et al, 2006). Since there are no analytical solutions for (8), two very accurate approximations for w_{min} and w_{max} that are obtained by using optimization techniques are proposed here. These approximations are

$$\begin{aligned}
\hat{w}_{\min}(d,\tau_I) &= f_{w_{\min}}(d,\tau_I)\sqrt{\frac{1}{\tau_I - d(1+\tau_I)}} \\
\hat{w}_{\max}(d,\tau_I) &= f_{w_{\max}}(d,\tau_I)\frac{\pi}{2d}\frac{(\tau_I - 0.9463(\tau_I+1)d)}{(\tau_I - 0.5609(\tau_I+1)d)}
\end{aligned} \tag{21}$$

$$f_{w_{\min}}(d,\tau_I) = 1 + \frac{(0.006+0.03\text{d}/(1.14\text{-d}))\hat{\tau}_{I,\min}}{(0.973+0.05/(1-d))\tau_I - \hat{\tau}_{I,\min}}$$

$$f_{w_{\max}}(d,\tau_I)=\left(1+0.22d^4\right)\left[1+\left(0.1-0.3\sqrt{d}\right)\left(\hat{\tau}_{I,\min}/\tau_I\right)^2\right] \quad (22)$$

where $\hat{\tau}_{I,\min}$ is an approximation of $\tau_{I,min}$, given by

$$\hat{\tau}_{I,\min}(d)=\left(0.0029\text{-}0.0682\sqrt{\mathrm{d}}+1.4941\mathrm{d}\right)/(1.003\text{-d})^2 \quad (23)$$

The normalized errors of the ultimate gains, defined by $\tilde{K}_{C,\min}=(K_{C,\min}-\hat{K}_{C,\min})/K_{C,\min}$ and $\tilde{K}_{C,\max}=(K_{C,\max}-\hat{K}_{C,\max})/K_{C,\max}$, where $\hat{K}_{C,\max}$ and $\hat{K}_{C,\min}$ are the approximations of $K_{C,max}$ and $K_{C,min}$, respectively, obtained using (21), never exceed 2.2% for $d\leq 0.9$ and $\tau_I > 1.2\,\hat{\tau}_{I,\min}$. Moreover the normalized error relative to $\hat{\tau}_{I,\min}$ never exceeds 1.4% for $d\leq 0.9$.

Since, here $\tau_D=\tau_S$, and K_C is obtained according to relations (19)-(23) as a function of τ_I, in order to tune a PID-like controller it only remains to specify τ_I. In the present Section, we propose to select the controller parameter τ_I, in order to maximize the real part of the slowest dominant pole (i.e. the pole with the smallest real part). This way the resulting closed loop system will have a very fast settling time and, at the same time, a very smooth (non-oscillatory) response.

In order to obtain a pole-zero description of (18), the exponential term in (18) is approximated by the relation

$$\exp(-d\hat{s})=\lim_{n\to\infty}\left[(d/n)\hat{s}+1\right]^{-n} \quad (24)$$

From (24), it can be easily recognized that the exponential term $exp(-d\hat{s})$ is equivalent to an infinite number of poles at $\hat{s}=-n/d+j0$. A typical example of the root locus of (18) is shown in Fig. 2 (for $d=0.5$, $n=25$, K_C given by (19) and $1.1\tau_{I,min}<\tau_I<10\tau_{I,min}$). From this figure, it becomes clear that, there exist three dominant poles that are responsible for the shape of the closed-loop system response. The rest of the poles contribute only to the delay of the response. Extensive simulation analysis (for $0<d<0.9$, $\tau_I>\tau_{I,min}$ and $K_{C,min}<K_C<K_{C,max}$) shows that the step response of an USOPDT system controlled by a PID-like controller (when $\tau_D=\tau_S$) cannot be easily distinguished from that of a 3rd order system with the same dominant poles and the same initial delay, when $n>20$ in (24). This fact is illustrated in Fig. 3.

Range of d	Estimated $\tau_I(d)$	M.N.E.
$0<d<0.17$	$3.06\sqrt{d}+4.19d-12.66d^2$	1.5%
$0.17<d<0.9$	$\left(3.47\sqrt{d}\text{-}2.9d+8.37d^2+18.28d^5\right)(0.95-d)^{-1}$	2%

Table 2. Approximate expressions of $\tau_I(d)$ for the DPC method.

In order to solve the tuning problem presented above, MATLAB® control toolbox was used to estimate the poles of a 27th order closed loop system (n=25 in (24)). Moreover, a simple algorithm based on the dissection method was used to find the value of τ_I that maximizes the real part of the slowest dominant pole. Since this procedure cannot be applied on-line due to its computational burden, the function $\tau_I(d)$ obtained by the DPC method has been approximated by analytical functions $\hat{\tau}_I(d)$. The parameters involved in these functions have been estimated using the optimization toolbox of MATLAB®, in order to minimize the maximum normalized error (M.N.E.), defined by $\tilde{\tau}_I = (\tau_I - \hat{\tau}_I)/\tau_I$. These approximate expressions are given in Table 2, together with their maximum normalized error. The response obtained by the DPC method can be distinguished as follows: For *d<0.157* the method gives three real dominant poles (the two slowest are identical) and the response approximates that of a critical second-order system response. For *d>0.157* the method gives two complex and one real poles all with the same real part (see also Fig. 4).

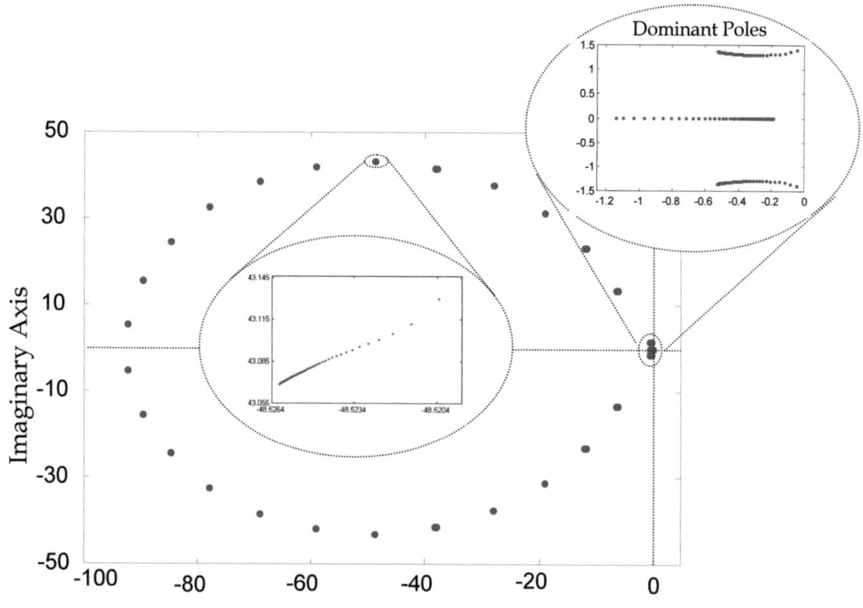

Fig. 2. A typical root locus of (18) for $d=0.5$, $n=25$, $1.1\tau_{I,min}<\tau_I<10\tau_{I,min}$ and K_C given by (19).

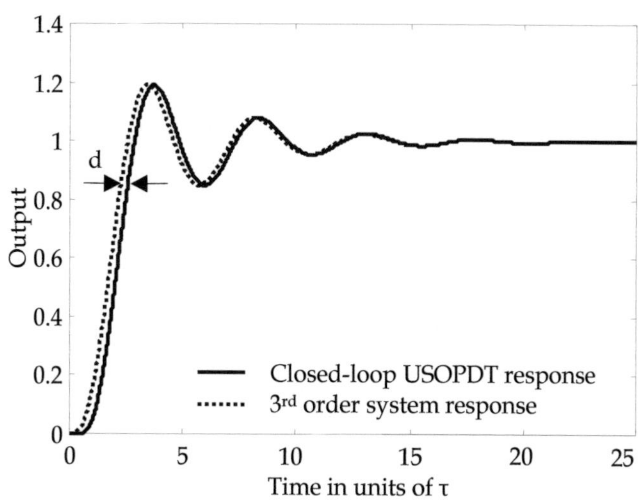

Fig. 3. A typical closed loop set-point step response of the USOPDT process and the response of the 3rd order system.

5. Controller tuning based on closed-loop time-response criteria

In this Section, we consider again that $\tau_D=\tau_S$ as well as that K_C is obtained through (19)-(23), and we present three alternative methods for the selection of the parameter τ_I. These methods are based on some very useful closed-loop set-point step response criteria.
A first, widely used, criterion for tuning PID-like controllers is the fastest settling time (FST) method. In the case of an oscillatory response, the settling time is usually estimated from the envelope of the response. Since for systems with time delay the closed-loop response is not known in analytical form, to estimate here the envelope of the response, we use the response of a third-order system having the dominant poles of the closed-loop USOPDT system. In particular, the response of a third order system, with two complex poles ($p_{I,1}=a+jb$ and $p_{I,2}=a-jb$) and one real pole (p_R), is given by

$$y(t)=1-\left[e^{-\zeta w_0 t}\left(A\cos(w_n t)+B\sin(w_n t)\right)+Ce^{-p_R t}\right] \tag{25}$$

where $w_0=\sqrt{a^2+b^2}$, $\zeta=a/w_0$, $w_n=w_0\sqrt{1-\zeta^2}$, $A=p_R(-p_R+2\zeta w_0)/D$, $B=p_R w_0(-\zeta p_R+2\zeta^2 w_0-w_0)/(D w_n)$, $C=-w_0^2/D$ and $D=-p_R^2+2p_R\zeta w_0-w_0^2$. The two envelopes (top and bottom) of (25) are given by

$$y_{g1,2}(t)=1\pm\left[e^{-\zeta w_0 t}\left(\sqrt{A^2+B^2}\right)+Ce^{-p_R t}\right] \tag{26}$$

Therefore, for the application of the FST method, a simple algorithm based on the dissection method, is used to estimate the value of parameter τ_I that minimizes the time t_{stl} required for obtaining $\left|1-y_{g1}(t_{stl})\right|=0.01$.
A second criterion, on the basis of which the tuning of the PID-like controller is performed, stems from the need to provide the fastest possible set-point step response of the closed loop system with a maximum overshoot of 1% (OPOS method). Also in this case a search algorithm is used to estimate the smallest value of the parameter τ_I (and hence the fastest response) for which the maximum of $y(t)$, given by (25), is smaller than 1.01 for all $t>0$.
Finally, the third method is based on the minimization of the integral of squared errors due to a unit step change in the set point (ISE-Sp method). The first part of the response, for $t<d$, can not be affected by the controller. Hence, for the optimization problem of minimizing the integral of squared errors, one can use the response obtained by (25). The integral of $(1-y(t))^2$ can then be calculated analytically, and it is given by

$$\begin{aligned} ISE_{Sp}&=\int_0^{\infty}\left(1-y(t)\right)^2 dt=\frac{C^2}{2p_R}+2C\frac{A(\zeta w_0+p_R)+Bw_n}{p_R^2+w_0^2+2\zeta w_0 p_R}\\ &+\left[A^2(1+\zeta^2)+B^2(1-\zeta^2)+2AB\zeta\sqrt{1-\zeta^2}\right](4w_0\zeta)^{-1} \end{aligned} \tag{27}$$

Then, using (27) in combination with a simple search algorithm, the parameter τ_I that minimizes the value of ISE_{Sp} can be estimated.
All three methods presented above cannot be applied on-line because of the excessive computational burden required to calculate the values of the three dominant poles. For this reason, the parameter τ_I obtained by the application of these methods, is next calculated for

all values of $d<0.9$ and the function $\tau_I(d)$ is approximated using the optimization toolbox of MATLAB®. The resulting approximations $\hat{\tau}_I(d)$ are given in Table 3. The M.N.E. in the estimation of the function $\tau_I(d)$ is less than 2.8%, for all these approximations. This error in τ_I does not produce a significant change in the response of the closed loop system.

Method	Range of d	Estimated $\tau_I(d)$	M.N.E.
FST	$0<d<0.17$	$0.017+0.42\sqrt{d}+8.08d$	1.5%
	$0.17<d<0.9$	$\frac{3.26\sqrt{d}-1.96d+5.55d^2+15.47d^5}{0.96-d}$	2%
OPOS	$0<d<0.9$	$\frac{2.29\sqrt{d}+0.69d+2.29d^2+15.07d^5}{0.96-d}$	2.8%
ISE-Sp	$0<d<0.9$	$\frac{0.1\sqrt{d}+2.47d+2.78d^2+5.59d^5}{0.95-d}$	2.7%

Table 3. Estimates of $\tau_I(d)$ for the tuning methods based on closed-loop time-domain criteria.

Method	$d=0.1$	$d=0.5$	$d=0.9$
DPC	-12.61, -2.502±j0.175	-0.425, -0.412±j1.312	-0.0377, -0.0377±j0.412
FST	-12.949, -2.326±j1.641	-0.516, -0.368±j1.302	-0.0550, -0.0291±j0.411
OPOS	-12.964, -2.318±j1.675	-0.556, -0.349±j1.299	-0.0609, -0.0262±j0.410
ISE-Sp	-14.765, -1.378±j4.231	-0.785, -0.237±j1.298	-0.0883, -0.0129±j0.409

Table 4. Locations of dominant poles for some typical examples.

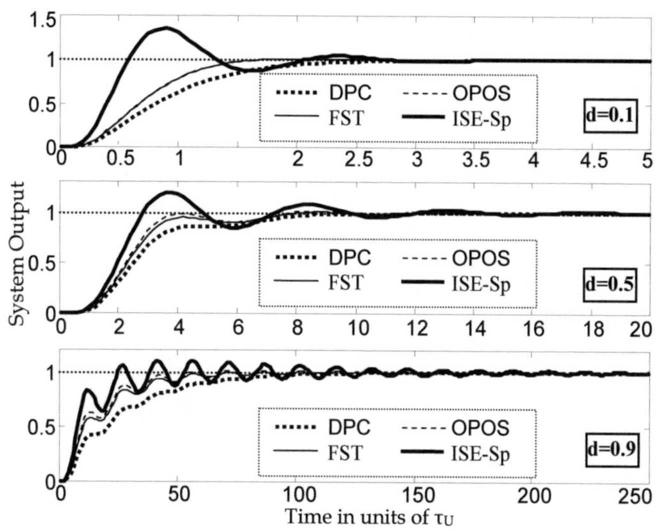

Fig. 4. Characteristic set-point step responses obtained by the proposed tuning methods.

For example, when $\hat{\tau}_I(d)$ is used instead of τ_I, to apply the FST method, the maximum normalized error in the settling time is less than 0.5%.

In Table 4, the locations of the three dominant poles of the closed loop system are given in the case where the normalized dead time takes the values 0.1, 0.5 and 0.9, for all methods presented above. The corresponding closed loop responses obtained from a unit change in the set-point are illustrated in Fig. 4. From these responses and the locations of the dominant poles reported in Table 4, one can easily recognize that the FST and the OPOS methods provide us controllers with similar performance. Moreover, the response obtained when the ISE_Sp method is used is the fastest, although very oscillatory. Finally, in the case where the DPC method is used, the response obtained is sluggish and smooth. Moreover, since this method yields a large value of τ_I, it provides a very robust controller.
Table 5 presents a stability robustness comparison with other existing PID tuning methods. In particular, the tuning methods presented in Sections 4 and 5 are compared with the R&L method with $\lambda=2.2$ (Rotstein & Lewin, 1991), the P&M method (De Paor and O'Malley, 1989), the H&X method with specifications $A_m=1.3$ and $\varphi_m=10^o$ (Ho & Xu, 1998), the P&P method based on the ITAE criterion (Poulin & Pomerleau, 1997) and the J&C method based on the IMC tuning rule with $\lambda=2.5$ (Jacob & Chidambaram, 1996), in the special case where $d=0.5$ and $\tau_S=1$. Table 5 presents the increasing and decreasing gain margins GM_{inc} and GM_{dec} as well as the phase margin PM. Moreover, it presents the maximum simultaneous multiplicative uncertainty A_a of all system parameters (i.e. when the system parameters d, τ_S, K are increased by A_a and τ_U is decreased by A_a) and the maximum multiplicative uncertainty A_d of the time delay (i.e. when only d is increased by A_d), for which the closed loop system remains stable. The results presented in Table 5 show that the DPC method provides more robust controllers than most other methods (except the J&C method with $\lambda=2.5$, that gives a significantly slower response in both set point tracking and regulatory control). The aim of the other three methods, presented in this Section, is to provide faster responses and hence they provide lesser robustness. Finally, it is worth noticing that all the other methods used in robustness comparison are not applicable in cases where $d>0.7$.

Method	K_C	τ_I	τ_D	$PM_{(rad)}$	GM_{inc}	GM_{dec}	a_a	a_d
DPC	1.618	8.150	1	0.172	1.469	1.462	1.101	1.268
FST	1.622	6.948	1	0.155	1.446	1.436	1.091	1.240
OPOS	1.623	6.539	1	0.148	1.436	1.425	1.088	1.225
ISE-SP	1.632	4.834	1	0.107	1.372	1.353	1.064	1.163
R&L (λ=2.2)	2.116	10.24	0.902	0.087	1.173	1.860	1.043	1.103
P&M	1.357	6.960	1	0.133	1.729	1.202	1.103	1.288
H&X	1.518	6.543	1	0.148	1.536	1.332	1.095	1.255
P&P	1.798	8.431	1	0.154	1.325	1.631	1.082	1.204
J&C (λ=2.5)	1.573	9.495	1	0.188	1.528	1.443	1.113	1.307

Table 5. Robustness performance comparison with other existing tuning methods.

6. Controller tuning based on closed-loop stability margins specifications

When a PID-like controller is used to control an USOPDT process, it is possible, in some cases, to simultaneously satisfy the design specifications GM_{dec}, GM_{inc}, and PM exactly. The PID-like controller sought can be found from the solution of the system of equations (8)-(14). Unfortunately, this system of equations is too complicated to be solved on-line and it is not always solvable. Furthermore, the solution might not be appropriate or useful, especially if the derivative term is too large. For this reason, we propose here, to select a priori the derivative term τ_D of the controller, on the basis of the designer's knowledge relative to the process. If there are no restrictions imposed by the process, then it is recommended to select τ_D as large as possible in the range proposed by (16). This way, the resulting closed-loop system has the fastest possible response, for both, the set-point tracking and the load attenuation case, a well as the smallest possible maximum error in the case of regulatory control. Having selected τ_D, as previously mentioned, three methods are then proposed, in order to tune the rest of the controller parameters.

6.1 The Phase Margin (PM) tuning method

In the case where, the only specification for the closed loop system is the desired phase margin PM^{des}, then it is recommended to tune the PID-like controller in such a way that this single specification is achieved at the maximum phase margin corresponding to the frequency w_p, namely, when $w_G = w_p$. This way, the integral reset time τ_I is the smallest possible that satisfies the specification and, hence, the obtained controller provides the fastest possible response, for both set-point tracking and regulatory control. The main steps of this tuning method are the following:

6.1.1 The PM algorithm

Step 1. Given the system parameters d, τ_S, the controller derivative term τ_D and the phase margin specification PM^{des}, set initially $\tau_I = 0$.
Step 2. With this value of τ_I, calculate w_p as the maximum real root of (14).
Step 3. Select the new value of τ_I from the solution of $PM^{des} = \varphi_L(w_p) + \pi$, with respect to τ_I, which is given by

$$\tau_I = w_p^{-1} \tan\left[PM^{des} + \frac{\pi}{2} + dw_p + a\tan(\tau_S w_p) - a\tan(w_p) - a\tan(\tau_D w_p) \right] \qquad (28)$$

Step 4. Repeat Steps 2 and 3 until convergence.
Step 5. With known τ_I, calculate the corresponding frequency w_p from (14) and the controller gain K_C from (13). This completes the method.
The above algorithm converges to the correct solution, if such a solution exists, i.e. if for given d, τ_S, τ_D there exists a value of τ_I for which $PM(d, \tau_S, \tau_D, \tau_I) = PM^{des}$.

6.2 The Gain Margin (GM) tuning method

This method is applicable in the case where the specifications of the closed loop system are described in the form of increasing and decreasing gain margins ($GM_{inc,des}$ and $GM_{dec,des}$). To present the method, two iterative algorithms for the calculation of the crossover frequencies w_{min} and w_{max} are first presented.

6.2.1 The w_{min} algorithm

Step 1. Start with an initial estimate for w_{min}. An appropriate value for fast convergence is

$$w_{\min}^{init} = \sqrt{\left[\tau_I - d(1-\tau_I)\right]^{-1}} \qquad (29)$$

Step 2. Calculate the error of this approximation using the relation

$$e_r = -\pi/2 - dw_{\min}^{init} + a\tan(w_{\min}^{init}) + a\tan(\tau_I w_{\min}^{init}) + a\tan(\tau_D w_{\min}^{init}) - a\tan(\tau_S w_{\min}^{init}) \qquad (30)$$

Step 3. Take the new value of w_{min} as $w_{\min}^{new} = w_{\min}^{old}(1-e_r)$.
Step 4. Repeat Steps 2 and 3 until a convergence.

6.2.2 The w_{max} algorithm

Step 1. Start with a very large initial estimate of w_{max}, say $w_{\max}^{init}=10^4$.
Step 2. Using (8), calculate the new value of w_{max} as

$$w_{\max}^{new} = d^{-1}\left[-\frac{\pi}{2} + a\tan(w_{\max}^{old}) + a\tan(\tau_I w_{\max}^{old}) + a\tan(\tau_D w_{\max}^{old}) - a\tan(\tau_S w_{\max}^{old})\right] \qquad (30)$$

Step 3. Repeat Steps 2 and 3 until convergence.
These two algorithms always converge to the correct values of w_{min} and w_{max}, if for given d, τ_S, τ_D and τ_I there exists a solution of (8), with respect to w_C, when the *atan* function takes values in the range $(-\pi/2,\pi/2)$. We are now able to present the main steps of proposed GM tuning method.

6.2.3 The GM algorithm

Step 1. Given the system parameters d, τ_S, the controller derivative term τ_D and the desired gain matrix product $GM_{prod,des}$, solve $max(PM(d,\tau_I,\tau_D,\tau_S))=0$ to obtain $\tau_{I,min}$.
Step 2. Set $\tau_{I,1}=\tau_{I,min}$ and $\tau_{I,2}=10^3\tau_{I,min}$.
Step 3. Take the new value of τ_I as the average of $\tau_{I,1}$ and $\tau_{I,2}$, i.e. $\tau_I=(\tau_{I,1}+\tau_{I,2})/2$.
Step 4. Calculate the values of w_{min} and w_{max} using the w_{min} Algorithm and the w_{max} Algorithm, respectively, for the obtained τ_I, and obtain $K_{C,min}$ and $K_{C,max}$ from (9).
Step 5. Calculate the value of GM_{prod} from (11).
Step 6. If $GM_{prod}<GM_{prod,des}$ or $w_{min}\leq 0$ or $w_{max}\leq 0$, then $\tau_{I,1}=\tau_I$ or else $\tau_{I,2}=\tau_I$.
Step 7. Repeat Steps 3 to 6 until convergence.
Step 8. The controller gain is evaluated from either $K_C=K_{C,max}/G_{inc,des}$ or $K_C=K_{C,min}G_{dec,des}$. This completes the algorithm.
The above algorithm converges to the correct solution, if such a solution exists, i.e. if for given d, τ_S, τ_D there exists a value of τ_I for which $GM_{prod}(d,\tau_S,\tau_D,\tau_I)=GM_{prod,des}$.

6.3 The Phase and Gain Margin (PGM) tuning method

If the derivative term is a priori selected, then it is not possible, in general, to simultaneously satisfy the specifications on GM_{dec}, GM_{inc}, and PM exactly, with the remaining two free controller parameters. This is due to the fact that, it is not possible to assign three independent specifications with only two independent controller parameters, namely K_C and τ_I. Indeed, with the controller parameters K_C and τ_I obtained from the GM Algorithm, in order to satisfy GM_{dec} and GM_{inc}, then a specific value of the phase margin $PM(d,K_C,\tau_I,\tau_D)$ is obtained, and, hence, in this case the phase margin cannot be selected independently. Keeping these in mind, we propose here a tuning method, in order to achieve simultaneous, although not exact, satisfaction of all three specifications PM, GM_{dec} and GM_{inc}. This method is based on the tuning methods presented in the previous two subsections. The basic steps, for the selection of the parameters of a PID-like controller that satisfy all three specifications, are the following:

6.3.1 The PGM algorithm

Step 1. For the selected value of τ_D, check if there exists a value of K_C that is able satisfy all three specifications, when $\tau_I \rightarrow \infty$.

Step 2. Calculate the two controllers obtained by the PM and the GM methods. If the controller with the largest value of τ_I satisfies all three specifications, then this is the controller sought. In the opposite case continue with *Step 3.*

Step 3. Assume that $K_{C,PM}$ and $\tau_{I,PM}$ are the controller parameters obtained form the application of the PM tuning method and $K_{C,GM}$ and $\tau_{I,GM}$ are the controller parameters obtained from the GM tuning method. Then, if none of these two controllers satisfy all specifications, check which controller gives the largest gain K_C, and distinguish the following two cases:

1. If $K_{C,PM}>K_{C,GM}$, then in order to satisfy all specifications with the smallest value of τ_I, gradually increase τ_I (starting from the $max(\tau_{I,GM},\tau_{I,PM})$), while maintaining the same increasing gain margin GM_{inc} (by selecting $K_C=K_{C,max}(d,\tau_S,\tau_I,\tau_D)/GM_{inc,des}$), until the phase margin specification is also satisfied.
2. If $K_{C,PM}<K_{C,GM}$, then gradually increase τ_I (starting from the $max(\tau_{I,GM},\tau_{I,PM})$), while maintaining the same decreasing gain margin GM_{dec} (by selecting $K_C= K_{C,min}(d,\tau_S,\tau_I,\tau_D)GM_{dec,des}$), until the phase margin specification is also satisfied.

This completes the algorithm.

Although there are several ways to select the controller parameters in order to satisfy all three specifications (although not exactly), the method presented here is preferred, because it requires the smallest computational effort, since for a given τ_I, the phase margin can be calculated exactly without the use of iterative algorithms (using (12) and $PM=\varphi_L(w_G)+\pi$). It is noted here that, in all PID tuning methods presented above, if the response obtained is too oscillatory (due to the small value of τ_I), then, by increasing the value of τ_I, the damping of the closed-loop system increases. From the analysis presented in Section 3, it becomes clear that, when τ_I is increased, the resulting closed-loop system is more robust, and hence all the stability robustness specifications are still satisfied (although not exactly).

6.4 Simplification of the tuning rules for on-line tuning

The tuning rules presented in the previous sections can significantly be simplified, in the case where $\tau_D=\tau_S$. In this case, the loop transfer function is given by (17), and the solutions of the algorithms presented in Subsections 6.1.1 and 6.2.1-6.2.3, can easily be approximated with satisfactory accuracy for all systems with $0<d<0.9$. In particular, the solutions for w_{min} and w_{max}, can be approximated by relations (21)-(23). Note that, here, $\hat{\tau}_{I,\min}(d)$ is an accurate approximation of the smallest value of the integral term τ_I, for which (8) has a solution, when $\tau_D=\tau_S$, and when the atan function takes values in the range $(-\pi/2, \pi/2)$. Table 6 summarizes useful approximations of some other parameters involved in the aforementio-ned algorithms. Note that the maximum normalized errors for the parameters $K_{C,min}$ and $K_{C,max}$, when their estimates are obtained by (20), using $\hat{w}_{\min}$ and $\hat{w}_{\max}$ as given by (21), never exceed 2.2% for $d\leq 0.9$ and $\tau_I>1.2\,\hat{\tau}_{I,\min}$.

In Table 7, numerical applications of the PM, GM and PGM tuning methods are presented for three processes with normalized dead time 0.1, 0.5 and 0.9. The controller parameters obtained from the application of these tuning methods are presented in the left section of Table 7 for both the exact (K_C, τ_I) and the approximated controller parameters ($\hat{K}_C, \hat{\tau}_I$). In the right section of Table 7, the polar plots of the resulting closed-loop systems are presented. Solid and dashed lines are used for the exact and the approximate controller, respectively. The gain margin specifications are indicated by the symbol '**o**' and the point on the unit circle which determines the phase margin specification is indicated by the symbol '⊗'. From all these polar plots, it becomes obvious that the approximate solution is very accurate and in most cases cannot be distinguished from the exact solution. Note that, since the proposed tuning methods provide a controller that satisfies the required stability robustness specifications with significant accuracy, it is possible to design a closed loop system with any desired design specifications. The most robust (but slow) closed loop system possible (when $\tau_D=\tau_S$) can be obtained when $PM_{des}\to PM_{max}$ or when $GM_{prod,des}\to GM_{pred,max}$ (i.e. $\tau_I\to\infty$), while it is possible to design a faster but less robust system with less conservative stability margins specifications.

Function	Approximation	MNE	Valid Range $d<0.9$ and
$GM_{prod,\max}(d)$	$\dfrac{\pi}{2d[1+0.4085d/(1-0.2864d)]}$	3%	
$\tau_I(d,PM^{des})$	$\hat{\tau}_{I,\min}(d)\left[1+f_{PM}(d)\dfrac{PM^{des}/PM_{\max}(d)}{1-\left[PM^{des}/PM_{\max}(d)\right]}\right]$	5%	$PM^{des}>0.2PM_{max}$
$\tau_I(d,GM_{prod,des})$	$\hat{\tau}_{I,\min}(d)\left(1+0.65\dfrac{A^{d+1}}{1-A}\right)+g(d)$	3%	$GM_{prod,des}>1+0.2\times(GM_{prod,max}-1)$
$f_{PM}(d)=(-0.0153+0.436\sqrt{d}+0.632d)/d$, $A=\dfrac{\sqrt{GM_{prod,des}}-1}{\sqrt{GM_{prod,\max}(d)}-1}$ $g(d)=10^{-2}[-0.18+5\sqrt{d}-32d+75d^2-51d^3+(-2.3d^2+3d^4)/(1-d)^3]$			

Table 6. Approximations of parameters involved in the PM, GM and PGM algorithms, when $\tau_D=\tau_S$.

Specifications			Controller parameters		Nyquist Plots
d=0.1 , $\tau_D=\tau_S$	PM=0.3, GM_{inc}=4, GM_{dec}=2	PM Method	$K_C=5.2293$ $\hat{K}_C=5.2170$	$\tau_I=0.3010$ $\hat{\tau}_I=0.2980$	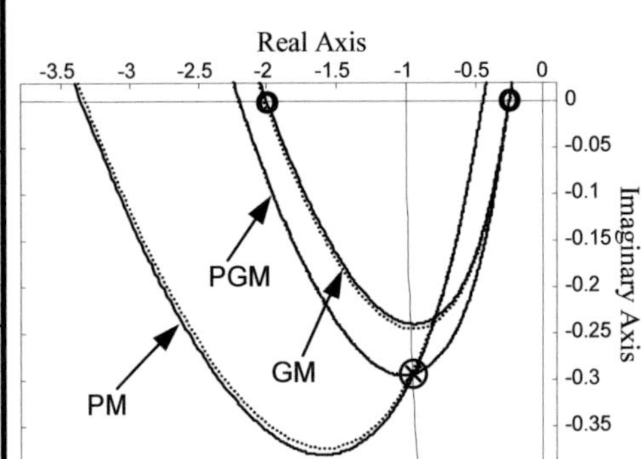
		GM Method	$K_C=3.0225$ $\hat{K}_C=3.0400$	$\tau_I=0.3184$ $\hat{\tau}_I=0.3216$	
		PGM Method	$K_C=3.1333$ $\hat{K}_C=3.1411$	$\tau_I=0.3598$ $\hat{\tau}_I=0.3597$	

Specifications			Controller parameters		Nyquist Plots
d=0.5 , $\tau_D=\tau_S$	PM=0.15, GM_{inc}=1.3, GM_{dec}=1.5	PM Method	$K_C = 1.5690$ $\hat{K}_C = 1.5688$	$\tau_I = 6.5667$ $\hat{\tau}_I = 6.5160$	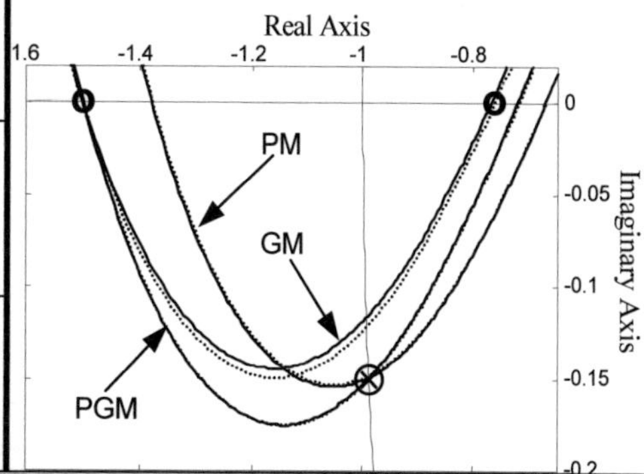
		GM Method	$K_C = 1.7581$ $\hat{K}_C = 1.7505$	$\tau_I = 5.5286$ $\hat{\tau}_I = 5.7115$	
		PGM Method	$K_C = 1.6933$ $\hat{K}_C = 1.6916$	$\tau_I = 6.6907$ $\hat{\tau}_I = 6.9608$	
d=0.9 , $\tau_D=\tau_S$	PM=0.018, GM_{inc}=1.07, GM_{dec}=1.07	PM Method	$K_C = 1.0602$ $\hat{K}_C = 1.0602$	$\tau_I = 777.17$ $\hat{\tau}_I = 744.56$	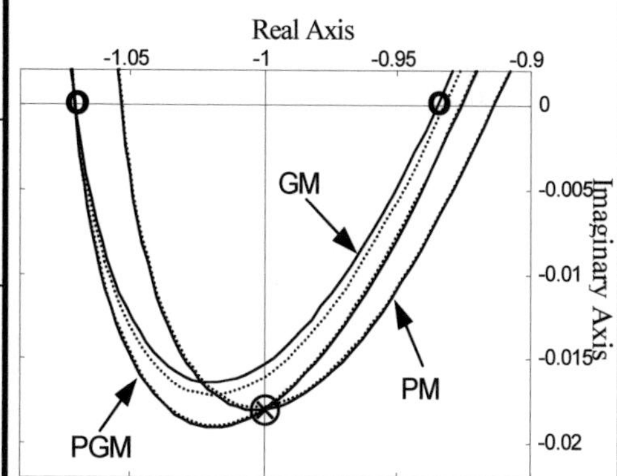
		GM Method	$K_C = 1.0811$ $\hat{K}_C = 1.0795$	$\tau_I = 511.24$ $\hat{\tau}_I = 590.60$	
		PGM Method	$K_C = 1.0756$ $\hat{K}_C = 1.0759$	$\tau_I = 971.4$ $\hat{\tau}_I = 930.70$	

Table 7. Some characteristic numerical examples of the proposed tuning methods reported in Section 6.

7. Application to an experimental magnetic levitation system

In this section the tuning methods presented above will be applied to the experimental magnetic levitation system shown in Figure 5. This experimental system is a popular gravity-biased one degree of freedom magnetic levitation system in which an electromagnet exerts attractive force to levitate a steel ball. The dynamics of the MagLev system can be described by the following simplified state space model (Yang & Tateishi, 2001)

$$dx / dt = v \ , \quad dv / dt = g - (c / M)\left[i^2 / (x_\infty + x)^2 \right] \tag{31}$$

where x, v and M are the air gap (vertical position), the velocity and the mass of the steel ball respectively, g is the gravity acceleration, i is the coil current, c and x_∞ are constants that are determined by the magnetic properties of the electromagnet and the steel ball. Moreover the coil of the electromagnet has an inductance L and a total resistance R.

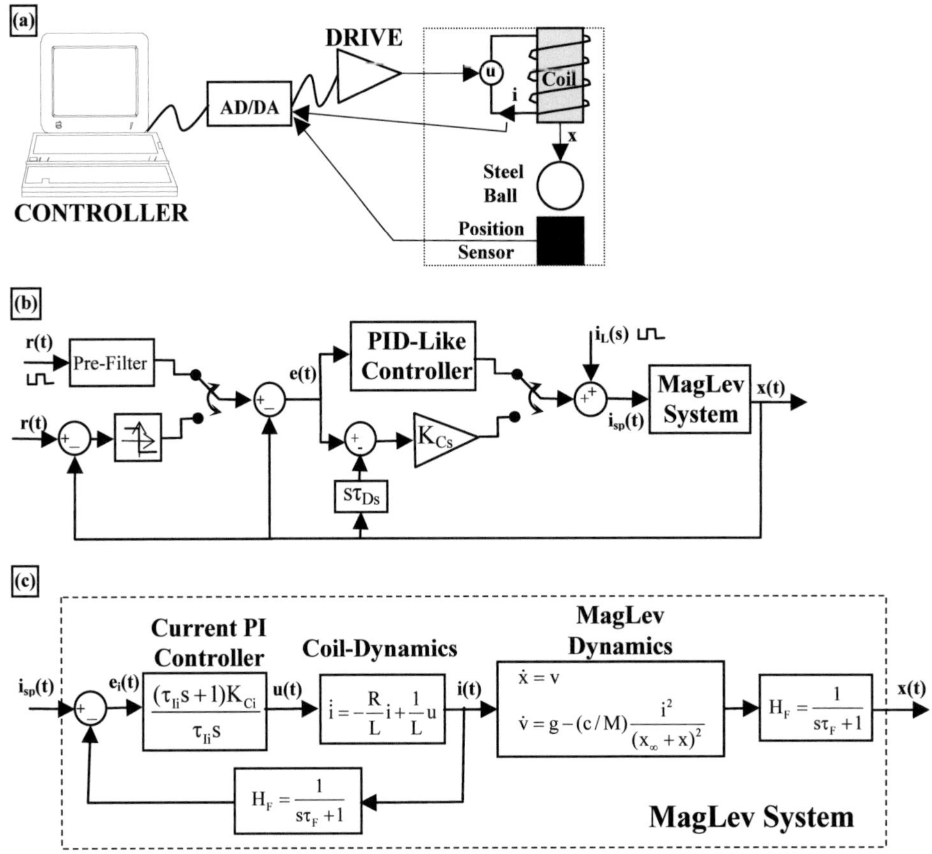

Fig. 5. MagLev system diagrams: (a) Schematic diagram, (b) Control diagram and (c) Block diagram.

Linearizing (31) about an operating point x_0, the following second order transfer function for the MagLev system is obtained

$$H_1^M(s) = \frac{K_m}{(\tau_{Um}s - 1)(\tau_{Sm}s + 1)} \tag{32}$$

where, K_m, τ_{Um}, and τ_{Sm} are the gain, the unstable and the stable time constants of the system given by

$$K_m = \sqrt{c/(Mg)} \quad , \quad \tau_{Um} = \tau_{Sm} = \sqrt{0.5(x_\infty + x_0)/g} \tag{33}$$

For the MagLev system used in the following experiments the current i is controlled by a PI controller (see Figure 5c). Moreover, to reduce measurement noise additional first order filters with time constants τ_F are used for the measurement of x and i (Figure 5c). The unmodelled dynamics of the current control loop, the measurement filters and the dynamics of the electrical circuitry (amplifiers, drivers etc.) is modelled here as a time delay d_m. Therefore, the complete transfer function of the linearized MagLev system is given by

$$H_1^M(s) = \frac{K_m \exp(-d_m s)}{(\tau_{Um}s-1)(\tau_{Sm}s+1)} \tag{34}$$

The model parameters c and x_∞ are obtained from measurements of the steady state value of the coil current (which is given by $i_0 = \sqrt{g(M/c)(x_\infty + x_0)^2}$), for several values of x_0 (3mm<x_0<11mm), using a stabilizing PID controller. Since the model parameters K_m, τ_{Um} and τ_{Sm} can be obtained from (33), to identify the time delay d_m of the system, a single closed loop relay-feedback experiment can be used. The control diagram for this experiment is shown in Figure 5b. Using a PD stabilizing controller with derivative time $\tau_{Ds}=\tau_{Um}$, one can easily verify that d_m is given by

$$d_m = \omega_C^{-1} a\sin\left[(\tau_{Um}\omega_C K_m K_{Cs}) / (\tau_{Um}^2\omega_C^2 + 1) \right] \tag{35}$$

where ω_C is the ultimate frequency of the closed loop system, which is measured by the relay experiment. The values of the model parameters for the linearized system given by (33), about the operating point x_0=7mm, are listed in Table 8 together with the parameters of the PI-current controller and the time constant of the two measurement filters used. It is noted here that the selection of the filter time constant τ_F and the gains of the PI current controller are performed intentionally in order to produce a significant time delay to the MagLev system. Finally, it is mentioned that the sampling intervals for all experiments is chosen as τ_{st}=0.5ms, which is fast enough to assume a continuous-time system.

Physical parameters	
M=0.068 Kg , g=9.81 m/sec^2 , c=8.068 $\cdot 10^{-5}$Hm, x_∞=0.00215m , L=0.4125 , R=11Ω	
Linearized Model parameters (around x_0=0.007m)	
K_m=0.008474 m/A , τ_{Um}= τ_{Sm} =0.0216 sec, d_m=0.01037 sec , i_0=1.08 A	
Current controller and measurement filter parameters	
K_{Ci}=200 , τ_{Ii}=1 , τ_F=0.005	
Parameters of the designed PID controller	
OPOS	K_{Cm}=196.7 , τ_{Im} =0.1273 , τ_{Dm} =0.0216
ISE-Sp	K_{Cm}=197.9 , τ_{Im}=0.0936 , τ_{Dm} =0.0216
DPC	K_{Cm}=196.1 , τ_{Im}=0.1565 , τ_{Dm}=0.0216
FST	K_{Cm}=196.5 , τ_{Im}=0.1346 , τ_{Dm}=0.0216
GM	K_{Cm}=118.5 , τ_{Im}= 0.428 , τ_{Dm} =0.0216
PM	K_{Cm}= 147.5 , τ_{Im}=0.1162 , τ_{Dm}=0.0216

Table 8. System and controller parameters for the experiments in the MagLev system.

A series of experiments have been performed by applying all four methods reported in Sections 4 and 5 to the MagLev system. In Fig. 6, the set-point and load step responses around the operating point x_0=7mm are presented. In particular, in Figs 6a and 6b, the response of the MagLev to a pulse waveform with amplitude 1 mm and period 5 sec is shown in the case where the PID controller is tuned using the OPOS and ISE-Sp methods, respectively. Fig. 6c shows the tracking response in the case where the DPC method is used. In this case the amplitude of the pulse waveform used as reference input is 7mm (from 3.5mm to 10.5mm). Finally, Fig. 6d shows the regulatory control response, in the case where the FST method is used with a change in the system input (current set-point) produced by a pulse waveform with amplitude 0.2A (i.e. 20% change in the steady state value of the coil current). Fig. 6 verifies the efficiency and good performance of the proposed methods. As expected, the ISE-Sp method provides the fastest response, but with an overshoot of about 20%. The FST and OPOS methods produce very smooth and fast regulatory and set-point tracking responses. Finally, the DPC method provides a very robust controller that can control the MagLev system in a large operating region. However, this controller provides a rather sluggish response.

As a second application of the proposed tuning methods, a robust PID controller is designed in order to guarantee a stable closed loop system in a wide operating region (3.5mm< x<10.5mm) and in the case of ±20% uncertainty in the parameters c, x_∞ and 10% uncertainty in the time delay d_m. The problem of converting the parametric uncertainties into gain and phase margin specifications is a very complicated problem that remains unsolved, in the general case. Here, in order to select appropriate specifications for the design of the controller, the following observations are made: (a) From (8), it is clear that the uncertainty in the model parameters τ_{Um} and τ_{Sm} (which depend on x_∞ and x_0) does not affect the argument of the loop transfer function. The only term which influences the phase uncertainty is the uncertainty in the identification of the time delay. (b) Assuming that τ_I>5$\tau_{I,min}$, (this assumption is in accordance with our desire to design a very robust controller as suggested in the work (Paraskevopoulos et al, 2006)) one can easily verify from $\hat{\tau}_{I,\min}(d)$ (given in Table 6), that a ±10% change in d_m produces a change in ω_{min} and ω_{max} smaller than

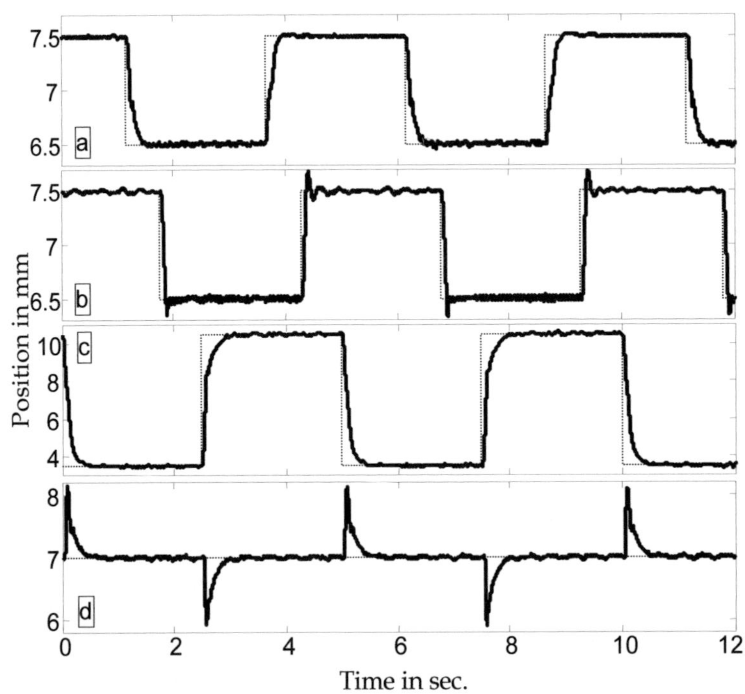

Fig. 6. Experimental MagLev position responses. Set-point tracking response: (a) using OPOS method, (b) using ISE-Sp method, (c) using DPC method. (d) Regulatory control step response using FST method (current load disturbance amplitude 20% or 0.2 A).

5% and 15%, respectively. (c) The magnitude of the loop transfer function is affected by all parameters, as well as the operating point. The two extreme worst cases are obtained when d and c are maximized and x_0, x_∞ are minimized (scenario A) and when d, x_0, x_∞ are maximized and c is minimized (scenario B). From scenario A, we obtain the smallest maximum ultimate gain $min(K_{C,max})$, while from scenario B, we obtain the largest minimum ultimate gain $max(K_{C,min})$. Obviously, for the closed loop system to be stable under the assumed uncertainty and for the whole desired operating region, there must be $min(K_{C,max})> max(K_{C,min})$. Based on the above observation one can easily verify that for $\tau_I>5\tau_{I,min}$, the inequalities $min(K_{C,max})>0.53K_{C,max,0}$ and $max(K_{C,min})<1.2K_{C,min,0}$, must hold, where $K_{C,max,0}$ and $K_{C,min,0}$ are the nominal values of $K_{C,max}$ and $K_{C,min}$ at the operating point x_0=0.007m, i.e. the case where there is no uncertainty. To guarantee stability, the increasing and decreasing gain margins must be selected grater than 1/0.53 and 1.2, respectively.

Based on the above results and observations, in order to tune the PID controller, the GM tuning method is next applied with specifications GM_{inc}=2 and GM_{dec}=1.25. The obtained controller gains are listed in Table 8. The Nyquist plots for the two extreme scenarios A and B and for the nominal system, using the obtained robust controller, are shown in Fig. 7, which verifies that the closed loop system is always stable.

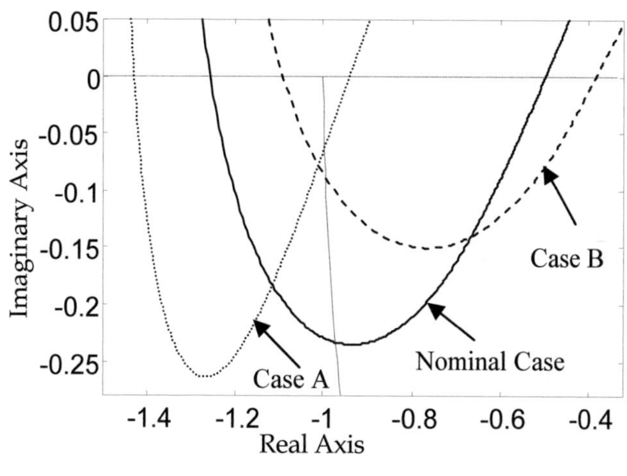

Fig. 7. Nyquist plots of the MagLev system using the robust controller designed with the GM method.

For the experimental application, a pre-filter with transfer function $G_{F,PID}(s)=1/(s\tau_I+1)$ is used in order to cancel the zero introduced by the PID controller. With this filter excessive overshoots in the set-point step response of the system are avoided. The experimental results obtained are presented in Figure 8. The set-point step response from 3.5mm to 10.5mm is shown in Figure 8a. This response is rather slow due to the small value of K_{Cm} and the very large value of τ_{Im} (τ_{Im}/τ_{Um}=19.81=9.728$\tau_{I,min}$). This is more evident in the regulatory control case, around the operating point x_0=7mm, shown in Figure 8b. This response is obtained from a change in the system input (current set-point) produced by a pulse waveform with amplitude 0.2A (or 20% change in i).

A faster controller can be designed if the desired operating region is smaller under the assumption of the same parameter uncertainties as in the previous application. In this case, the PM tuning method is used with a specification PM_{des}=0.15 rad. The obtained controller is presented in Table 6. Figures 9a and 9b, show the set-point step response from 6.5mm to 7.5mm and the regulatory control around the operating point x_0=7mm using the new controller. Clearly, the obtained responses are significantly faster, as it was expected from the design of the PID controller (smaller τ_{Im}, larger K_{Cm}). Moreover, in the case of regulatory control the maximum error produced in the present case is significantly smaller (at least three times smaller) than the maximum error produced when the robust controller is used.

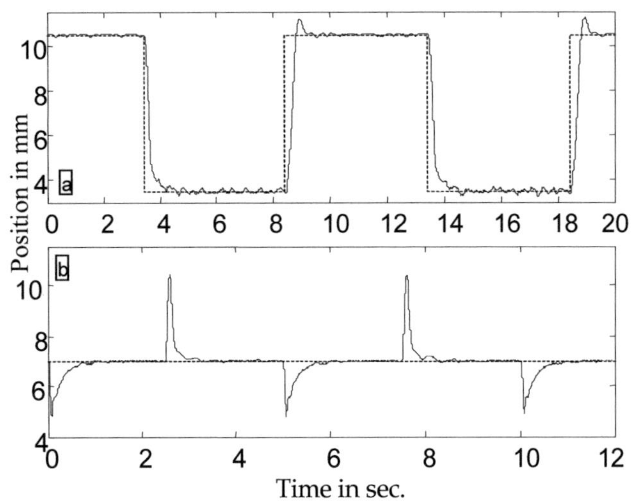

Fig. 8. Position response of MagLev system using the robust controller designed with the GM-method: (a) Set-point response and (b) Load step response (current load disturbance amplitude 20% or 0.2 A).

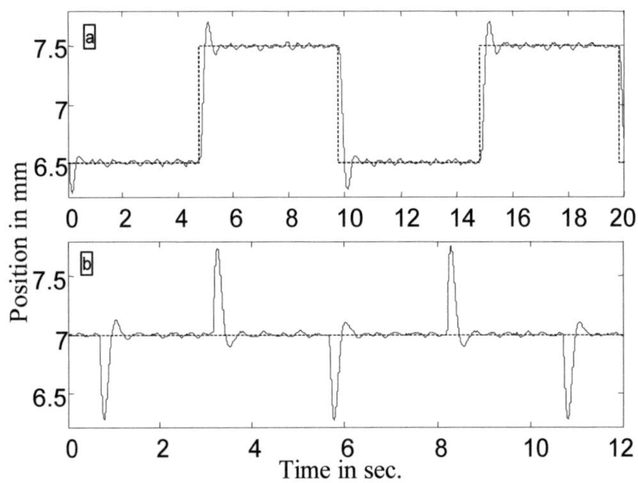

Fig. 9. Position response of MagLev system using a fast controller designed with the PM-method. Other legend as in Fig. 8.

8. Conclusions

New methods for tuning PID-like controllers for USOPDT systems have been developed in this work. These methods are based on various criteria, such as the appropriate assignment of the dominant poles of the closed-loop system, the attainment of various time-domain closed-loop characteristics, as well as the satisfaction of gain and phase margins specifications of the closed-loop system. In the general case, where the derivative action of the controller is selected arbitrarily, the tuning methods require the use of iterative algorithms for the solution of nonlinear systems of equations. In the special case where the controller derivative time constant is selected equal to the stable time constant of the system, the solutions of the nonlinear system of equations involved in the tuning methods are given in the form of quite accurate analytic approximations and, thus, the iterative algorithms can be avoided. In this case the tuning methods can readily be used for on-line applications. The proposed tuning methods have successfully been applied to the control of an experimental magnetic levitation system that is modelled as an USOPDT process. The obtained experimental results verify the efficiency of the proposed tuning methods that provide a very satisfactory performance of the closed-loop system.

9. References

Clement, C.V. & Chidambaram, M. (1997). PID Controller for Unstable Time Delay Systems. *Chemical Engineering Communications,* Vol.162, pp. 63-74.

De Paor, A. M. & O'Malley, M. (1989). Controllers of Ziegler-Nichols Type for Unstable Process with Time Delay. *International Journal of Control.,* Vol.49, No.4, pp. 1273–1284.

Fung, H.-W.; Wang, Q.-G. & Lee, T.-H. (1998). PI Tuning in Terms of Gain and Phase Margins. *Automatica,* Vol.34, No.9, pp. 1145–1149.

Ho, E.K. & Xu, W. (1998). PID Tuning for Unstable Processes based on Gain and Phase Margin Specifications. *Proceedings of the Institute of Electrical Engineers-Part D: Control Theory and Applications,* Vol.145, No.5, pp. 392–396.

Huang, H.P. & Chen, C.C. (1997). Control-System Synthesis for Open-Loop Unstable Process with Time Delay. *Proceedings of the Institute of Electrical Engineers – Part D: Control Theory and Applications,* Vol.144, pp. 334–346.

Jacob, E.F. & Chidambaram, M. (1996). Design of Controllers for Unstable First-Order plus Time Delay Systems. *Computers and Chemical Engineering,* Vol.20, No.5, pp. 579–584.

Jhunjhunwala, M.K. & Chidambaram, M. (2001). PID Controller Tuning for Unstable Systems by Optimization Method. *Chemical Engineering Communications,* Vol. 185, pp. 91-113.

Jung, C.S.; Song, H.K. & Hyun, J.C. (1999). A Direct Synthesis Tuning Method of Unstable First-Order-plus-Time-Delay-Processes. *Journal of Process Control,* Vol.9, No.3, pp. 265–269.

Lee, C.H. & Teng, C.C. (2002). Tuning of PID Controllers for Stable and Unstable Processes based on Gain and Phase Margin Specifications: A Fuzzy Neural Approach. *International Journal of Fuzzy Systems,* Vol.128, No.1, pp. 95–106.

Lee, Y.; Lee, J. & Park, S. (2000). PID Controller Tuning for Integrating and Unstable Processes with Time Delay. *Chemical Engineering Science,* Vol.55, No.17, pp. 3481–3493.

Liu, T., Zhang, W. & Gu, D. (2004). Analytical Design of Two-Degree-of-Freedom Control Scheme for Open-Loop Unstable Process with Time Delay. *Journal of Process Control,* Vol.15, pp. 559-572.

O'Dwyer, A. (2009). *Handbook of PI and PID Controller Tuning Rules.* World Scientific, Singapore.

Padma Sree, R. & Chidambaram, M. (2004). A Simple Method of Tuning PID Controllers for Stable and Unstable FOPDT systems. *Computers and Chemical Engineering,* Vol.28, pp. 2201-2218.

Paraskevopoulos, P.N.; Pasgianos, G.D. & Arvanitis, K.G. (2004). New Tuning and Identification Methods for Unstable First Order plus Dead Time Processes Based on Pseudo-Derivative Feedback Control. *IEEE Transactions on Control Systems Technology,* Vol.12, No.3, pp. 455-464.

Paraskevopoulos, P.N.; Pasgianos, G.D. & Arvanitis, K.G. (2006). PID-Type Controller Tuning for Unstable First-Order plus Dead-Time processes based on Gain and Phase Margin Specifications. *IEEE Transactions on Control Systems Technology,* Vol.14, No.5, pp. 926-936.

Park, J.H.; Sung, S.W. & Lee, I.B. (1998). An Enhanced PID Control Strategy for Unstable Processes. *Automatica,* Vol.34, No.6, pp. 751–756.

Poulin, E. & Pomerleau, A. (1997). Unified PID Design Method based on Maximum Peak Resonance Specification. *Proceedings of the Institute of Electrical Engineers-Part D: Control Theory and Applications,* Vol.144, No.6, pp. 566–574.

Prashanti, G. & Chidambaram, M. (2000). Set-Point Weighted PID Controllers for Unstable Systems. *Journal of the Franklin Institute,* Vol.337, pp. 201-215.

Rao, A.S. & Chidamabaram M. (2006). Enhanced Two-Degrees-of-Freedom Control Strategy for Second-Order Unstable Processes with Time Delay. *Industrial Engineering Chemistry Research,* Vo.45, pp. 3604–3614.

Rotstein, G.E. & Lewin, D.R. (1991). Simple PI and PID Tuning for Open-Loop Unstable Systems. *Industrial Engineering Chemistry Research,* Vol.30, No.8, pp. 1864–1869.

Shamsuzzoha, M., Yoon, M.K. & Lee, M. (2007). Analytical Controller Design of Integrating and First Order Unstable Time Delay Process. *Proceedings of the 8th International IFAC Symposium on Dynamics and Control of Process Systems,* pp. 397-402, Cancun, Mexico, June 6-8, 2007.

Tan, W., Marquez, H.J. & Chen, T. (2003). IMC Design for Unstable Processes with Time Delays. *Journal of Process Control,* Vol.13, pp. 203–213.

Venkatashankar, V. & Chidambaram, M. (1994). Design of P and PI Controllers for Unstable First-Order plus Time Delay Systems. *International Journal of Control,* Vol.60, pp. 137-144.

Visioli, A. (2001). Optimal Tuning of PID Controllers for Integral and Unstable processes. *Proceedings of the Institute of Electrical Engineers – Part D: Control Theory and Applications,* Vol.148, pp. 180–184.

Wang, Y.G. & Cai, W.J. (2002). Advanced Proportional-Integral-Derivative Tuning for Integrating and Unstable Processes with Gain and Phase Margin Specifications. *Industrial Engineering Chemistry Research,* Vol.41, No.12, pp. 2910–2914.

Yang, X.P., Wang, Q.G., Hang, C.C. & Lin, C. (2002). IMC-based Control System Design for Unstable Processes. *Industrial and Engineering Chemistry Research,* Vol.41, pp. 4288–4294.

Yang, Z.J. & Tateishi, M. (2001). Adaptive Robust Nonlinear Control of a Magnetic Levitation Systems, *Automatica,* Vol.31, pp. 1125-1131.

4

Magnitude Optimum Techniques for PID Controllers

Damir Vrančić
Jožef Stefan Institute
Slovenia

1. Introduction

Today, most tuning rules for PID controllers are based either on the process step response or else on relay-excitation experiments. Tuning methods based on the process step response are usually based on the estimated process gain and process lag and rise times (Åström & Hägglund, 1995). The relay-excitation method is keeping the process in the closed-loop configuration during experiment by using the on/off (relay) controller. The measured data is the amplitude of input and output signals and the oscillation period.

The experiments mentioned are popular in practice due to their simplicity. Namely, it is easy to perform them and get the required data either from manual or from automatic experiments on the process. However, the reduction of process time-response measurement into two or three parameters may lead to improperly tuned controller parameters.

Therefore, more sophisticated tuning approaches have been suggested. They are usually based on more demanding process identification methods (Åström et al., 1998; Gorez, 1997; Huba, 2006). One such method is a magnitude optimum method (MO) (Whiteley, 1946). The MO method results in a very good closed-loop response for a large class of process models frequently encountered in the process and chemical industries (Vrančić, 1995; Vrančić et al., 1999). However, the method is very demanding since it requires a reliable estimation of quite a large number of process parameters, even for relatively simple controller structures (like a PID controller). This is one of the main reasons why the method is not frequently used in practice.

Recently, the applicability of the MO method has been improved by using the concept of 'moments', which originated in identification theory (Ba Hli, 1954; Strejc, 1960; Rake, 1987). In particular, the process can be parameterised by subsequent (multiple) integrals of its input and output time-responses. Instead of using an explicit process model, the new tuning method employs the mentioned multiple integrals for the calculation of the PID controller parameters and is, therefore, called the "Magnitude Optimum Multiple Integration" (MOMI) tuning method (Vrančić, 1995; Vrančić et al., 1999). The proposed approach therefore uses information from a relatively simple experiment in a time-domain while retaining all the advantages of the MO method.

The deficiency of the MO (and consequently of the MOMI) tuning method is that it is designed for optimising tracking performance. This can lead to the poor attenuation of load disturbances (Åström & Hägglund, 1995). Disturbance rejection performance is particularly

decreased for lower-order processes. This is one of the most serious disadvantages of the MO method, since in process control disturbance rejection performance is often more important than tracking performance.
The mentioned deficiency has been recently solved by modifying the original MO criteria (Vrančić et al., 2004b; Vrančić et al., 2010). The modified criteria successfully optimised the disturbance rejection response instead of the tracking response. Hence, the concept of moments (multiple integrations) has been applied to the modified MO criteria as well, and the new tuning method has been called the "Disturbance Rejection Magnitude Optimum" (DRMO) method (Vrančić et al., 2004b; Vrančić et al., 2010).
The MOMI and DRMO tuning methods are not only limited to the self-regulating processes. They can also be applied to integrating processes (Vrančić, 2008) and to unstable processes (Vrančić & Huba, 2011). The methods can also be applied to different controller structures, such as Smith predictors (Vrečko et al., 2001) and multivariable controllers (Vrančić et al., 2001b). However, due to the limited space and scope of this book, they will not be considered further.

2. System description

A stable process may be described by the following process transfer function:

$$G_P(s) = K_{PR} \frac{1 + b_1 s + b_2 s^2 + \cdots + b_m s^m}{1 + a_1 s + a_2 s^2 + \cdots + a_n s^n} e^{-sT_{delay}}, \tag{1}$$

where K_{PR} denotes the process steady-state gain, and a_1 to a_n and b_1 to b_m are the corresponding parameters ($m \leq n$) of the process transfer function, whereby n can be an arbitrary positive integer value and T_{delay} represents the process pure time delay. Note that the denominator in (1) contains only stable poles.
The PID controller is defined as follows:

$$U(s) = G_R(s)R(s) - G_C(s)Y(s), \tag{2}$$

where U, R and Y denote the Laplace transforms of the controller output, the reference and the process output, respectively. The transfer functions $G_R(s)$ and $G_C(s)$ are the feed-forward and the feedback controller paths, respectively:

$$\begin{aligned} G_R(s) &= \frac{K_I + bK_P s + cK_D s^2}{s(1 + sT_F)} \\ G_C(s) &= \frac{K_I + K_P s + K_D s^2}{s(1 + sT_F)} \end{aligned} \tag{3}$$

The PID controller parameters are proportional gain K_P, integral gain K_I, derivative gain K_D, filter time constant T_F, proportional reference weighting factor b and derivative reference weighting factor c (Åström & Hägglund, 1995). Note that the first-order filter is applied to all three controller terms instead of only the D term in order to reduce noise amplitude at the controller output and to simplify the derivation of the PID controller parameters. The range of parameters b and c is usually between 0 and 1. Since the feed-

forward and the feedback paths are generally different, the PID controller (2) is a two-degrees-of-freedom (2-DOF) controller. Note that controller (2) becomes a 1-DOF controller when choosing b=c=1.
The PID controller in a closed-loop configuration with the process is shown in Figure 1.

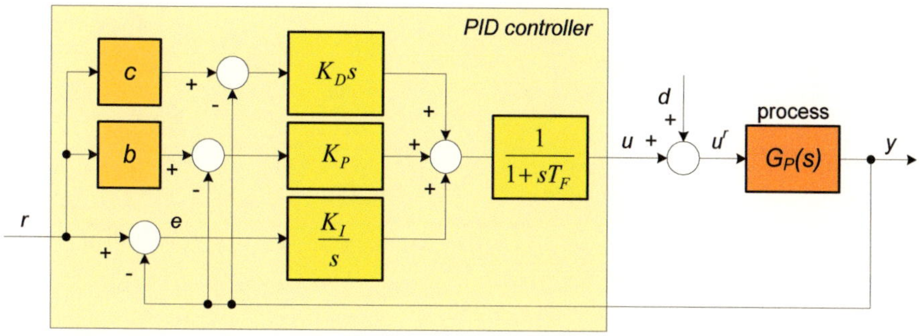

Fig. 1. The closed-loop system with the PID controller

Signals e, d and u^r denote the control error, disturbance and process input, respectively. The closed-loop transfer function with the PID controller is defined as follows:

$$G_{CL}(s) = \frac{Y(s)}{R(s)} = \frac{G_R(s)G_P(s)}{1+G_C(s)G_P(s)}. \tag{4}$$

For the 1-DOF PID controller (b=c=1), the closed-loop transfer function becomes:

$$G_{CL}(s) = \frac{Y(s)}{R(s)} = \frac{G_C(s)G_P(s)}{1+G_C(s)G_P(s)}. \tag{5}$$

The deficiency of 1-DOF controllers is that they usually cannot achieve optimal tracking and disturbance rejection performance simultaneously. 2-DOF controllers may achieve better overall performance by keeping the optimal disturbance rejection performance while improving tracking performance.

3. Magnitude Optimum (MO) criteria

One possible means of control system design is to ensure that the process output (y) follows the reference (r). The ideal case is that of perfect tracking without delay (y=r). In the frequency domain, the closed-loop system should have an infinite bandwidth and zero phase shift. However, this is not possible in practice, since every system features some time delay and dynamics while the controller gain is limited due to physical restrictions.

The new design objective would be to maintain the closed-loop magnitude (amplitude) frequency response (G_{CL}) from the reference to the process output as flat and as close to unity as possible for a large bandwidth (see Figure 2) (Whiteley, 1946; Hanus, 1975; Åström & Hägglund, 1995; Umland & Safiuddin, 1990). Therefore, the idea is to find a controller that makes the frequency response of the closed-loop amplitude as close as possible to unity for lower frequencies.

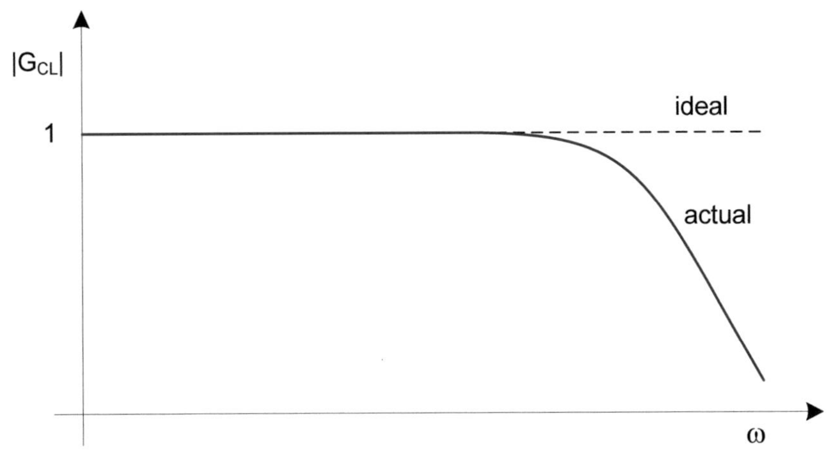

Fig. 2. The amplitude (magnitude) frequency response of the closed-loop system

These requirements can be expressed in the following way:

$$G_{CL}(0) = 1 , \tag{6}$$

$$\left. \frac{d^{2k} \left| G_{CL}(j\omega) \right|^2}{d\omega^{2k}} \right|_{\omega=0} = 0 \ ; \ k = 1,2,\cdots,k_{\max} \tag{7}$$

for as many k as possible (Åström & Hägglund, 1995).
This technique is called "Magnitude Optimum" (MO) (Umland & Safiuddin, 1990), "Modulus Optimum" (Åström & Hägglund, 1995), or "Betragsoptimum" (Åström & Hägglund, 1995; Kessler, 1955), and it results in a fast and non-oscillatory closed-loop time response for a large class of process models.
If the closed-loop transfer function is described by the following equation:

$$G_{CL}(s) = \frac{f_0 + f_1 s + f_2 s^2 + \cdots}{e_0 + e_1 s + e_2 s^2 + \cdots} , \tag{8}$$

then expression (7) can be met by satisfying the following conditions (Vrančić et al., 2010):

$$\sum_{i=0}^{2n}(-1)^{i+n}\left(f_i f_{2n-i}e_0^2 - e_i e_{2n-i} f_0^2\right) = 0; \quad n = 1,2,\ldots \tag{9}$$

Before calculating the parameters of the 1-DOF PID controller, according to the given MO criteria, the pure time delay in expression (1) has to be developed into an infinite Taylor series:

$$e^{-sT_{delay}} = 1 - sT_{delay} + \frac{\left(sT_{delay}\right)^2}{2!} - \frac{\left(sT_{delay}\right)^3}{3!} + \cdots + \frac{(-1)^k\left(sT_{delay}\right)^k}{k!} + \cdots \tag{10}$$

or Padé series:

$$e^{-sT_{delay}} = \lim_{n\to\infty}\left(\frac{1-\frac{sT_{delay}}{2n}}{1+\frac{sT_{delay}}{2n}}\right)^n = \frac{1-\frac{sT_{delay}}{2}+\frac{s^2T_{delay}^2}{2^2 2!}-\ldots+(-1)^k\frac{s^kT_{delay}^k}{2^k k!}+\cdots}{1+\frac{sT_{delay}}{2}+\frac{s^2T_{delay}^2}{2^2 2!}+\ldots+\frac{s^kT_{delay}^k}{2^k k!}+\cdots}. \tag{11}$$

Then, the closed-loop transfer function (5) is calculated from expressions (1), (3) and (10) or else (11). The closed-loop parameters e_i and f_i can be obtained by comparing expressions (8) and (5). The PID controller parameters are then obtained by solving the first three equations (n=1, 2 and 3) in expression (9) (Vrančić et al., 1999):

$$K_P = f_1\left(K_{PR}, a_1, a_2, \ldots, a_5, b_1, b_2, \ldots, b_5, T_{delay}, T_F\right) \tag{12}$$

$$K_I = f_2\left(K_{PR}, a_1, a_2, \ldots, a_5, b_1, b_2, \ldots, b_5, T_{delay}, T_F\right) \tag{13}$$

$$K_D = f_3\left(K_{PR}, a_1, a_2, \ldots, a_5, b_1, b_2, \ldots, b_5, T_{delay}, T_F\right) \tag{14}$$

The expressions (12)-(14) are not explicitly given herein, since they would cover several pages. In order to calculate the three PID controller parameters – according to the given MO tuning criteria – *only* the parameters K_{PR}, a_1, a_2, a_3, a_4, a_5, b_1, b_2, b_3, b_4, b_5, and T_{delay} of the process transfer function (1) are required, even though the process transfer function can be of a higher-order. However, accurately estimating such a high number of process parameters from real measurements could be very problematic. Moreover, if one identifies the fifth-order process model from the actually higher-than-fifth-order process, a systematic error in the estimated process parameters would be obtained, therefore leading to the calculation of non-optimal controller parameters. Accordingly, the accuracy of the estimated process parameters in practice remains questionable.

Note that the actual expressions (12)-(14) remain exactly the same when the process with pure time-delay is developed into a Taylor (10) or Padé (11) series (Vrančić et al., 1999).

4. Magnitude Optimum Multiple Integration (MOMI) tuning method

The problems with original MO tuning method just mentioned can be avoided by using the concept of 'moments', known from identification theory (Ba Hli, 1954; Preuss, 1991).

Namely, the process transfer function (1) can be developed into an infinite Taylor series around s=0, as follows:

$$G_P(s) = A_0 - A_1 s + A_2 s^2 - A_3 s^3 + \cdots, \tag{15}$$

where parameters A_i (i=0, 1, 2, ...) represent time-weighted integrals of the process impulse response h(t) (Ba Hli, 1954; Preuss, 1991; Åström & Hägglund, 1995):

$$A_k = \frac{1}{k!}\int_0^\infty t^k h(t)dt. \tag{16}$$

However, the process impulse response cannot be obtained easily in practice since - due to several restrictions - we cannot apply an infinite impulse signal to the process input. Fortunately, the moments A_i can also be obtained by calculating repetitive (multiple) integrals of the process input (u) and output (y) signals during the change of the process steady-state (Strejc, 1960; Vrančić et al., 1999; Vrančić, 2008):

$$\begin{array}{ll} u_0(t) = \dfrac{u(t)-u(0)}{u(\infty)-u(0)} & y_0(t) = \dfrac{y(t)-y(0)}{u(\infty)-u(0)} \\ I_{U1}(t) = \int_0^t u_0(\tau)d\tau & I_{Y1}(t) = \int_0^t y_0(\tau)d\tau \\ I_{U2}(t) = \int_0^t I_{U1}(\tau)d\tau & I_{Y2}(t) = \int_0^t I_{Y1}(\tau)d\tau \\ \vdots & \vdots \end{array}. \tag{17}$$

The moments (integrals, areas) can be calculated as follows:

$$\begin{array}{l} A_0 = y_0(\infty)\ ;\ y_1 = A_0 I_{U1}(t) - I_{Y1}(t) \\ A_1 = y_1(\infty)\ ;\ y_2 = A_1 I_{U1}(t) - A_0 I_{U2}(t) + I_{Y2}(t) \\ A_2 = y_2(\infty)\ ;\ y_3 = A_2 I_{U1}(t) - A_1 I_{U2}(t) + A_0 I_{U3}(t) - I_{Y3}(t) \\ \vdots \end{array}. \tag{18}$$

It is assumed that:

$$\dot{y}(0) = \ddot{y}(0) = \dddot{y}(0) = \cdots = 0. \tag{19}$$

Given that in practice the integration horizon should be limited, there is no need to wait until $t=\infty$. It is enough to integrate until the transient of $y_0(t)$ in (17) dies out. Note that the first impulse (A_0) equals the steady-state process gain, K_{PR}.

In order to clarify the mathematical derivation, a graphical representation of the first moment (area) is shown in Figure 3. Note that u_0 and y_0 represent scaled process input and process output time responses, respectively.

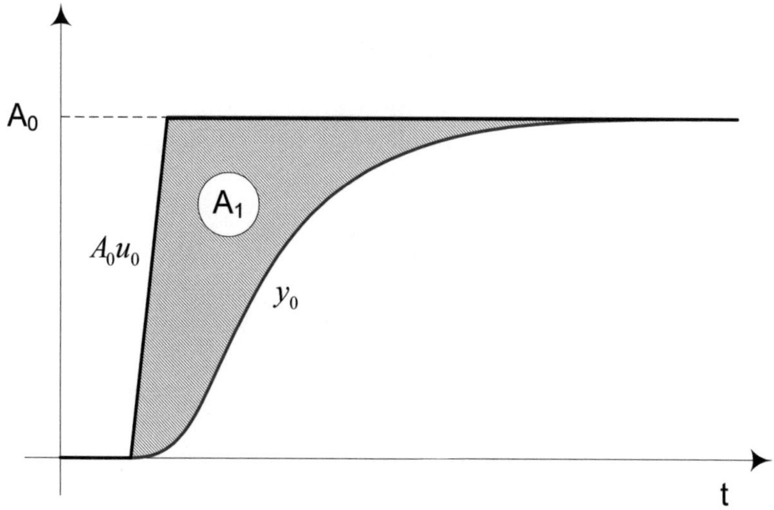

Fig. 3. Graphical representation of the moment (area) A_1 measured from the process steady-state change time response (see shadowed area).

Therefore, in practice the process can be easily parameterised by the moments A_i from the process step-response or else from any other change of the process steady-state.
On the other hand, the moments can also be obtained directly from the process transfer function (1), as follows (Vrančić et al., 1999; Vrančić et al., 2001a):

$$\begin{aligned}
A_0 &= K_{PR} \\
A_1 &= K_{PR}\left(a_1 - b_1 + T_{delay}\right) \\
A_2 &= K_{PR}\left[b_2 - a_2 - T_{delay}b_1 + \frac{T_{delay}{}^2}{2!}\right] + A_1 a_1 \\
&\vdots \\
A_k &= K_{PR}\left((-1)^{k+1}\left(a_k - b_k\right) + \sum_{i=1}^{k}(-1)^{k+i}\frac{T_{delay}{}^i b_{k-i}}{i!}\right) + \\
&+\sum_{i=1}^{k-1}(-1)^{k+i-1}A_i a_{k-i}
\end{aligned} \tag{20}$$

Let us now calculate the 1-DOF PID controller parameters by using the process transfer function parameterised by moments (15). In order to simplify derivation of the PID controller parameters, the filter within the PID controller (3) is considered to be a part of the process (1):

$$G_P^*(s) = \frac{G_P(s)}{1 + sT_F}. \tag{21}$$

Therefore, $G_C(s)$ (3) simplifies into the "schoolbook" PID controller without a filter:

$$G_C^*(s)=\left(K_I+K_Ps+K_Ds^2\right)/s. \tag{22}$$

Since a filter is considered as a part of the process, the measured moments (18) should be changed accordingly. One solution to calculate any new moments is to filter the process output signal:

$$Y_F(s)=\frac{Y(s)}{1+sT_F} \tag{23}$$

and use signal $y_F(t)$ instead of $y(t)$ in expression (17). However, a much simpler solution is to recalculate the moments as follows:

$$\begin{aligned} A_0^* &= A_0 \\ A_1^* &= A_1 + A_0T_F \\ A_2^* &= A_2 + A_1T_F + A_0T_F^2 \\ &\vdots \end{aligned}, \tag{24}$$

where A_i^* denote the moments of the process with included the filter (21).
The parameters e_i and f_i in expression (8) can be obtained by placing expressions (22) and (15) (by replacing moments A_i with A_i^*) into (5). By solving the first three equations in (9), the following PID controller parameters are obtained (Vrančić et al., 2001a):

$$\begin{bmatrix} K_I \\ K_P \\ K_D \end{bmatrix}=\begin{bmatrix} -A_1^* & A_0^* & 0 \\ -A_3^* & A_2^* & -A_1^* \\ -A_5^* & A_4^* & -A_3^* \end{bmatrix}^{-1}\begin{bmatrix} -0.5 \\ 0 \\ 0 \end{bmatrix}. \tag{25}$$

The expression for the PID controller parameters is now much simpler when compared to expressions (12)-(14). There are several other advantages to using expression (25) instead of expressions (12)-(14) for the calculation of the PID controller parameters.
First, only the steady-state process gain $A_0=K_{PR}$ and five moments (A_1 to A_5) instead of the 12 transfer function parameters (K_{PR}, $a_1..a_5$, $b_1..b_5$, and T_{delay}) are needed as input data.
Second, the expression for K_I, K_P, and K_D is simplified, which makes it more transparent and simpler to handle.
Third, the moments A_1 to A_5 can be calculated from the process time-response using numerical integration, whilst the gain $A_0=K_{PR}$ can be determined from the steady-state value of the process steady-state change in the usual way. This procedure replaces the much more demanding algorithm for the estimation of the transfer function parameters.
In addition, it is important to note that the mapping of expressions (12)-(14) into expression (25) results in exact (rather than approximate) controller parameters. This means that the frequency-domain control criterion can be achieved with a model parameterised in the time-domain. Thus the proposed tuning procedure is a simple and very effective way for controller tuning since no background in control theory is needed.

Note that the calculation of the filtered PID controller parameters is based on the fact that the filter time constant is given *a priori*. In practice this is often not entirely true, since the usual way is rather to define the ratio (N) between the derivative time constant ($T_D=K_D/K_P$) and the filter time constant:

$$N=\frac{T_D}{T_F}=\frac{K_D}{K_P T_F}. \qquad (26)$$

Typical values of N are 8 to 20 (Åström & Hägglund, 1995).
The controller parameters can be calculated iteratively by first choosing $T_F=0$ (or any relatively small positive value) and then calculating the controller parameters by using expression (25). In the second iteration, the filter time constant can be calculated from (26), as follows:

$$T_F=\frac{K_D}{K_P N}. \qquad (27)$$

The moments are recalculated according to expression (24) and the new controller parameters from (25). By performing a few more iterations, quite accurate results can be obtained for the *a priori* chosen ratio N.
The PI controller parameters can be calculated in a similar manner to those of the PID controller by choosing $K_D=0$. Since a filter is usually not needed in a PI controller ($T_F=0$), the original moments (A_i) are applied in the calculation. Repeating the same procedure as before and solving the first two equations in (9), the following PI controller parameters are obtained (Vrančić et al., 2001a):

$$\begin{bmatrix} K_I \\ K_P \end{bmatrix}=\begin{bmatrix} -A_1 & A_0 \\ -A_3 & A_2 \end{bmatrix}^{-1}\begin{bmatrix} -0.5 \\ 0 \end{bmatrix}. \qquad (28)$$

Note that the vectors and matrices in (28) are just sub-vectors and sub-matrices of expression (25). Similarly, the I (integral-term only) controller gain is the following:

$$K_I=\frac{0.5}{A_1}. \qquad (29)$$

The proportional (P) controller gain can be obtained by fixing $K_I=0$ and $K_D=0$, repeating the procedure and solving the first equation in (9):

$$K_P=\frac{2A_0A_2-A_1^2}{2A_0\left(A_1^2-A_0A_2\right)}. \qquad (30)$$

However, condition (6) is not satisfied, since proportional controllers cannot achieve closed-loop gain equal to one at lower frequencies. Therefore the proportional controller does not entirely fulfil the MO conditions and will not be used in any further derivations.
In some cases, the controller parameters have to be re-tuned for certain practical reasons. In particular, when tuning the PID controllers for the first-order or the second-order process, the controller gain is theoretically infinite. In practice (when there is process noise), the

calculated controller gain can have a very high positive or negative value. In this case, the controller gain should be limited to some acceptable value, which would depend on the controller and the process limitations (Vrančić et al., 1999). Note that the sign of the proportional gain is usually the same to the sign of the process gain:

$$\operatorname{sgn}(K_P) = \operatorname{sgn}(K_{PR}). \tag{31}$$

The recommended values of the proportional gain are:

$$\left|\frac{1}{A_0}\right| \leq |K_P| \leq \left|\frac{10}{A_0}\right|. \tag{32}$$

The remaining two controller parameters can now be calculated according to the limited (fixed) controller gain from expression (25). If the chosen controller gain is:

$$K_P > \frac{1}{\frac{2A_1^* A_2^*}{A_3^*} - 2A_0^*}, \tag{33}$$

then:

$$K_I = \frac{0.5 + K_P A_0^*}{A_1^*} \tag{34}$$

and:

$$K_D = \frac{A_3^*}{A_1^{*2}} \left[\frac{A_1^* A_2^* K_P}{A_3^*} - 0.5 - A_0^* K_P \right]. \tag{35}$$

If expression (33) is not true:

$$K_D = 0. \tag{36}$$

When limiting the proportional gain of the PI controller, only Eq. (34) is used. Note that proposed re-tuning can also be used in cases when a slower and more robust controller should be designed (by decreasing K_P), or if a faster but more oscillatory response is required (by increasing K_P).

The PID controller tuning procedure, according to the MOMI method, can therefore proceed as follows:

- If the process model is not known *a priori*, modify the steady-state process by changing the process input signal.
- Find the steady-state process gain $K_{PR}=A_0$ and moments A_1-A_5 by using numerical integration (summation) from the beginning to the end of the process time response according to expressions (17) and (18). If the process model is known, calculate the moments from expression (20).
- Fix the filter time constant T_F to some desired value and calculate the PID controller parameters from (25). If needed, change the filter time constant and recalculate the PID controller parameters. If the proportional gain K_P is too high or has a different sign to

the process gain ($K_{PR}=A_0$), set K_P manually to some desired value (32) and recalculating remaining parameters according to expressions (33)-(36).

- The PI or I parameters can be calculated from expressions (28) or (29), respectively.

The proposed tuning procedure will be illustrated by the following process models:

$$G_{P1}(s)=\frac{1}{(1+2s)^2(1+s)^2}$$
$$G_{P2}(s)=\frac{1}{(1+s)^6}$$
$$G_{P3}(s)=\frac{1-4s}{(1+s)^2} \qquad (37)$$
$$G_{P4}(s)=\frac{e^{-5s}}{1+s}$$

The process models have been chosen in order to cover a range of different processes, including higher-order processes, highly non-minimum phase processes and dominantly delayed processes. The models have the same process gain (A_0=1) and the first moment A_1=6. If the process transfer function is not known in advance, the moments (areas) can be calculated according to the time-domain approach given above. The ramp-like input signal has been applied to the process inputs. The process open-loop responses are shown in Figure 4.

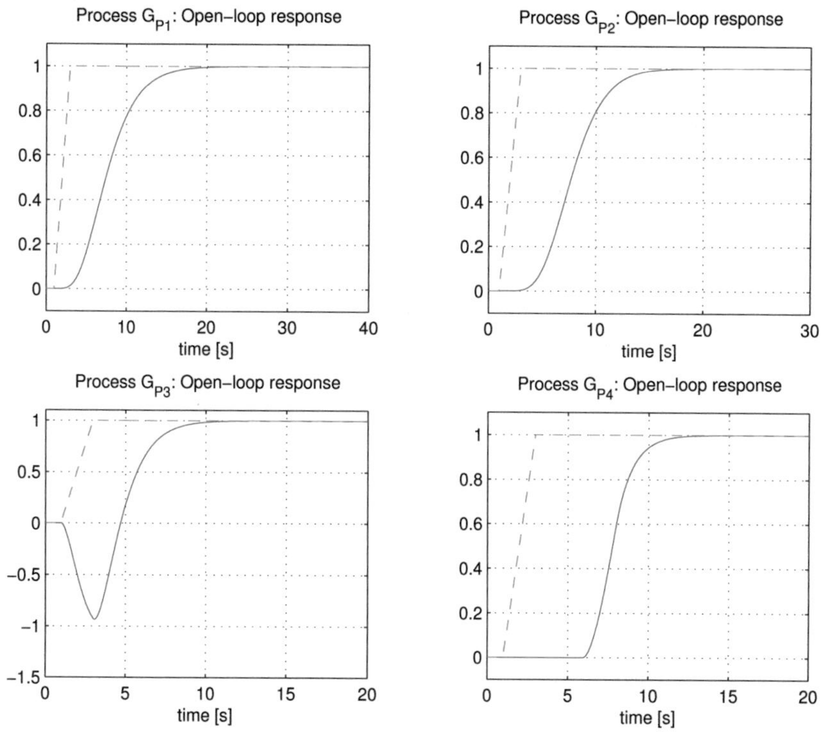

Fig. 4. The process input (--) and the process output (__) signals during an open-loop experiment for processes G_{P1} to G_{P4}.

The moments are calculated by using expressions (17) and (18) and the controller parameters by using expressions (25), (28) and (29). The calculated parameters are given in Table 1.

	Moments (areas)					PID				PI		I
	A_1	A_2	A_3	A_4	A_5	K_I	K_P	K_D	T_F	K_I	K_P	K_I
G_{P1}	6	23	72	201	521	0.31	1.45	1.76	0.2	0.17	0.55	0.08
G_{P2}	6	21	56	126	252	0.22	0.87	0.96	0.2	0.15	0.4	0.08
G_{P3}	6	11	16	21	26	0.12	0.25	0.13	0.2	0.11	0.16	0.08
G_{P4}	6	18.5	39.3	65.4	91.3	0.16	0.49	0.45	0.2	0.13	0.27	0.08

Table 1. The values of moments and controller parameters for processes (37) when using a time-domain approach (by applying multiple integration of the process time-response).

The closed-loop responses for all the processes, when using different types of controllers tuned by the MOMI method, are shown in Figure 5. As can be seen, the responses are stable and relatively fast, all according to the MO tuning criteria.

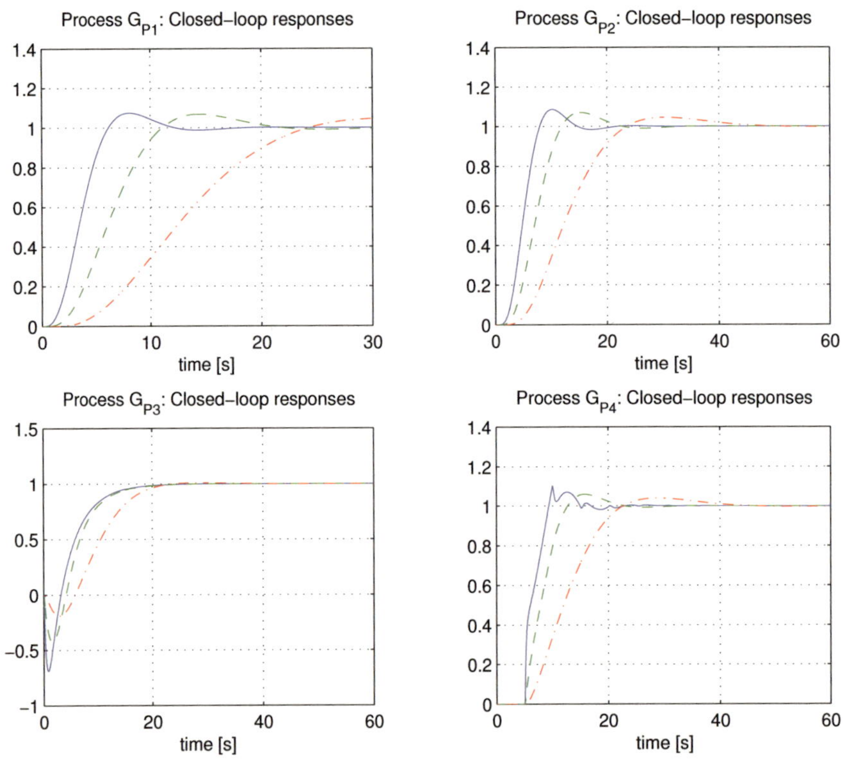

Fig. 5. Closed-loop responses for processes G_{P1} to G_{P4} when using PID controller (__), PI controller (--) and I controller (-.-) tuned by the MOMI method.

The results can be verified by calculating the moments and controller parameters directly from the process transfer functions (37). The moments can be calculated from expression (20). The controller parameters are calculated as before. The obtained parameters are given in Table 2. It can be seen that the values are practically equivalent, so the closed-loop responses are the same to those shown in Figure 5.

	Moments (areas)					PID				PI		I
	A_1	A_2	A_3	A_4	A_5	K_I	K_P	K_D	T_F	K_I	K_P	K_I
G_{P1}	6	23	72	201	522	0.31	1.44	1.76	0.2	0.17	0.55	0.08
G_{P2}	6	21	56	126	252	0.22	0.87	0.96	0.2	0.15	0.4	0.08
G_{P3}	6	11	16	21	26	0.12	0.25	0.13	0.2	0.11	0.16	0.08
G_{P4}	6	18.5	39.3	65.4	91.4	0.16	0.49	0.45	0.2	0.13	0.27	0.08

Table 2. The values of moments and controller parameters for processes (37) by using direct calculation from the process model.

The MOMI tuning method will be illustrated by the three-water-column laboratory setup shown in Figure 6. It consists of two water pumps, a reservoir and three water columns. The water columns can be connected by means of electronic valves. In our setup, two water columns have been used (R_1 and R_2), as depicted in the block diagram shown in Figure 7.

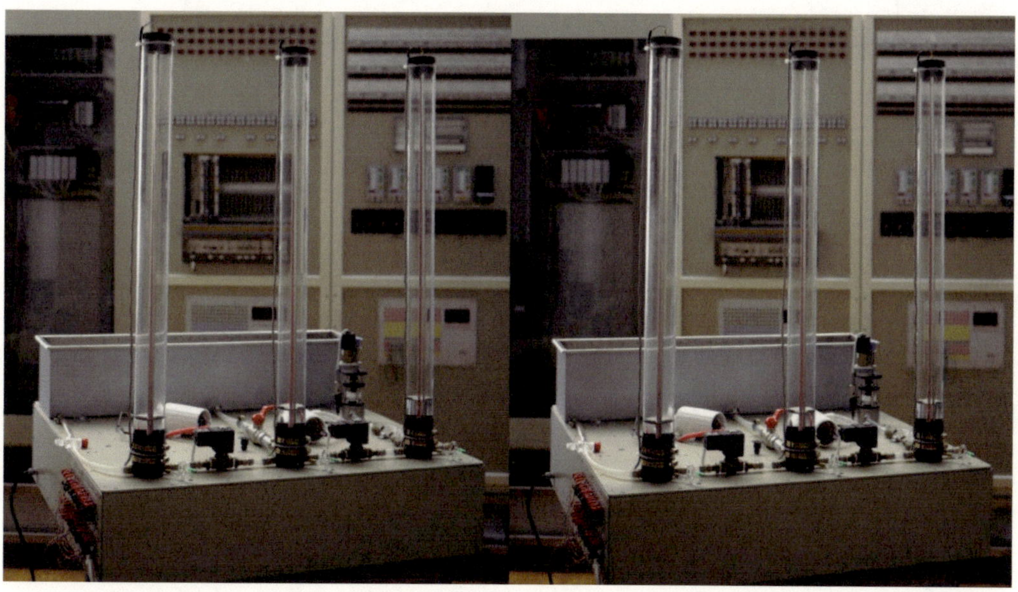

Fig. 6. Picture of the laboratory hydraulic setup (taken in stereoscopic side-by-side format).

The selected control loop consists of the reservoir R_0, the pump P_1, an electronic valve V_1 (open), a valve V_3 (partially open) and water columns R_1 and R_2. The valve V_2 is closed and the pump P_2 is switched off. The process input is the voltage on pump P_1 and the process output is the water level in the second tank (h_2), measured by the pressure to voltage transducer. The actual process input and output signals are voltages measured by an A/D and a D/A converter (NI USB 6215) via real-time blocks in Simulink (Matlab).

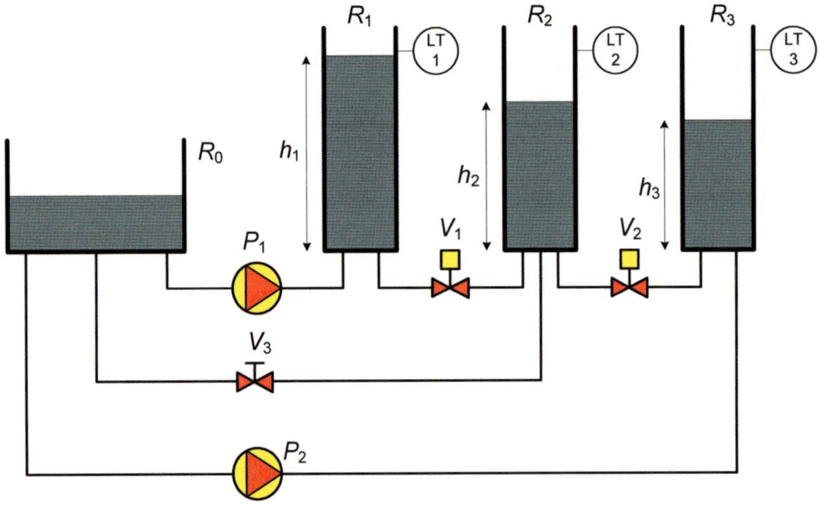

Fig. 7. Block diagram of the laboratory hydraulic setup.

First, the linearity of the system was checked by applying several steps at the process input. The process input and output responses are shown in Figure 8. It can be seen that both - the process steady-state gain and the time-constants - change according to the working point. In order to partially linearise the process, the square-root function has been placed between the controller output (u) and the process input (u^r) signals:

$$u^r = \sqrt{10 \cdot u}\ , \qquad (38)$$

The control output signal u is limited between values 0 and 10. The pump actually starts working when signal u^r becomes higher than 1V.
Note that artificially added non-linearity cannot ideally linearise the non-linearity of the process gain. Moreover, the process time constants still differ significantly at different working points.
After applying the non-linear function (38), the open-loop process response has been measured (see Figure 9). The moments (areas) have been calculated by using expressions (17) and (18):

$$A_0 = 0.507,\ A_1 = 33.9,\ A_2 = 1.76 \cdot 10^3,\ A_3 = 8.44 \cdot 10^4,\ A_4 = 3.9 \cdot 10^6,\ A_5 = 1.78 \cdot 10^8 \qquad (39)$$

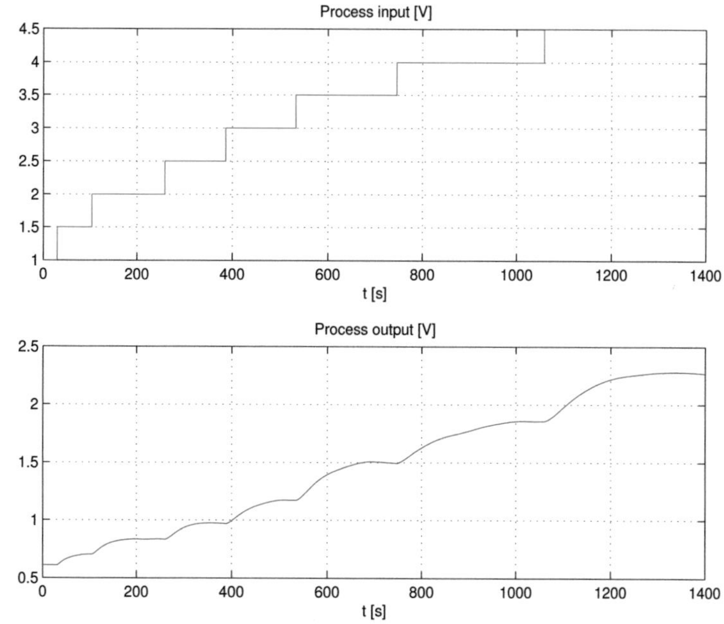

Fig. 8. The process input and process output responses over the entire working region.

The calculated PID controller parameters, for an *a priori* chosen filter parameter T_F=1s, were the following (the proportional gain has been limited to the value $K_P=10/A_0$):

$$K_I = 0.305,\ K_P = 19.7,\ K_D = 264 \tag{40}$$

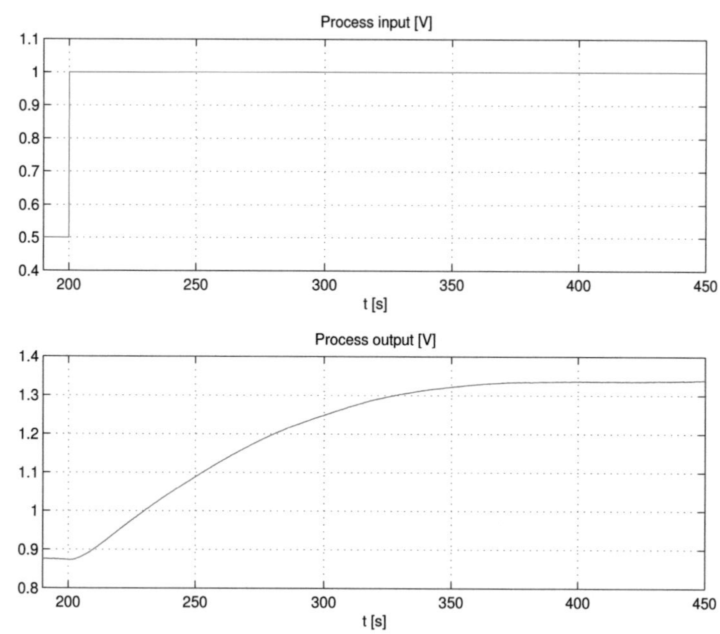

Fig. 9. Process open-loop response.

The closed-loop response of the process with the controller was calculated in the previous step, as shown in Figure 10. At t=300s, the set-point has been changed from 1.2 to 1.5 and at t=900s it is returned back to 1.2. A step-like disturbance has been added to the process input at t=700s and t=1300s. It can be seen that the closed-loop response is relatively fast (when compared to the open-loop response) and without oscillations.

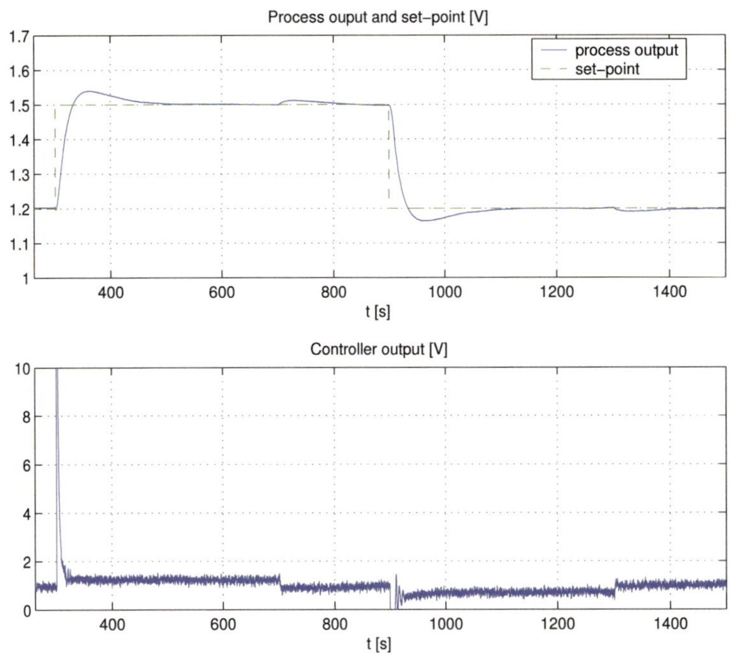

Fig. 10. The process closed-loop response in the hydraulic setup when using the PID controller tuned by the MOMI method.

5. Disturbance-Rejection Magnitude Optimum (DRMO) tuning method

The efficiency of the MOMI method has been demonstrated on several process models (Vrančić, 1995). The MO criteria, according to expressions (6) and (7), optimises the closed-loop transfer function between the reference (r) and the process output (y). However, this may lead to the poor attenuation of load disturbances (Åström & Hägglund, 1995). The disturbance-rejection performance is particularly degraded when controlling lower-order processes.

Let us observe the disturbance-rejection performance of the following process models:

$$\begin{aligned} G_{P1}(s) &= \frac{1}{(1+6s)} \\ G_{P2}(s) &= \frac{1}{(1+3s)^2} \\ G_{P3}(s) &= \frac{1}{(1+s)^6} \\ G_{P4}(s) &= \frac{e^{-5s}}{1+s} \end{aligned} \tag{41}$$

Two of them (G_{P3} and G_{P4}) are the same as in the previous section (37) while we added two lower-order processes in order to clearly show the degraded disturbance-rejection performance. The moments and controller parameters for the chosen processes are given in Table 3. Note that the proportional gain has been limited to 10 for G_{P1} and G_{P2}.

	Moments (areas)					PID				PI		I
	A_1	A_2	A_3	A_4	A_5	K_I	K_P	K_D	T_F	K_I	K_P	K_I
G_{P1}	6	36	216	1296	7776	1.75	10	0	0	1.75	10	0.08
G_{P2}	6	27	108	405	1458	1.69	10	14.5	0.2	0.25	1	0.08
G_{P3}	6	21	56	126	252	0.22	0.87	0.96	0.2	0.15	0.4	0.08
G_{P4}	6	18.5	39.3	65.4	91.4	0.16	0.49	0.45	0.2	0.13	0.27	0.08

Table 3. The values of the moments and controller parameters for processes (41) using the MOMI method.

A step-like disturbance (d) has been applied to the process input (see Figure 1). The process output responses are shown in Figure 11. It is clearly seen that the closed-loop responses of the processes G_{P1} and G_{P2}, when using the PI and the PID controllers, are relatively slow with visible "long tails" (exponential approaching to the reference).

It is obvious that the MO criteria should be modified in order to achieve a more optimal disturbance rejection. The closed-loop transfer function between the disturbance (d) and the process output (y) is the following:

$$G_{CLD}(s) = \frac{Y(s)}{D(s)} = \frac{G_P(s)}{1+G_C(s)G_P(s)} \tag{42}$$

However, the function G_{CLD} (42) cannot be applied instead of G_{CL} in expressions (6) and (7), since G_{CLD} has zero gain in the steady-state (s=0). However, by adding integrator to function (42) and multiplying it with K_I, it complies with the MO requirements (Vrančić et al., 2004b; 2010):

$$G_{CLI}(s) = \frac{K_I}{s} G_{CLD}(s) = \frac{K_I G_P(s)}{s\left(1+G_C(s)G_P(s)\right)} \tag{43}$$

Therefore, in order to achieve optimal disturbance-rejection properties, the function G_{CLI} should be applied instead of G_{CL} in the MO criteria (6) and (7).

However, the expression for the PID controller parameters - due to higher-order equations - is not analytic and the optimisation procedure should be used (Vrančić et al., 2010). Initially, the derivative gain K_D is calculated from expression (25). As such, the proportional and integral term gains are calculated as follows (Vrančić et al., 2010):

$$\begin{aligned} K_P &= \frac{\beta - \sqrt{\beta^2 - \alpha\gamma}}{\alpha} \\ K_I &= \frac{\left(1 + K_P A_0^*\right)^2}{2\left(K_D A_0^{*2} + A_1^*\right)} \end{aligned}, \tag{44}$$

where

$$\begin{aligned} \alpha &= A_1^{*3} + A_0^{*2} A_3^* - 2A_0^* A_1^* A_2^* \\ \beta &= A_1^* A_2^* - A_0^* A_3^* + K_D\left(A_0^* A_1^{*2} - A_0^{*2} A_2^*\right) \\ \gamma &= K_D^3 A_0^{*4} + 3K_D^2 A_0^{*2} A_1^* + K_D\left(2A_0^* A_2^* + A_1^{*2}\right) + A_3^* \end{aligned} . \tag{45}$$

The optimisation iteration steps consist of modifying the derivative gain K_D and re-calculating the remaining two parameters from (44) until the following expression becomes true (Vrančić et al., 2010):

$$\begin{aligned} &-4A_0A_4K_IK_D - 2A_3K_D + 2A_4K_P - 2A_5K_I + 2A_0A_4K_P^2 - 2A_0A_2K_D^2 - \\ &-2A_1A_3K_P^2 - 2A_2^2K_IK_D + A_1^2K_D^2 + A_2^2K_P^2 + 4A_1A_3K_DK_I = 0 \end{aligned} . \tag{46}$$

Any method that employs an iterative search for a numeric solution - that solves the system of nonlinear equations - can be applied. However, in Vrančić et al. (2004a) it was shown that the initially calculated parameters of the PID controller are usually very close to optimal ones. Therefore, a simplified (sub-optimal) solution is to use only the initial PID parameters. In the following text, the simplified version will be applied and denoted as the DRMO tuning method.

Note that the PI controller parameters do not require any optimisation procedure. The derivative gain is fixed at $K_D=0$ and the PI controller parameters are then calculated from expression (44).

The PID controller tuning procedure, according to the DRMO method, can therefore proceed as follows:

- If the process model is not known *a priori*, modify the process steady-state by changing the process input signal.
- Find the steady-state process gain $K_{PR}=A_0$ and moments A_1-A_5 by using numerical integration (summation) from the beginning to the end of the process step response according to expressions (17) and (18). If the process model is defined, calculate the gain and moments from expression (20).
- Fix the filter time constant T_F to some desired value and calculate moments and the derivative gain K_D from (24) and (25). Calculate the remaining controller parameters from expression (44). If the value $\alpha=0$ or if the proportional gain K_P is too high or has a

different sign to the process gain ($K_{PR}=A_0$), set K_P manually to some more suitable value and then recalculate K_I from (44).

- The PI controller parameters can be calculated by fixing $K_D=0$ and using expression (44). If the value $\alpha=0$ or if the proportional gain K_P is too high or has a different sign to the process gain ($K_{PR}=A_0$), set K_P manually to some more suitable value and then recalculate K_I from (44).

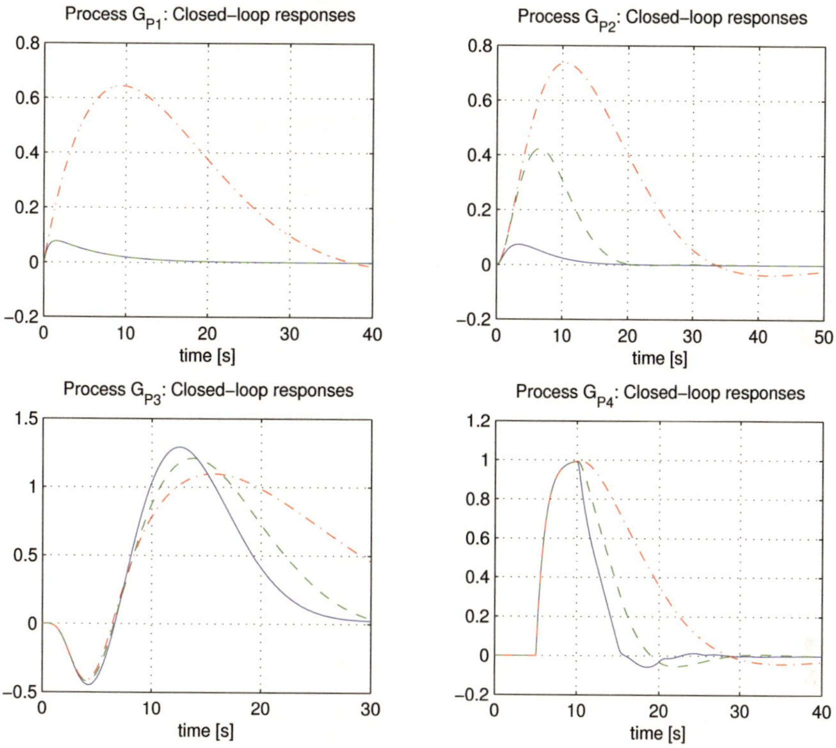

Fig. 11. Closed-loop responses to step-like input disturbance (d) for processes G_{P1} to G_{P4} when using a PID controller (__), a PI controller (--) and an I controller (-.-) tuned by the MOMI method.

The proposed DRMO tuning procedure will be illustrated by the same four process models (41), as before. The PID and PI controllers' parameters are calculated by the procedure given above. Note that the I controller parameters remain the same as with the MOMI method (29). The parameters for all of the controllers are given in Table 4.

	Moments (areas)					PID				PI		I
	A_1	A_2	A_3	A_4	A_5	K_I	K_P	K_D	T_F	K_I	K_P	K_I
G_{P1}	6	36	216	1296	7776	10.1	10	0	0	1.75	10	0.08
G_{P2}	6	27	108	405	1458	2.92	10	14.5	0.2	0.25	1	0.08
G_{P3}	6	21	56	126	252	0.27	0.97	0.96	0.2	0.17	0.43	0.08
G_{P4}	6	18.5	39.3	65.4	91.4	0.18	0.52	0.45	0.2	0.14	0.29	0.08

Table 4. The values of moments and controller parameters for processes (41) using the DRMO method.

A step-like disturbance (d) has been applied to the process input. The process output responses, when using the PID and the PI controllers, are shown in Figures 12 and 13. It can be clearly seen that the closed-loop performance for processes G_{P1} and G_{P2} is now improved when compared with the original MOMI method.

However, improved disturbance-rejection has its price. Namely, the optimal controller parameters for disturbance-rejection are usually not optimal for reference following. Deterioration in tracking performance, in the form of larger overshoots, can be expected for the lower-order processes. A possible solution for improving deteriorated tracking performance, while retaining the obtained disturbance-rejection performance, is to use a 2-DOF PID controller, as shown in Figure 1. Namely, it has been shown that tracking performance can be optimised by choosing b=c=0 (Vrančić et al., 2010). The closed-loop responses on a step-wise reference changes and input disturbances (at the mid-point of the experiment) are shown in Figures 14 and 15. It can be seen that the overshoots are reduced when using b=c=0 while retaining disturbance-rejection responses.

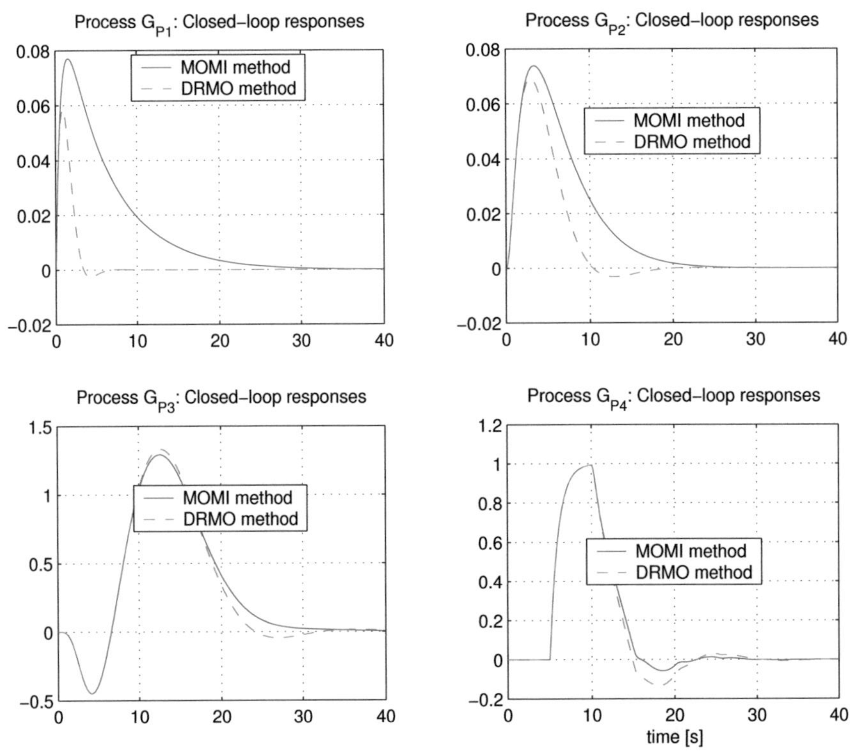

Fig. 12. A comparison of process output disturbance-rejection performance for processes G_{P1} to G_{P4} when using a PID controller tuned by the MOMI (__) and DRMO (--) tuning methods.

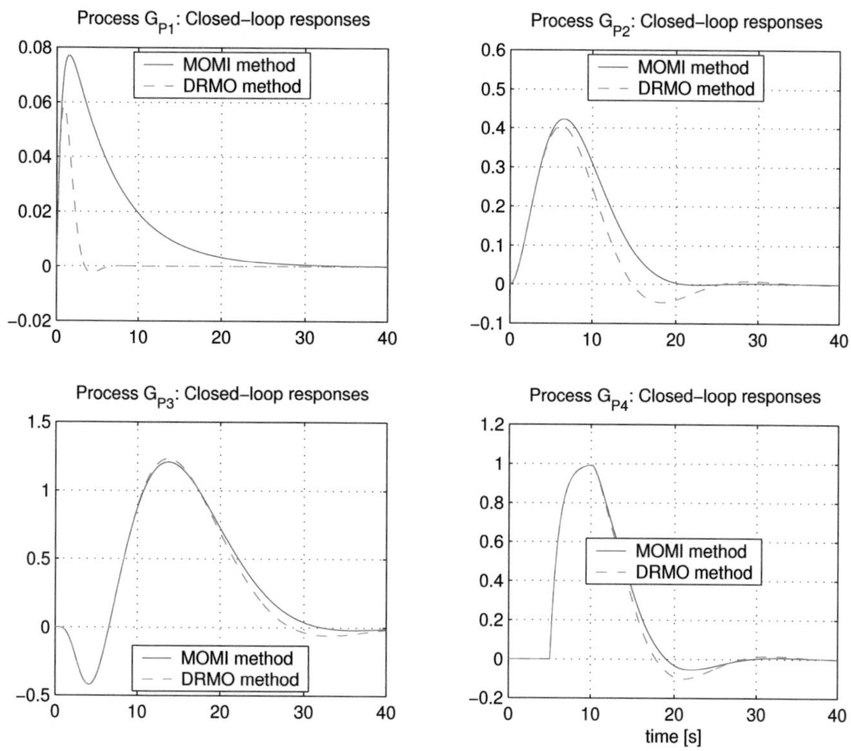

Fig. 13. A comparison of process output disturbance rejection performance for processes G_{P1} to G_{P4} when using a PI controller tuned the by MOMI (__) and DRMO (--) tuning methods.

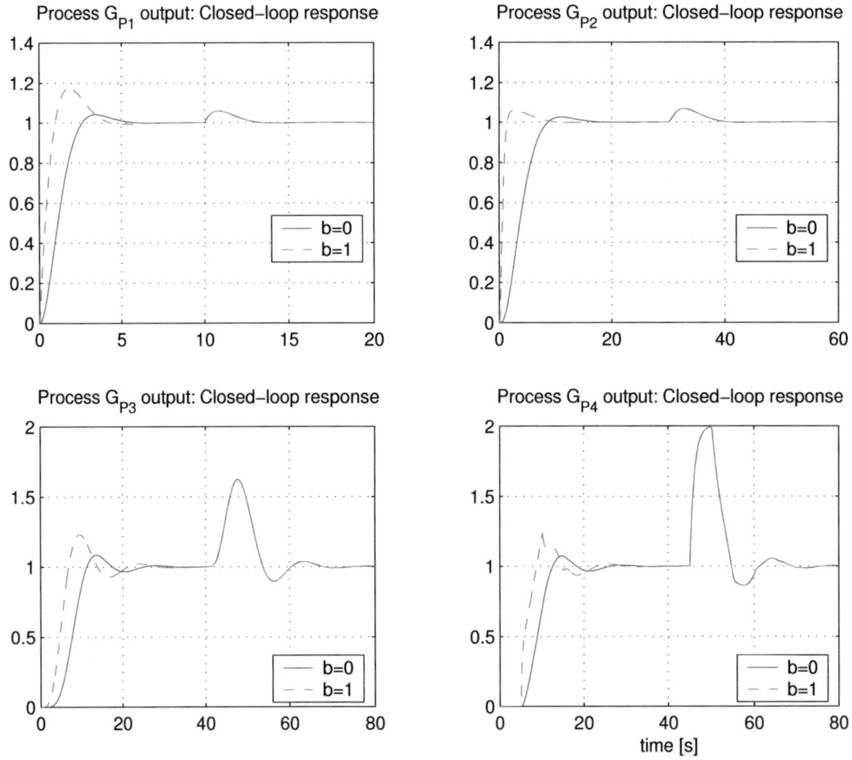

Fig. 14. Process output tracking and disturbance-rejection performance for processes G_{P1} to G_{P4} when using a PID controller tuned by the DRMO tuning method for the controller parameters b=c=0 and b=c=1.

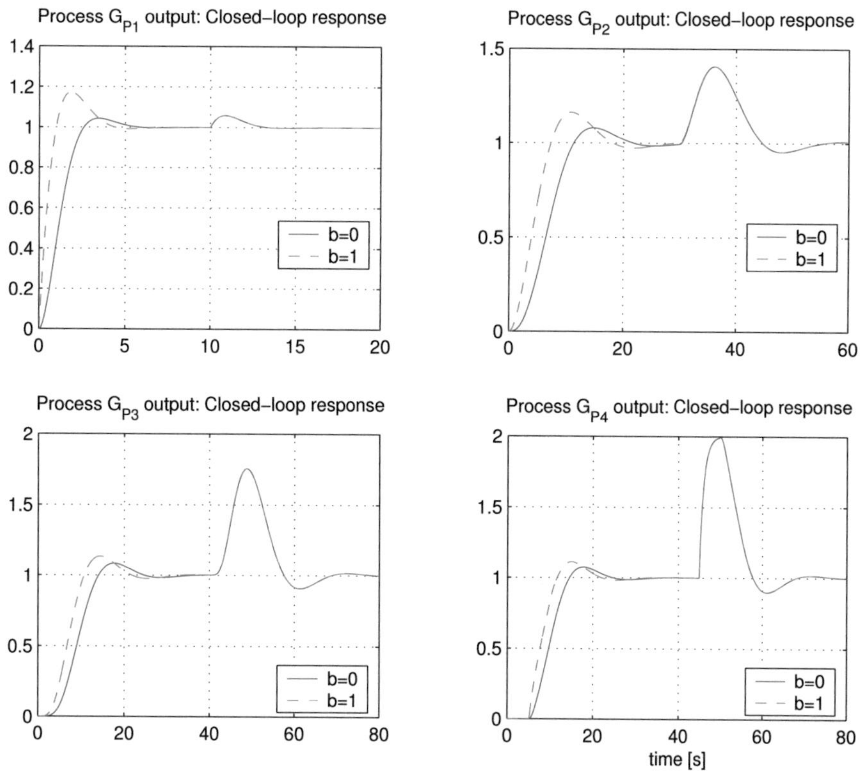

Fig. 15. Process output tracking and disturbance-rejection performance for processes G_{P1} to G_{P4} when using a PI controller tuned by the DRMO tuning method for the controller parameters b=c=0 and b=c=1.

The DRMO tuning method will be illustrated on the same three-water-column laboratory setup, described in the previous section. According to the previously calculated values of moments (39), the PID controller parameters are the following (the proportional gain has been limited to value $K_P=10/A_0$) for the chosen $T_F=1$s:

$$K_I = 0.59,\ K_P = 19.7,\ K_D = 264 \tag{47}$$

The closed-loop responses, when setting the parameter b=c=0.1, are shown in Figure 16. Similarly, as with the MOMI method, the set-point has been changed from 1.2 to 1.5 at t=300s and is returned to 1.2 at t=900s. A step-like disturbance has been added to the process input at t=700s and t=1300s. The disturbance rejection performance is now improved when compared with Figure 10. A comparison of responses obtained by the MOMI and the DRMO methods with PID controllers is shown in Figure 17. It is clear that the tracking response is slower and with a smaller overshoot, while the disturbance-rejection is significantly improved.

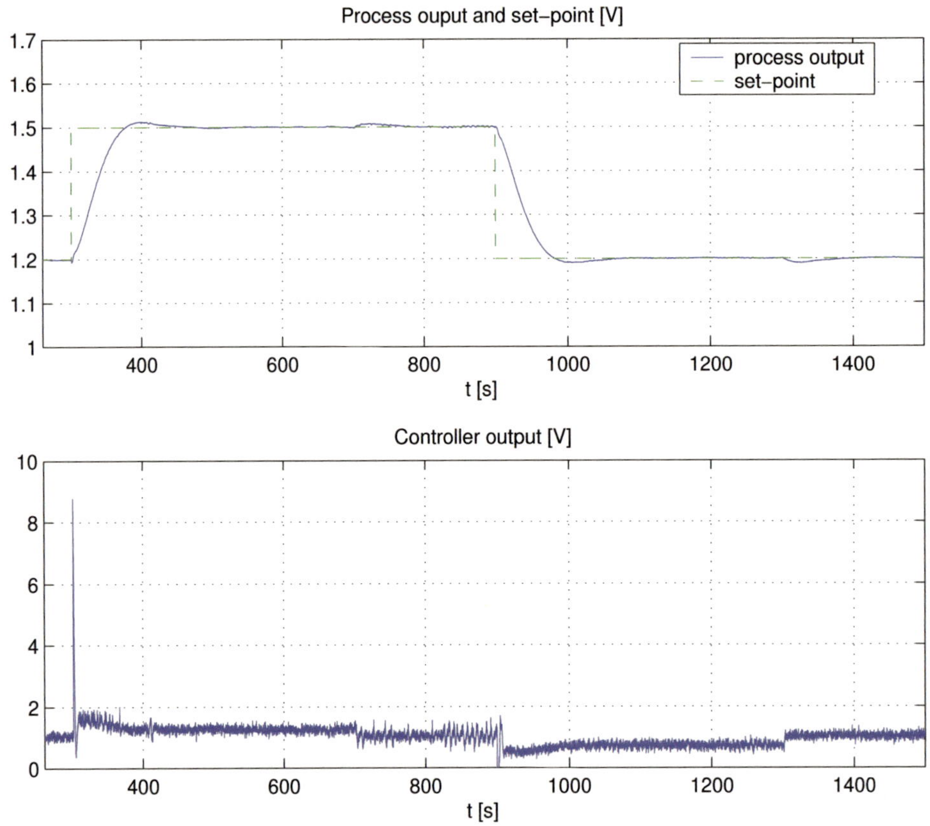

Fig. 16. The process closed-loop response in the hydraulic setup when using the PID controller tuned by the DRMO method.

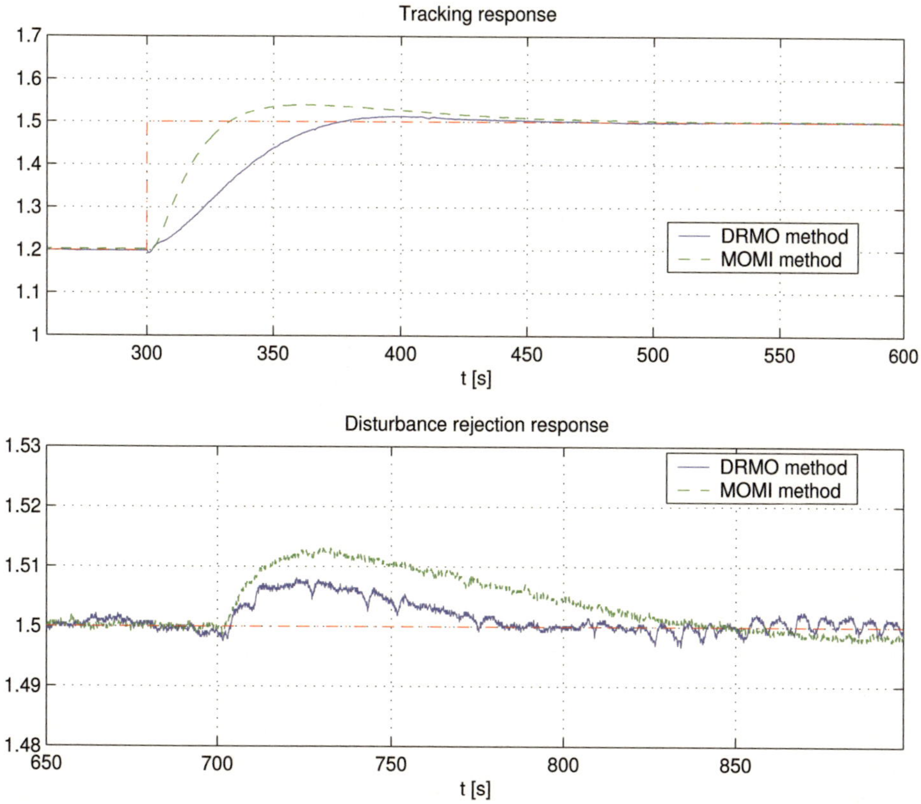

Fig. 17. A comparison of the process closed-loop responses in the hydraulic setup with PID controllers tuned by the MOMI and DRMO methods.

6. Conclusion

The purpose of this Chapter is to present tuning methods for PID controllers which are based on the Magnitude Optimum (MO) method. The MO method usually results in fast and stable closed-loop responses. However, it is based on demanding criteria in the frequency domain, which requires the reliable estimation of a large number of the process parameters. In practice, such high demands cannot often be satisfied.

It was shown that the same MO criteria can be satisfied by performing simple time-domain experiments on the process (steady-state change of the process). Namely, the process can be parameterised by the moments (areas) which can be simply calculated from the process steady-state change by means of repetitive integrations of time responses. Hence, the method is called the "Magnitude Optimum Multiple Integration" (MOMI) method. The measured moments can be directly used in the calculation of the PID controller parameters without making any error in comparison with the original MO method. Besides this, from the time domain responses, the process moments can also be calculated from the process transfer function (if available). Therefore, the MOMI method can be considered to be a universal method which can be used either with the process model or the process time-responses.

The MO (and therefore the MOMI) method optimises the closed-loop tracking performance (from the reference to the process output). This may lead to a degraded disturbance-rejection performance, especially for lower-order processes. In order to improve the disturbance-rejection performance, the MO criteria have been modified. The modification was based on optimising the integral of the closed-loop transfer function from the process input (load disturbance) to the process output. Hence, the method is called the "Disturbance-Rejection Magnitude Optimum" (DRMO) method.

The MOMI and the DRMO tuning methods have been tested on several process models and on one hydraulic laboratory setup. The results of the experiments have shown that both methods give stable and fast closed-loop responses. The MOMI method optimises tracking performance while the DRMO method improves disturbance-rejection performance. By using a two-degrees-of-freedom (2-DOF) PID controller structure, the optimal disturbance-rejection and improved tracking performance have been obtained simultaneously.

The MOMI and DRMO methods are not limited to just PID controller structures or stable (self-regulatory) processes. The reader can find more information about different controller structures and types of processes in Vrančić (2008), Vrančić & Huba (2011), Vrečko et al., (2001), Vrančić et al., (2001b) and in the references therein.

The drawback of the MO method (and therefore the MOMI method and, to an extent, the DRMO method) is that stability is not guaranteed if the controller is of a lower-order than the process. Therefore, unstable closed-loop responses may be obtained on some processes containing stronger zeros or else complex poles. Although the time-domain implementation of the method is not very sensitive to high-frequency process noise (due to multiple integrations of the process responses), the method might give sub-optimal results if low-frequency disturbances are present during the measurement of the process steady-state change.

7. Acknowledgments

The author gratefully acknowledges the contribution of the Ministry of Higher Education, Science and Technology of the Republic of Slovenia, Grant No. P2-0001.

8. References

Åström, K. J., & Hägglund, T. (1995). PID controllers: Theory, design, and tuning. Instrument Society of America Research Triangle Park (2nd ed.).

Åström, K. J., Panagopoulos, H. & Hägglund, T. (1998). Design of PI Controllers based on Non-Convex Optimization. Automatica, 34 (5), pp. 585-601.

Ba Hli, F. (1954). A General Method for Time Domain Network Synthesis. IRE Transactions - Circuit Theory, 1 (3), pp. 21-28.

Gorez, R. (1997). A survey of PID auto-tuning methods. Journal A. Vol. 38, No. 1, pp. 3-10.

Hanus, R. (1975). Determination of controllers parameters in the frequency domain. Journal A, XVI (3).

Huba, M. (2006). Constrained pole assignment control. Current Trends in Nonlinear Systems and Control, L. Menini, L. Zaccarian, Ch. T. Abdallah, Edts., Boston: Birkhäuser, pp. 163-183.

Kessler, C. (1955). Über die Vorausberechnung optimal abgestimmter Regelkreise Teil III. Die optimale Einstellung des Reglers nach dem Betragsoptimum. Regelungstechnik, Jahrg. 3, pp. 40-49.

Preuss, H. P. (1991). Model-free PID-controller design by means of the method of gain optimum (in German). Automatisierungstechnik, Vol. 39, pp. 15-22.

Rake, H. (1987). Identification: Transient- and frequency-response methods. In M. G. Singh (Ed.), Systems & control encyclopedia; Theory, technology, applications. Oxford: Pergamon Press.

Strejc, V. (1960). Auswertung der dynamischen Eigenschaften von Regelstrecken bei gemessenen Ein- und Ausgangssignalen allgemeiner Art. Z. Messen, Steuern, Regeln, 3(1), pp. 7-10

Umland, J. W. & M. Safiuddin (1990). Magnitude and symmetric optimum criterion for the design of linear control systems: what is it and how does it compare with the others? IEEE Transactions on Industry Applications, 26 (3), pp. 489-497.

Vrančić, D. (1995). A new PI(D) tuning method based on the process reaction curve, Report DP-7298, J. Stefan Institute, Ljubljana.
Available on http://dsc.ijs.si/Damir.Vrancic/bibliography.html

Vrančić, D. (2008). MOMI Tuning Method for Integral Processes. Proceedings of the 8Th Portuguese Conference on Automatic Control, Vila Real, July 21-23, pp. 595-600.

Vrančić, D. & Huba, M. (2011). Design of feedback control for unstable processes with time delay. Proceedings of the 18th International Conference on Process Control. June 14-17, Tatranska Lomnica, Slovakia, pp. 100-105.

Vrančić, D. Kocijan, J. & Strmčnik, S. (2004a). Simplified Disturbance Rejection Tuning Method for PID Controllers. Proceedings of the 5th Asian Control Conference. July 20-23, Melbourne, pp. 491-496.

Vrančić, D., Lieslehto, J. & Strmčnik, S. (2001b). Designing a MIMO PI controller using the multiple integration approach. Process Control and Quality, Vol. 11, No. 6, pp. 455-468.

Vrančić, D., Peng, Y. & Strmčnik, S. (1999). A new PID controller tuning method based on multiple integrations. Control Engineering Practice, Vol. 7, pp. 623-633.

Vrančić, D., Strmčnik S. & Kocijan J. (2004b). Improving disturbance rejection of PI controllers by means of the magnitude optimum method. ISA Transactions. Vol. 43, pp. 73-84.

Vrančić, D., Strmčnik S., Kocijan J. & Moura Oliveira, P. B. (2010). Improving disturbance rejection of PID controllers by means of the magnitude optimum method. ISA Transactions. Vol. 49, pp. 47-56.

Vrančić, D., Strmčnik S. & Juričić Đ. (2001a). A magnitude optimum multiple integration method for filtered PID controller. Automatica. Vol. 37, pp. 1473-1479.

Vrečko, D., Vrančić, D., Juričić Đ. & Strmčnik S. (2001). A new modified Smith predictor: the concept, design and tuning. ISA Transactions. Vol. 40, pp. 111-121.

Whiteley, A. L. (1946). Theory of servo systems, with particular reference to stabilization. The Journal of IEE, Part II, 93(34), pp. 353-372.

Part 3

Multivariable Systems – Automatic Tuning and Adaptation

5

Identification and Control of Multivariable Systems – Role of Relay Feedback

Rames C. Panda* and V. Sujatha
Department of Chemical Engineering, CLRI(CSIR), Adyar, Chennai, India

1. Introduction

Batch and continuous systems are of multivariable in nature. A multivariable system is one in which one input not only affects its own outputs but also one or more other outputs in the plant. Multivariable processes are difficult to control due to the presence of the interactions. Increase in complexity and interactions between inputs and outputs yield degraded process behavior. Such processes are found in process industries as they arise from the design of plants that are subject to rigid product quality specifications, are more energy efficient, have more material integration, and have better environmental performance. Most of the unit operations in process industry require control over product rate and quality by adjusting one/more inputs to the process; thus making multivariable systems. For example, chemical reactors, distillation column, heat exchanger, fermenters are typical multivariable processes in industry. In case of chemical reactor, the output variables are product composition and temperature of reaction mass. The input variables are reactant or feed flow rate and energy added to the system by heating/ cooling through jackets. Product composition can be controlled by manipulating feed rate whereas rate of reaction (thereby temperature) can be controlled by changing addition/ removal rate of energy. But, while controlling product composition, temperature is affected; similarly, while controlling temperature of reaction mass, the composition gets affected, thus, exhibiting interactions between input and output variables. Distillation is widely used for separating components from mixture in refineries. Composition of top and bottom products are controlled by adjusting energy input to the column. A common scheme is to use reflux flow to control top product composition whilst heat input is used to control bottom product composition. However, changes in reflux also affect bottom product composition and component fractions in the top product stream are also affected by changes in heat input. Hence, loop interactions occur in composition control of distillation column. Thus, unless proper precautions are taken in terms of control system design, loop interactions can cause performance degradation and instability. Control system design needs availability of linear models for the multivariable system.
The basic and minimum process model for multivariable system is considered here as 2x2 system. The outputs of the loops are given by

---

* Corresponding Author

$$\begin{aligned} y_1 &= u_1 G_{P11} + u_2 G_{P12} \\ y_2 &= u_2 G_{P21} + u_2 G_{P22} \end{aligned} \tag{1}$$

where y_i are system outputs and u_i are the system inputs, G is system transfer functions. Eqn (1) can be expressed as $y = G_P u$ where

$$y = [y_1 \;\; y_2]^T \; ; \quad u = [u_1 \;\; u_2]^T \; ; \quad G_P = \begin{bmatrix} G_{P11} & G_{P12} \\ G_{P21} & G_{P22} \end{bmatrix} \tag{2}$$

In order to achieve desired quality, specified output characteristics at the cost of spending optimum inputs one needs to design a controller and run the plant under closed loop so that optimal production of product under safe operation. The first thing we need is to select input-output pairs, i.e., which output should be controlled by which input? This needs knowledge in control structure selection or interaction analysis. In the next section, a brief state of art on interaction analysis is presented.

Relative gain array (RGA) (Bristol 1966) is the most discussed method for analyzing interactions and it is based on steady state gain information of MIMO processes. Control loops should have input-output pairs which give positive relative gains that have values which are as close as unity as possible. It is dependent on process models, independent of scaling of inputs and outputs and can include all ways of pairing in a single matrix. Niederlinski index (NI) is a useful tool to analyse interactions and stability of the control loop pairings determined using process gain matrix. NI is found by the following formula, $NI = \frac{|G_P|}{\prod_{i-1}^{n} g_{ii}} |_{SS}$ where each element of G_P is rational and is openloop stable. The values of NI need to be positive. A negative value of NI will imply that the system is un-stable. Ni is used to check if the system (more than 2x2) is unstable or not. NI will detect instability introduced by closing the other control loops. Generally, NI is not used for systems with time delays. Any loop pairing is unacceptable if it leads to a control system configuration for which the NI is negative. But both RGA & NI do not provide dynamic information on the process transients. They do not give information on change in in/op pairing for instances when there is a sudden load disturbance. Singular value decomposition (SVD) is a useful tool to determine whether a system will be prone to control loop interactions resulting in sensitivity problems that rises from model mismatch in process gains. SVD considers directional changes in the disturbances. SVD is applied to steady state gain matrix that is decomposed into product of three matrices,

$S = U\Sigma V^T$ where U is matrix of normalized eigen vectors of GG_P^T, Σ is diagonal matrix of eigenvalues and V is matrix of normalized eigenvectors of $G_P^T G_P$ The condition number (CN) is defined as ratio between maximum and minimum eigenvalues. Generally if the CN < 50 then the system is not prone to sensitivity problems (a small error in process gain will not cause a large error in the controller's reactions). The greater the CN value, the harder it is for the system in question to be decoupled. An ideal system would have a CN number of one, where each control variable controls a single distinct output variable. CN value tells us how easy it is to decouple a system. Though SVD has good geometric interpretation in terms of selection of measurement and pairing of variables, SVD depends on input-output scaling.

Moreover, with weak interactions and with large dimensional systems they induce to go for more criteria for selection of pairs. Morari resiliency index (MRI) is also used to select in/out pairs. $MRI = \sigma\left(G_P(j\omega)\right)$ where σ is eigenvalue. The MRI is the minimum singular value *(g)* of the plant transfer function matrix G(iw). The set of manipulated variables that gives the largest minimum singular value over the frequency range of interest is the best. The MRI is a measure of the inherent ability of the process (control structure) to handle disturbances, model plant mismatches, changes in operating conditions, etc. The larger the value of MRI, the more resilient the control structure. Dynamic Relative Gain Array (DRGA) is defined to extend the RGA notion to non-zero frequencies. The RGA provides only limited knowledge about when to use multivariable controllers and gives no indication of how to choose multivariable controller structures. A somewhat different approach for investigating channel interaction was therefore employed by Conley and Salgado (2000) and Salgado and Conley (2004) when considering observability and controllability gramians in so called Participation Matrices (PM). In a similar approach Wittenmark and Salgado (2002) introduced the Hankel Interaction Index Array (HIIA). These gramian based interaction measures seem to overcome most of the disadvantages of the RGA. One key property of these is that the whole frequency range is taken into account in one single measure. Furthermore, these measures seem to give appropriate suggestions for controller structures selection. The use of the system H_2 norm as a base for an interaction measure has been proposed by Birk and Medvedev (2003) as an alternative to the HIIA. But, dynamic simulation is a powerful tool to be used to test the viability of a control scheme during various process disturbances. Controllers for MIMO systems can be of either multiloop (controllers are designed only for diagonal elements of process TF) or multivariable (controllers are designed for all the elements of the MIMO TF). Multiloop control scheme has an edge over multivariable as the former can work even if a single loop fails. In presence of interactions between input/output, the process need to be decoupled and then multiloop controllers can be designed. When interaction effects produce a significant deterioration in control system performance, decoupling control should be considered. One of the most powerful and simplest ways of reducing or eliminating interaction is by altering manipulated and / or controlled variables. Improvement of closed-loop performance needs proper tuning of controller parameters that requires process model structure and estimation of respective parameters. There are many methods to select input/output pairs or to design control structures, design control strategy (either PID or IMC or predictive or heuristics etc.) and tuning of controller parameters in literature. But because of hazy pictures on above selections, till today, it is difficult to choose correct pairs, carryout interaction analysis and choose tuning rules. Thus the aim of this chapter is to bring out a clear picture of identifying process parameters and designing controller for MIMO systems. The rest of the chapter is carried out as follows: section 2 discusses identification methods of multivariable systems. Interaction analysis is explained in section 3. Control structure selection and determination of input/output pairs are given in section 4. Tuning of controllers is presented in section 5. Stability analysis for multivariable systems is provided in section 6. At the end, conclusion is drawn.

2. System identification

Most of the chemical and bio-chemical processes are multivariable in nature, having more than one input and outputs. Estimation of process parameters is a key element in multivariable controller design. Thus, as better performance is achieved by model based tuning algorithms, estimation of model structures are necessary from either open-loop or closed-loop data. This is due to the fact that tuning rules are based on model structures & parameters. As their exist advantages and disadvantages in both of these identification strategies, for example, open-loop responses may show unstable behavior with certain inputs, whereas, closed-loop strategy needs more excitation to yield observable response. Here we use mostly used methods of identification for multivariable systems. Least square method (Tungnait 1998) is an old but reliable technique that was in use to estimate multivariable parameters of open-loop systems. But, MIMO systems with interactions may not yield satisfactory transfer function estimates with these techniques. Overschee and Moor (1994) proposed subspace method of identification that mostly applies to identification of multivariable state space models. This method involves more computational time. Practical industrial plants are easy to identify in closed-loop using relay feedback method (Astrom and Hagguland 1984) and Yu (1999) explains advances in autotuning using sequential identification. System identification is the method of estimating parameters from system's input/output data using numerical techniques:

2.1 Transfer function identification

Model structures and parameters of transfer function are constructed from observed plant input output data. Transfer function models are developed using three schemes: (a) Least square (b) subspace and (c) sequential identification method. These approximations made out through each of the methods carry errors that propagate to controller tuning and in turn deteriorates the overall performance.

2.1.1 Least-squares method

Least-squares method, used to reduce the mean square error, is very simple and more numerically stable and can be used to identify the unknown parameters of the 2x2 MIMO transfer function model from the input (u) and output (y) data. Though any type of forcing function (step, pulses or a sequence of positive and negative pulses) can be used, a very popular sequence of inputs, "Pseudo-random binary sequence" (PRBS) is made use of in the present work.

Let us consider a process with continuous transfer function

$$\frac{y(s)}{u(s)} = G(s) = \frac{K_p e^{-Ds}}{\tau s + 1} \tag{2.1}$$

The pulse transfer function of this process with a zero-order hold is

$$\frac{y(z)}{u(z)} = HG(z) = \frac{K_p(1-b)}{z^{nk}(z-b)} = \frac{K_p(1-b)z^{-1}}{z^{nk}\left(1-bz^{-1}\right)} = \frac{z^{-nk}\left(b_0 + b_1 z^{-1}\right)}{1 + a_1 z^{-1}} \tag{2.2}$$

where $n_k = D/T_s$; T_s=sampling period;

$$b = e^{-T_s/\tau} ; b_o = 0; b_1 = k_p(1-b); a_1 = -b$$

The discrete transfer function has three parameters that need to be identified: dead time (D) contained in n_k, and other two parameters of the model (k_p and τ) contained in b_1, and a_1. The discrete output can be represented in the following form:

$$\overline{y_n} = b_1 u_{n-1} + b_2 u_{n-2} + ... + b_{nb} u_{n-nb} - a_1 y_{n-1} - a_2 y_{n-2} - .... - a_{na} y_{n-na} \tag{2.3}$$

where $\overline{y_n}$ is the predicted value of the current output of the process. For a FOPDT process, equation (2.3) can be written as

$$y(k) + a_1 y(k-1) = b_1 u(k-1) \tag{2.4}$$

which can be written in matrix form as

$$y = \phi\theta + e \tag{2.5}$$

where

$$\phi = \begin{bmatrix} -y(0) & u(0) \\ -y(1) & u(1) \\ .... & .... \\ -y(N-1) & u(N-1) \end{bmatrix} \quad \text{and } \theta = \begin{bmatrix} a_1 \\ b_1 \end{bmatrix}$$

The parameters a_1 and b_1 are calculated using

$$\theta = \left(\phi^T \phi\right)^{-1} \phi^T y \tag{2.6}$$

where θ is the parameter vector ϕ is state matrix and y is outputs.

2.2 State-space model

In the state space form the relationship between the input, noise and output signals are written as a system of first-order differential or difference equations using auxillary state vectors. Transfer function in laplace domain is converted to state space form using a sampling period of 0.1s

2.2.1 Subspace method

The beginning of the 1990s witnesses the birth of a new type of linear system identification algorithms, called subspace method. Subspace identification methods are indeed attractive since a state-space realization can be directly estimated from input/output data without nonlinear optimization. Furthermore, these techniques are characterized by the use of robust numerical tools such as RQ factorization and the singular values decomposition (SVD). Interesting from numerical point of view, the batch subspace model identification (SMI) algorithms are not usable for online implementation because of the SVD computational complexity. Indeed, in many online identification scenarios, it is important to update the model as time goes on with a reduced computational cost.

Linear subspace identification methods are concerned with systems and models of the form

$$x_{k+1} = Ax_k + Bu_k + w_k \tag{2.7}$$

$$y_k = Cx_k + Du_k + v_k \tag{2.8}$$

with

$$E\left[\begin{pmatrix} w_p \\ v_p \end{pmatrix}\begin{pmatrix} w_q^{\,T} & v_q^{\,T} \end{pmatrix}\right] = \begin{pmatrix} Q & S \\ S^T & R \end{pmatrix}\delta_{pq} \tag{2.9}$$

The vectors $u_k \in R^{mx1}$ and $y_k \in R^{lx1}$ are the measurements at time instant k of, respectively, the m inputs and l outputs of the process. The vector x_k is the state vector of the process at discrete time instant k, $v_k \in R^{lx1}$ and $w_k \in R^{nx1}$ are unobserved vector signals, v_k is called the measurement noise and w_k is called the process noise. It is assumed that they are zero mean, stationary white noise vector sequences and uncorrelated with the inputs u_k. $A \in R^{nxn}$ is the system matrix, $B \in R^{nxm}$ is the input matrix, $C \in R^{lxn}$ is the output matrix while $D \in R^{lxm}$ is the direct feed-through matrix. The matrices $Q \in R^{nxn}$, $S \in R^{nxl}$ and $R \in R^{lxl}$ are the covariance matrices of the noise sequences w_k and v_k.

In subspace identification it is typically assumed that the number of available data points goes to infinity, and that the data is ergodic. The main problem of identification is arranged as follows:

Given a large number of measurements of the input u_k and the output y_k generated by the unknown system described by equations (2.7)-(2.9). The task is to determine the order n of the unknown system, the system matrices A, B, C, D up to within a similarity transformation and an estimate of the matrices Q, S and R.

Subspace identification algorithms always consist of two steps:

Step 1: Make a weighted projection of certain subspace generated from the data, to find an estimate of the extended observability matrix, Γ_i and/or an estimate $\tilde{X}_i$ of the state sequence X_i of the unknown system

Step 2: Retrieve the system matrices (A, B, C, D and Q, S, R) and from either this extended observability matrix (Γ_i) or the estimated states.

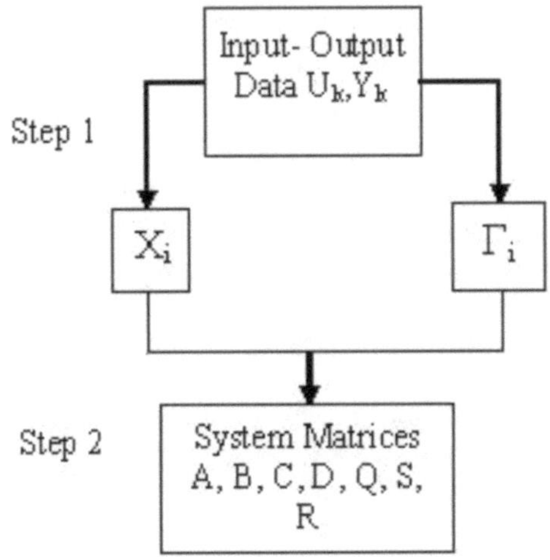

Fig. 1. Flow chart of subspace algorithm.

All the above identification methods involve more computations and many offline methods. These difficulties can be avoided easily by using another method of estimation technique, namely, relay feedback method as explained below:

2.3 Sequential identification

Based on the concept of sequential auto tuning (Shen & Yu, 1994) method each controller is designed in sequence. Let's consider a 2-by-2 MIMO system with a known pairing $(y_1 - u_1)$ and $(y_2 - u_2)$ under decentralized PI control (Figure 1). Initially, an ideal / biased relay is placed between y_1 and u_1, while loop 2 is on manual (Figure 2a). Following the relay-feedback test, a controller can be designed from the ultimate gain and ultimate frequency. The next step is to perform relay-feedback test between y_2 and u_2 while loop 1 is on automatic (Figure 2b). A controller can also be designed for loop 2 following the relay-feedback test. Once the controller on the loop 2 is put on automatic, another relay-feedback experiment is performed between y_1 and u_1, (Figure 2c). Generally, a new set of tuning constants is found for the controller in loop 1. This procedure is repeated until the controller parameters converge. Typically, the controller parameters converge in 3 - 4 relay-feedback tests for 2 x 2 systems.

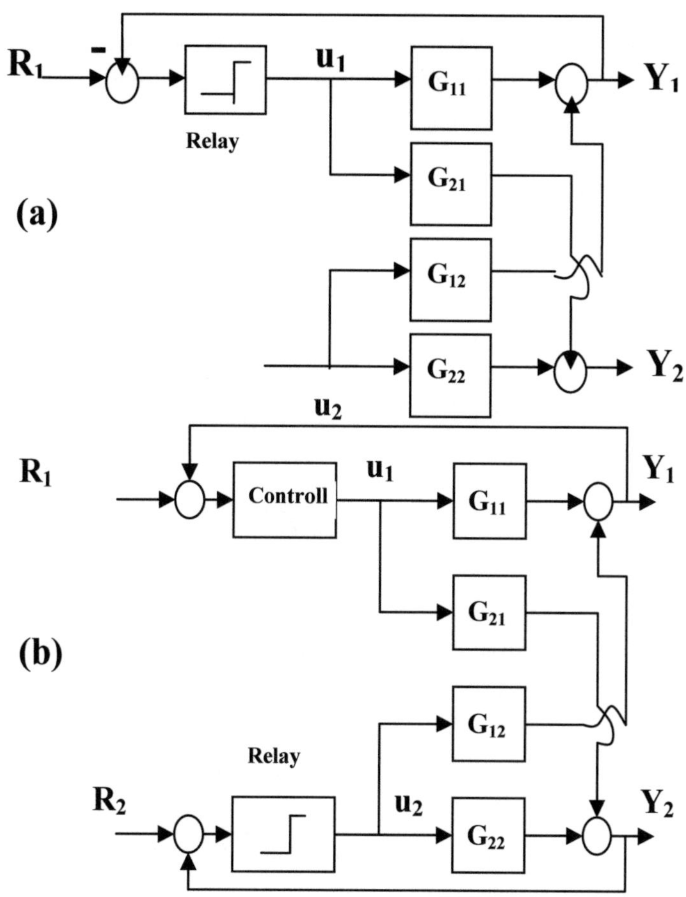

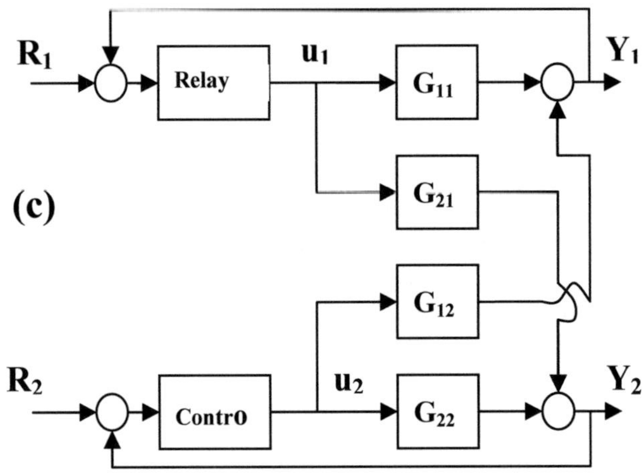

Fig. 2. Sequential method of tuning for 2x2 multivariable system. Steps are: (a) followed by (b) and followed by (c).

In order to proceed with sequential identification, it is necessary to derive closed-loop transfer functions for the above mentioned schemes. The following notations will be used for 2-by- 2 MIMO system:

$$g_p(s)=\begin{pmatrix} g_{p11} & g_{p12} \\ g_{p21} & g_{p22} \end{pmatrix} \quad G_c(s)=\begin{pmatrix} G_{c1} & 0 \\ 0 & G_{c2} \end{pmatrix} \quad y(s)=\begin{pmatrix} y_1 \\ y_2 \end{pmatrix} \text{ and } u(s)=\begin{pmatrix} u_1 \\ u_2 \end{pmatrix} \tag{2.10}$$

Thus, when perturbation is introduced in the second input u_2, transfer functions for the input $u_2(s)$ are

$$y_1 = G_{p12,CL}(s)u_2(s) = -\frac{g_{p12}(s)}{1+g_{p11}(s)G_{c1}(s)}u_2(s) \tag{2.11}$$

$$y_2(s) = G_{p22,CL}(s)u_2(s) = \left[g_{p22}(s) - \frac{g_{p21}(s)G_{c1}(s)g_{p12}(s)}{1+g_{p11}(s)G_{c1}(s)} \right] u_2(s) \tag{2.12}$$

By applying the above identification method to the 2nd loop (by collecting output y_2 for the change in input u_1), we can obtain models for $Gp_{12,CL}(s)$ and $Gp_{22,CL}(s)$. Then, we have

$$y_2 = G_{p21,CL}(s)u_1(s) = -\frac{g_{p21}(s)}{1+g_{p22}(s)G_{c2}(s)}u_1(s) \tag{2.13}$$

$$y_1(s) = G_{p11,CL}(s)u_1(s) = \left[g_{p11}(s) - \frac{g_{p21}(s)G_{c2}(s)g_{p12}(s)}{1+g_{p22}(s)G_{c2}(s)} \right] u_1(s) \tag{2.14}$$

From the identified step response models of $Gp_{12,CL}(s)$ and $Gp_{22,CL}(s)$, we can obtain their frequency response data and, by fitting them, we can get approximate low order models. Time domain modeling is obtained using equations (2.15) and (2.16) for 2x2 and 3x3 MIMO process with FOPDT models using relay feedback test as:

$$y_n = k_{21}\left(1-e^{-\frac{t}{\tau_{21}}}\left(\frac{2}{1+e^{-\frac{p_u}{2\tau_{21}}}}\right)\right)-\frac{k_{22}k_{c2}}{\tau_{i2}}\left(1+(\tau_{i2}-\tau_{22})e^{-\frac{t}{\tau_{22}}}\left(\frac{2}{1+e^{-\frac{p_u}{2\tau_{22}}}}\right)\right)y_n(t-1) \quad (2.15)$$

$$\begin{aligned} y_n = {} & k_{21}\left(1-e^{-\frac{t}{\tau_{21}}}\left(\frac{2}{1+e^{-\frac{p_u}{2\tau_{21}}}}\right)\right)-\frac{k_{22}k_{c2}}{\tau_{i2}}\left(1+(\tau_{i2}-\tau_{22})e^{-\frac{t}{\tau_{22}}}\left(\frac{2}{1+e^{-\frac{p_u}{2\tau_{21}}}}\right)\right) \\ & -\left(\left[\left[1-\left(a_1 e^{-t/\tau_{i3}}\left(\frac{2}{1+e^{-\frac{p_u}{2\tau_{i3}}}}\right)+a_2 e^{-t/\tau_{23}}\left(\frac{2}{1+e^{-\frac{p_u}{2\tau_{23}}}}\right)+a_3 e^{-t/\tau_{33}}\left(\frac{2}{1+e^{-\frac{p_u}{2\tau_{33}}}}\right)\right)\right]\right]-\right. \\ & \left.\frac{k_{33}k_{c3}}{\tau_{i3}}\left[1+(\tau_{i3}-\tau_{33})e^{-t/\tau_{33}}\left(\frac{2}{1+e^{-\frac{p_u}{2\tau_{33}}}}\right)\right]\right) \end{aligned} \quad (2.16)$$

2.4 Process dynamics of example under study

Wood and Berry (1973) (WB) reported a column for methanol-water separation with transfer function as given below

$$\begin{bmatrix} x_D \\ x_B \end{bmatrix} = \begin{bmatrix} \dfrac{12.8e^{-s}}{16.7s+1} & \dfrac{-18.9e^{-3s}}{21s+1} \\ \dfrac{6.6e^{-7s}}{10.9s+1} & \dfrac{-19.4e^{-3s}}{14.4s+1} \end{bmatrix} \begin{bmatrix} L \\ V \end{bmatrix} \quad (2.17)$$

The compositions of top (x_D) and bottom (x_B) products expressed in wt% of methanol are controlled variables. The reflux (L) and the reboiler (V) steam flow rates are the manipulated inputs are expressed in lb/min. time constants are in minutes. Feed flow rate is disturbance. Here the input variables are liquid (L) and vapour (V) flow rates (where as feed (F) flow rate is the load); outputs are distillate (x_D) and bottom (x_B) compositions. This plant given by Eq.(2.15) is considered as actual or real plant-model in present work.

On applying least square algorithms to individual transfer function elements of an unknown 2x2 MIMO process (WB column) the estimated transfer function is obtained as shown in Table 1.The output (y) and input data (to original WB plant transfer function) are used to form matrix. The parameters a_1 and b_1 were calculated using Eq.(2.6).

On applying subspace algorithms to an unknown 2x2 MIMO process (WB column) the following steps are followed

Step 1: From the transfer function matrix State space representation matrices are calculated.

Step 2: A, B, C and D matrices are simulate to get output data for a random input signal.

Step 3: From the output and input data Henkel matrix are formed and LQ decomposition method is used to spilt the matrix
Step 4: Then Singular value decomposition method is used to estimate A, B, C and D matrices.
Step 5: From estimated matrices the transfer function were found.

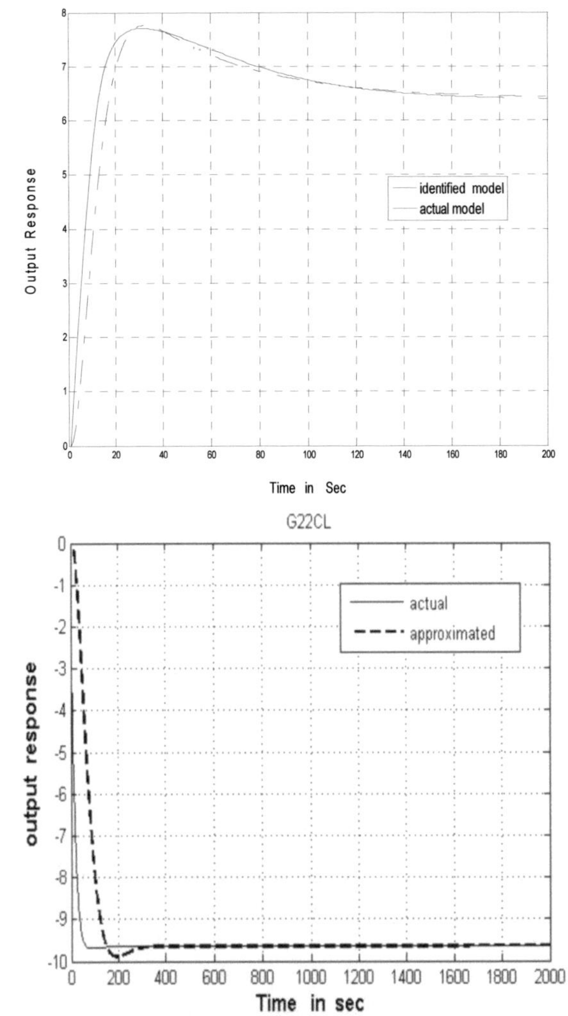

Fig. 3. Comparison of responses between actual (solid) and identified (Sequential identification, dashed line) models of WB column

Mostly, the purpose of identification of transfer functions is to design controller for the system in order to achieve desired performance. Three methods of identifications (two in openloop mode and the other in closed-loop mode) are used to identify the two-input-two-output process, WB column. Least square and subspace methods have been used to identify the process in openloop and sequential identification technique is used to estimate the process in closedloop.

The identified models and actual plant model are compared (Table-2.1). It is found that subspace identification method gives better result/

ACTUAL WB COLUMN	ESTIMATED TRANSFER FUNCTION		
	LEAST SQUARE	SUBSPACE	SEQUENTIAL
$\begin{bmatrix} \frac{12.8e^{-s}}{16.7s+1} & \frac{-18.9e^{-3s}}{21s+1} \\ \frac{6.6e^{-7s}}{10.9s+1} & \frac{-19.4e^{-3s}}{14.4s+1} \end{bmatrix}$	$\begin{bmatrix} \frac{12.692e^{-s}}{16.41s+1} & \frac{-18.89e^{-3s}}{21.46s+1} \\ \frac{6.4e^{-7s}}{10.95s+1} & \frac{-19.6e^{-3s}}{14.481s+1} \end{bmatrix}$	$\begin{bmatrix} \frac{12.799e^{-s}}{16.7541s+1} & \frac{-18.9e^{-3s}}{21.054s+1} \\ \frac{6.600e^{-7s}}{10.9505s+1} & \frac{-19.398e^{-3s}}{14.448s+1} \end{bmatrix}$	$G_{11,CL} = \frac{6.4e^{-s}}{42.25s^2+11.7s+1}\left(\frac{60s+1}{44s+1}\right)$ $G_{22,CL} = \frac{-9.655e^{-3s}}{1453s^2+58.15s+1}\left(\frac{0.4774s+1}{2.741s+1}\right)$

Table 2.1. Actual and estimated multivariable transfer functions using different methods

After identifying the model structures and estimating process parameters of the models, next work is to select a suitable control strategy for the process.

3. Different control strategies

MIMO systems came into use in chemical industries as the processes were redesigned to improve efficiency. Multivariable control involves the objective of maintaining several controlled variables at independent set points. Interaction between inputs and output cause a manipulated variable to affect more than one controlled variable. The various control schemes studied here are the decentralized, centralized and decoupled systems. In decentralized structure, diagonal controllers are used. Hence they result in systems having n controllers. The centralized control systems have n x n controllers. In decoupled systems the process interactions are decoupled before they can actually reach and affect the processes.

3.1 Centralized structure

Centralized control scheme is a full multivariable controller where the controller matrix is not a diagonal one. The decentralized control scheme is preferred over the centralized control scheme mainly because the control system has only n controlling n output variables, and the operator can easily understand the control loops. However, the design methods of such decentralized controllers require first pairing of input-output variables, and tuning of controllers requires trial and error steps. The centralized control system requires n x n controllers for controlling n output variables using n manipulated variables. But if we are calculating the control action using a computer, then this problem of requiring n x n controllers does not exist. The advantage of the centralized controller is easy to tune even with the knowledge of the steady state gain matrix alone, multivariable PI controllers can be easily designed.

For the centralized structure, Internal model control-proportional integral tuning is adopted, based on studies on the studies and recommendations of Reddy et al (1997) on the design of centralized PI controllers for a Multi-stage flash desalination plant using Davison, Maciejowski and Tanttu-Lieslehto methods.

The IMC-PID tuning relations are used in tuning the controller. For a first order system of the form $\frac{k_p e^{-Ds}}{(\tau s+1)}$, the PI controller settings are as follows:

$$k_c = \frac{\tau}{k_p \lambda} \tag{3.1}$$

$$\tau_i = \tau \tag{3.2}$$

where $\lambda = \max(1/0.7D, 0.2\tau)$

These tuning relations are derived by comparing IMC control with the conventional PID controller and solving to determine the proportional gain and integral time.

3.2 Decentralized structure

In spite of developments of advanced controller synthesis for multivariable controllers, decentralized controller remain popular in industries because of the following:

1. Decentralized controllers are easy to implement.
2. They are easy for operators to understand.
3. The operators can easily retune the controllers to take into account the change in process conditions.
4. Some manipulated variables may fail. Tolerances to such failures are more easily incorporated into the design of decentralized controllers than full controllers.
5. The control system can be bought gradually into service during process start up and taken gradually out of service during shut down.

The design of a decentralized control system consists of two main steps:
Step 1 is control structure selection and step 2 is the design of a SISO controller for each loop.
In decentralized control of multivariable systems, the system is decomposed into a number of subsystems and individual controllers are designed for each subsystem.
For tuning the controller, Biggest Log Modulus Tuning (BLT) method (Lubed 1986) is used, which is an extension of the Multivariable Nyquist Criterion and gives a satisfactory response. A detuning factor F (typical values are said to vary between 2 and 5) is chosen so that closed-loop log modulus, L_{cm}^{max} >= 2n,

$$L_{cm} = 20\log\left|\frac{w}{1+w}\right| \tag{3.3}$$

$$w = -1 + \det\left(I + G_p G_c\right) \tag{3.4}$$

where G_c is an n x n diagonal matrix of PI controller transfer functions, G_p is an n x n matrix containing the process transfer functions relating the n controlled variables to n manipulated variables.
Now the PI controller parameters are given as,

$$k_{ci} = {k_{ciZ-N}}/{F} \tag{3.5}$$

$$\tau_{Ii} = F\tau_{IiZ-N} \tag{3.6}$$

where k_{ciZ-N} and τ_{IiZ-N} are Zeigler-Nichols tuning parameters which are calculated from the system perturbed in closed loop by a relay of amplitude h, reaches a limit cycle whose amplitude a and period of oscillation P, are correlated with the ultimate gain (k_u) and frequency (w_u) by the following relationships:

$$k_u = \frac{4h}{\pi a} \tag{3.7}$$

$$\omega_u = \frac{2\pi}{P_u} \tag{3.8}$$

Detuning factor F determines the stability of each loop. The larger the value of F, more stable the system is but set point and load responses are sluggish. This method yields settings that give a reasonable compromise between stability and performance in multivariable systems.

The decentralized scheme is more advantageous in the fact that the system remains stable even when one controller goes down and is easier to tune because of the less number of tuning parameters. But however pairing (interaction) analysis needs to be done as n! pairings between input/output are possible.

3.3 Decoupled structure

This structure has additional elements called decouplers to compensate for the interaction phenomenon. When Relative gain Array shows strong interaction then a decoupler is designed. But however decouplers are designed only for orders less than 3 as the design procedure becomes more complex as order increases.

The BLT (Luyben 1986) procedure of tuning the decentralized structure follows the generalized way for all n x n systems as mentioned above. The centralized controllers are tuned using the IMC-PI tuning relations which are appropriately selected for first order and second order systems.

The decoupled structure adopts the various methods like partial, static and dynamic decoupling to procedure the best results. The design equations for a general decoupler for n x n systems are conveniently summarized using matrix notations defined as follows:

$$G = \begin{bmatrix} G_{11}(s) & G_{1n}(s) \\ G_{n1}(s) & G_{nn}(s) \end{bmatrix}; D = \begin{bmatrix} D_{11}(s) & D_{1n}(s) \\ D_{n1}(s) & D_{nn}(s) \end{bmatrix}; H = \begin{bmatrix} H_{11}(s) & \dots & 0 \\ \dots & H_{22}(s) & \dots \\ 0 & \dots & H_{nn}(s) \end{bmatrix}$$

Transfer function matrix; Decoupler matrix; Diagonal matrix of decoupler

$$u = \begin{bmatrix} u_1 \\ \dots \\ u_n \end{bmatrix}; \qquad M = \begin{bmatrix} M_1 \\ \dots \\ M_n \end{bmatrix}; \qquad C = \begin{bmatrix} C_1 \\ \dots \\ C_n \end{bmatrix}$$

Manipulated variable (new) Manipulated variable (old) Output

For a decoupled multivariable system, output can be written as

$$C = GM \tag{3.9}$$

$$M = Du \tag{3.10}$$

The equation (3.10) becomes,

$$C = GDu \quad (3.11)$$

The equation (3.11) becomes,

$$C = Hu \quad (3.12)$$

where,

$$GD = H \quad (3.13)$$

or

$$D = G^{-1}H \quad (3.14)$$

which defines the decoupler

For a 2 x 2 system, equations are derived for decouplers, taking that loop and the other interacting loops into account.

3.4 Examples

3.4.1 Centralized controller

A first order plus dead time process with $k_p = 1$, $\tau_p = 1$ and $D_p = 0.25$ is chosen for simulation study. The controller is designed with a first order filter with $\lambda = 1.4286$, $k_c = 0.7$ and $\tau_I = 1$. Closed loop responses with the present controller are obtained. The results are shown below:

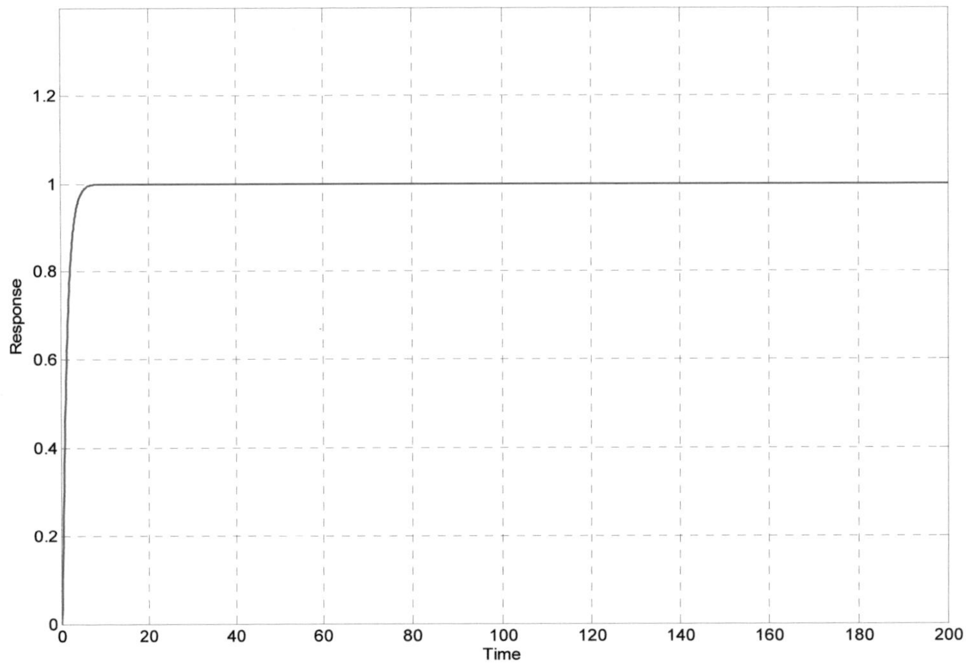

Fig. 4. Closed-loop response of example -processes using PID controller

3.4.2 Decentralized controller

The wood and berry distillation column process whose transfer function

$$\begin{bmatrix} \dfrac{12.8e^{-s}}{16.7s+1} & \dfrac{-18.9e^{-3s}}{21s+1} \\ \dfrac{6.6e^{-7s}}{10.9s+1} & \dfrac{-19.4e^{-3s}}{14.4s+1} \end{bmatrix}$$

is chosen for simulation study. The controller is designed using BLT method with F=2.55, $k_{c1} = 0.375$, $\tau_{I1} = 8.29$ (loop 1 controller settings) and $k_{c2} = -0.075$, $\tau_{I2} = 23.6$ (loop 2 controller settings). With these settings, the closed loop responses are obtained and are shown below.

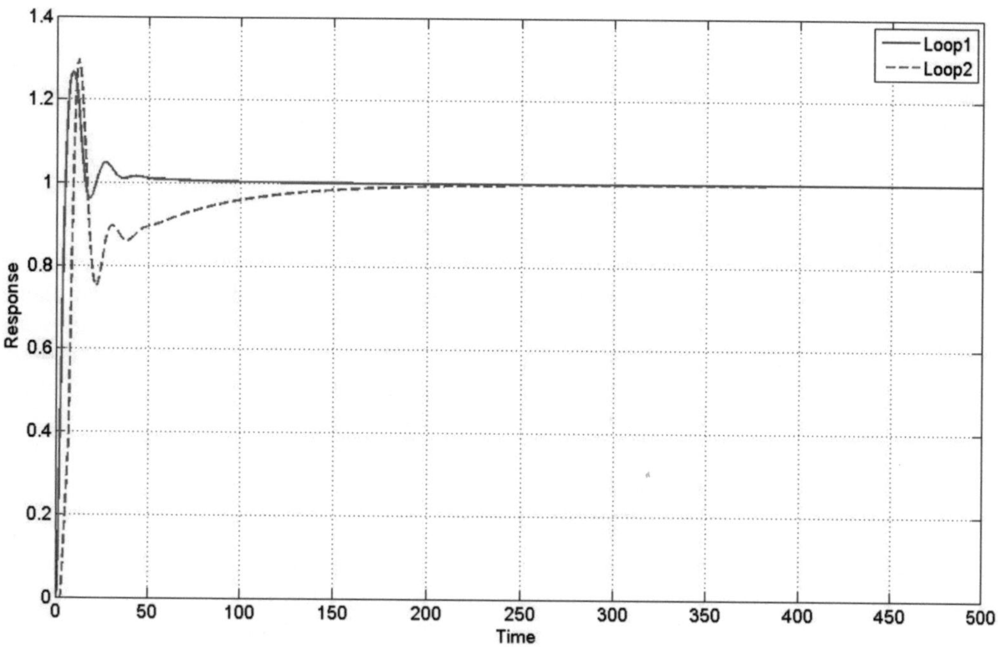

Fig. 5. Closed-loop response with BLT tuning for WB -Column using PID controller (solid line is loop 1 response and dashed line is loop 2 response)

3.4.3 Decoupled PID controller

The Wood and Berry binary distillation column is a multivariable system that has been studied extensively. The process has transfer function

$$\begin{bmatrix} \dfrac{12.8e^{-s}}{16.7s+1} & \dfrac{-18.9e^{-3s}}{21s+1} \\ \dfrac{6.6e^{-7s}}{10.9s+1} & \dfrac{-19.4e^{-3s}}{14.4s+1} \end{bmatrix}. \quad (3.15)$$

The decoupler is given by

$$D = G^{-1}(0) = \frac{1}{\det(C(0))}\begin{pmatrix} g_{22}(0) & -g_{12}(0) \\ -g_{21}(0) & g_{22}(0) \end{pmatrix} \tag{3.16}$$

$$D = G^{-1}(0) = \frac{1}{-123.58}\begin{pmatrix} -19.4 & -18.9 \\ 6.6 & 12.8 \end{pmatrix}$$

$$D = \begin{pmatrix} 0.15698 & 0.15293 \\ 0.0534 & -0.1035 \end{pmatrix}$$

The transfer function of the statistically decoupled system is given by

$$Q = GD \text{ or } Q = GG^{-1}(0) \tag{3.17}$$

$$Q = \begin{bmatrix} \dfrac{12.8e^{-s}}{16.7s+1} & \dfrac{-18.9e^{-3s}}{21s+1} \\ \dfrac{6.6e^{-7s}}{10.9s+1} & \dfrac{-19.4e^{-3s}}{14.4s+1} \end{bmatrix}\begin{pmatrix} 0.15698 & 0.15293 \\ 0.0534 & -0.1035 \end{pmatrix}$$

4. Input-output pairing

Many control systems are multivariable in nature. In such systems, each manipulated variable (input signal) may affect several controlled variables (output signals) causing interaction between the input/output loops. Due to these interactions, the system becomes more complex as well as the control of multivariable systems is typically much more difficult compared to the single-input single-output case.

4.1 The Relative Gain Array analysis

The RGA is a matrix of numbers. The i jth element in the array is called β_{ij}. It is the ratio of the steady-state gain between the ith controlled variable and the jth manipulated variable when all other manipulated variables are constant, divided by the steady-state gain between the same two variables when all other controlled variables are constant.

$$\beta_{ij} = \frac{\left[y_i / m_j \right]_{\overline{m_k}}}{\left[y_i / m_j \right]_{\overline{y_k}}} \tag{4.1}$$

For example, suppose we have a 2 **X** 2 system with the steady-state gains k_{pij}

$$y_1 = k_{p11}m_1 + k_{p12}m_2 \tag{4.2}$$

$$y_2 = k_{p21}m_1 + k_{p22}m_2$$

For this system, the gain between y_1 and m_1 when m_2 constant is

$$\left[\frac{y_1}{m_1}\right]_{\overline{m_2}} = k_{p11}$$

The gain between y_1 and m_1 when y_2 is constant ($y_2 = 0$) is found from solving the equations

$$y_1 = k_{p11}m_1 + k_{p12}m_2$$

$$0 = k_{p21}m_1 + k_{p22}m_2$$

$$y_1 = k_{p11}m_1 + k_{p12}\left[\frac{k_{p21}m_1}{k_{p22}}\right]$$

$$y_1 = \left[\frac{k_{p11}k_{p22} - k_{p12}k_{p21}}{k_{p22}}\right]m_1 \tag{4.3}$$

$$\left[\frac{y_1}{m_1}\right]_{\overline{y_2}} = \left[\frac{k_{p11}k_{p22} - k_{p12}k_{p21}}{k_{p22}}\right] \tag{4.4}$$

Therefore the term β_{11} in RGA is

$$\beta_{11} = \frac{1}{1 - \frac{k_{p12}k_{p21}}{k_{p11}k_{p22}}} \tag{4.5}$$

Example: Calculate β_{11} element of RGA for the wood and berry column

$$k_p = \begin{bmatrix} 12.8 & -18.9 \\ 6.6 & -19.4 \end{bmatrix}$$

$$\beta_{11} = \frac{1}{1 - \frac{k_{p12}k_{p21}}{k_{p11}k_{p22}}} = \frac{1}{1 - \frac{(-18.9)(6.6)}{(12.8)(-19.4)}} = 2.01$$

4.2 Singular Value Decomposition

SVD is a numerical algorithm developed to minimize computational errors involving large matrix operations. The singular value decomposition of matrix K results in three component matrices as follows:

$$k = U\Sigma V^T \tag{4.6}$$

where K is an n x m matrix. U is an n x n orthonormal matrix, the columns of which are called the 'left singular vectors'. V is an m x m orthonormal matrix, the columns of which are called the 'right singular vectors'. Σ is an n x m diagonal matrix of scalars called the "singular values"

SVD is designed to determine the rank and the condition of a matrix and to show geometrically the strengths and weaknesses of a set of equations so that the errors during computation can be avoided.

4.2.1 Example

Consider a very simple mixing example, a multivariable process whose gain matrix is as follows:

$$k = \begin{bmatrix} 0.7778 & -0.3889 \\ 1.0000 & 1.0000 \end{bmatrix}$$

which decomposes to

$$U = \begin{bmatrix} 0.2758 & -0.9612 \\ 0.9612 & 0.2758 \end{bmatrix}$$

$$V = \begin{bmatrix} 0.8091 & -0.5877 & 0.8091 \\ 0.5877 & 1.0000 & \end{bmatrix}$$

$$\Sigma = \begin{bmatrix} 1.4531 & 0 \\ 0 & 0.8029 \end{bmatrix}$$

At this point these singular values and vectors are merely numbers; however, consider the relationship between these values and an experimental procedure that could be applied to measure the steady-state process characteristics.

4.3 Niederlinski index

A fairly useful stability analysis method is the Niederlinski index. It can eliminate unworkable pairings of variables at an early stage in the design. The controller settings need not be known, but it applies only when integral action is used in all the loops. It utilizes only the steady state gains of the process transfer function matrix. The method is necessary but not the sufficient condition for stability of a closed loop system with integral action. If the index is negative, the system will be unstable for any controller settings. If the index is positive, the system may or may not be stable. Further analysis is necessary.

$$\text{Niederlinski index} = \text{NI} = \frac{Det\left[k_p\right]}{\prod_{j=1}^{N} k_{pjj}} \tag{4.7}$$

where, k_p is a matrix of steady state gains from the process openloop transfer function

k_{pjj} is the diagonal elements in steady state gain matrix

Example: Calculate the Niederlinski index for the wood and berry column:

$$k_p = \begin{bmatrix} 12.8 & -18.9 \\ 6.6 & -19.4 \end{bmatrix}$$

$$\text{NI} = \frac{Det\left[k_p\right]}{\prod_{j=1}^{N} k_{pjj}} = \frac{(12.8)(-19.4)-(-18.9)(6.6)}{(12.8)(-19.4)} = 0.498 \tag{4.8}$$

Since NI is positive, the closed loop system with the specified pairing may be stable.

4.4 Gramian based interaction measures

In 2004, Salgado and Conley investigated the channel interaction by considering controllability and observability gramians so called participation matrix. Similarly, Wittenmark and Salgado (2002) introduced Hankel Interaction Index array. These gramian measures namely HIIA, PM overcome the disadvantages of RGA. One key property of these is that the whole frequency range is taken into account in one single measure. Interaction measures recommend the input-output pairings that result in the largest sum when adding the corresponding elements in the measure. HIIA and PM give appropriate suggestions for decentralized multivariable controller.

The controllability Gramian, P, defined for stable time-invariant systems as

$$P = \int_0^\infty e^{A\tau} BB^T e^{A^T\tau} d\tau \tag{4.9}$$

If P has full rank, the system is state controllable.

A stable system will be *state observable* if the observability Gramian, Q, defined as

$$Q = \int_0^\infty e^{A\tau} CC^T e^{A^T\tau} d\tau \tag{4.10}$$

If Q has full rank, the system is state observable

These Gramians can be obtained by solving the following continuous time Lyapunov equations:

$$\left.\begin{aligned} AP + PA^T + BB^T &= 0 \\ A^T Q + QA + C^T C &= 0 \end{aligned}\right\} \tag{4.11}$$

Hankel singular values with controllability and observability gramians P and Q is given by

$$\sigma_H{}^{(i)} \triangleq \sqrt{\lambda_i} \qquad i = 1,2,.......n \tag{4.12}$$

The Hankel norm of the system with the transfer function G is

$$\|G\|_H \triangleq \sigma_H{}^{(1)} = \sqrt{\lambda_{\max}(PQ)} \tag{4.13}$$

Hankel interaction index array

The normalized version is the HIIA given by

$$[\Sigma_H]_{ij} = \frac{\|G_{ij}\|_H}{\sum_{kl}\|G_{kl}\|_H} \tag{4.14}$$

Participation matrix

Hankel norm is the largest singular values. For elementary SISO subsystems with several HSVs it can be argued that a more relevant way of quantifying the interaction is to take into account all of the HSVs, atleast if there are several HSVs that are of magnitudes close to maximum HSV.
Each element in PM is defined by

$$[\phi]_{ij} = \frac{tr\left(P_j Q_i\right)}{tr(PQ)} \tag{4.15}$$

$tr\left(P_j Q_i\right)$ is the sum of squared HSVs of the subsystems with input and output.

$tr(PQ)$ equals the sum of all $tr\left(P_j Q_i\right)$
Gramian based interaction measures are calculated and these values for benchmark 2-by-2 MIMO process is given in table 4.1.

2X2 MIMO PROCESS	HIIA	PM
WB	0.2218 0.3276 0.1144 0.3362	0.1741 0.3796 0.463 0.4000

Table 4.1. HIIA and PM for benchmark 2-by-2 MIMO process

5. Tuning of controller

Consider a process with transfer function $G_p(s) = \frac{k_p e^{-D_p s}}{\tau_p s + 1}$. This transfer function has two parts. One invertible: G_p^- and the other containing non-invertible part G_p^+ (time delay or right half plane zero that gives non-minimum phase behaviour). The IMC controller can be expressed as: $G_c^{IMC} = \frac{1}{G_p^-}$ where $G_p^- = \frac{k_p}{\tau_p s + 1}$ and $G_p^+ = e^{-D_p s}$.

Let us consider the desired closed loop response as $\frac{y}{R} = \frac{G_p^+}{(\lambda s + 1)} = \frac{e^{-D_p s}}{(\lambda s + 1)}$ which can be equated to complimentary sensitive function as $\frac{y}{R} = \frac{G_c^{true} G_p}{1 + G_c^{true} G_p}$. Thus the true controller can be expressed as:

$$G_c^{true} = \frac{G_c^{IMC}}{1 - \left(\frac{y}{R}\right)_d G_c^{IMC}} = \frac{1/G_p^-}{(\lambda s + 1) - G_p^+} \tag{5.1}$$

The right hand side of this equation can be written or rearranged to

$$G_c^{true} = \frac{1/G_p^-}{(\lambda s+1)-e^{-D_p s}} \tag{5.2}$$

In fact, the standard form of a PID controller can be given as

$$G_c^{true} = \frac{f(s)}{s} \quad \text{Or} \quad G_c^{true} = \frac{(\beta s+1)f(s)}{s(\beta s+1)} = \frac{\phi(s)}{s(\beta s+1)} \quad \text{where } \beta = \alpha\tau_D \tag{5.3}$$

This true controller can be expanded near the vicinity of s=0 using Laurent series as

$$G_c^{true}(s) = \frac{1}{s(\beta s+1)}\left[\sum_{j=-\infty}^{\infty} c_j(s)^j\right] = \frac{1}{s(\beta s+1)}\left[\ldots+\phi(0)+\phi'(0)s+\phi''(0)\frac{s^2}{2!}+\ldots\right] \tag{5.4}$$

By comparing the coefficients of s in equation (5.4) with the standard PID controller, we get

$$\begin{aligned} k_c &= a_0 = \phi'(0) = f'(0)+\beta f(0) \\ \frac{k_c}{\tau_I} &= b_1 = \phi(0) = f(0) \\ k_c\tau_D &= a_1 = \frac{\phi''(0)}{2!} = \frac{f'(0)+2\beta f'(0)}{2} \end{aligned} \tag{5.5}$$

where

$$\begin{aligned} G_c(s) &= \frac{\phi(s)}{(\beta s+1)} \\ \phi(s) &= (\beta s+1)f(s) \end{aligned} \tag{5.6}$$

The method described in earlier section is applied to some standard transfer functions and the comprehensive results are presented in Table 5.1 and selection of λ is given in Table 5.2. Detailed analysis on synthesis of PID tuning rules can be seen in Panda (2008 & 2009).

Example 5.1: The wood and berry binary distillation column is a multivariable system that has been studied extensively. The process has transfer function

$$\begin{bmatrix} \dfrac{12.8e^{-s}}{16.7s+1} & \dfrac{-18.9e^{-3s}}{21s+1} \\ \dfrac{6.6e^{-7s}}{10.9s+1} & \dfrac{-19.4e^{-3s}}{14.4s+1} \end{bmatrix}. \tag{5.7}$$

The closed loop response is given in Figure 5.1.

Example 5.2: The transfer function of multiproduct plant distillation column for the separation of binary mixture of ethanol-water (Ogunnaike-Ray (OR) column) is given by

$$\begin{bmatrix} y_1 \\ y_2 \\ y_3 \end{bmatrix} = \begin{bmatrix} \frac{0.66e^{-2.6s}}{6.7s+1} & \frac{-0.61e^{-3.5s}}{8.64s+1} & \frac{-0.0049e^{-s}}{9.06s+1} \\ \frac{-2.36e^{-3s}}{5s+1} & \frac{-2.3e^{-3s}}{5s+1} & \frac{-0.01e^{-1.2s}}{7.09s+1} \\ \frac{-34.68e^{-9.2s}}{8.15s+1} & \frac{46.2e^{-9.4s}}{10.9s+1} & \frac{0.87(11.61s+1)e^{-s}}{(3.89s+1)(18.8s+1)} \end{bmatrix} \begin{bmatrix} u_1 \\ u_2 \\ u_3 \end{bmatrix} \quad (5.8)$$

The closed loop response is given in Figure 5.2.

Transfer Function	PID-Tuning Rules
$\frac{K_p e^{-D_p s}}{\tau_p s+1}$	$k_c = \frac{\tau_I}{k_p(\lambda+D_p)}; \tau_I = (\tau_p+\beta)+\frac{D_p^2}{2(\lambda+D_p)}; \tau_D = \frac{D_p^2}{2(\lambda+D_p)\tau_I}\left[(\tau_I-\beta)-\frac{D_p}{3}\right]+\frac{\beta\tau_p}{\tau_I}$
$\frac{K_p e^{-D_p s}}{\tau_p^2 s^2+2\zeta\tau_p s+1}$	$k_c = \frac{\tau_I}{k_p(2\lambda+D_p)}; \tau_I = (2\zeta\tau_p+\beta)-\frac{2\lambda^2-D_p^2}{2(2\lambda+D_p)};$ $\tau_D = \frac{(\tau_p^2+2\zeta\tau_p\beta)}{\tau_I}-\frac{2\lambda^2-D_p^2}{2(2\lambda+D_p)\tau_I}\left[(\tau_I-\beta)+\frac{D_p^3}{3(2\lambda^2-D^2)}\right]$
$\frac{K_p(\tau_{pz}s+1)e^{-D_p s}}{(\tau_p^2 s^2+2\zeta\tau_p s+1)(\tau_{p1}s+1)}$	$k_c = \frac{\tau_I}{k_p(3\lambda+D_p-\tau_{pz})}; \tau_I = (\tau_{p1}+2\zeta\tau_p+2\beta)-\frac{\left(3\lambda^2+D_p\tau_{pz}-\frac{D_p^2}{2}\right)}{(3\lambda+D_p-\tau_{pz})}$ $\tau_D = \frac{\left\{(\tau_p^2+2\zeta\tau_p(\tau_{p1}+\beta)+\beta\tau_{p1})-\frac{c}{a}\right\}-\frac{b}{a}(\tau_I-\beta)}{\tau_I}$ $a = k_p(3\lambda+D_p-\tau_p z); b = 3\lambda^2+D_p\tau_{pz}-\frac{D_p^2}{2}; c = \lambda^3-\frac{D_p^2\tau_{pz}}{2}+\frac{D_p^3}{6}$

Table 5.1. Analytical expressions for PID controller parameters for standard transfer functions

	FOPDT	SOPDT	IPDT
PI	$\lambda = \max(1.7D_p, 0.2\tau_p)$	$\lambda = \max(0.25D_p, 0.2\tau_p)$	$\lambda = D_p\sqrt{10}$
PID	$\lambda = \max(0.25D_p, 0.2\tau_p)$	$\lambda = \max(0.25D_p, 0.2\tau_p)$	$\lambda = D_p\sqrt{10}$

Table 5.2. λ selection rule

6. Stability analysis

6.1 INA and DNA methods

Rosenbrock extended the nyquist stability and design concepts to MIMO systems containing significant interaction. The methods are known as the inverse and direct Nyquist array (INA and DNA) methods. As an extension from the SISO nyquist stability and design concepts, these methods use frequency response approach. These techniques are used because of their simplicity, high stability, and low noise sensitivity. In actual applications, there will be a region of uncertainty for interaction, as the process transfer function can be different from what was used in the controller design (due to modeling errors and process variations).

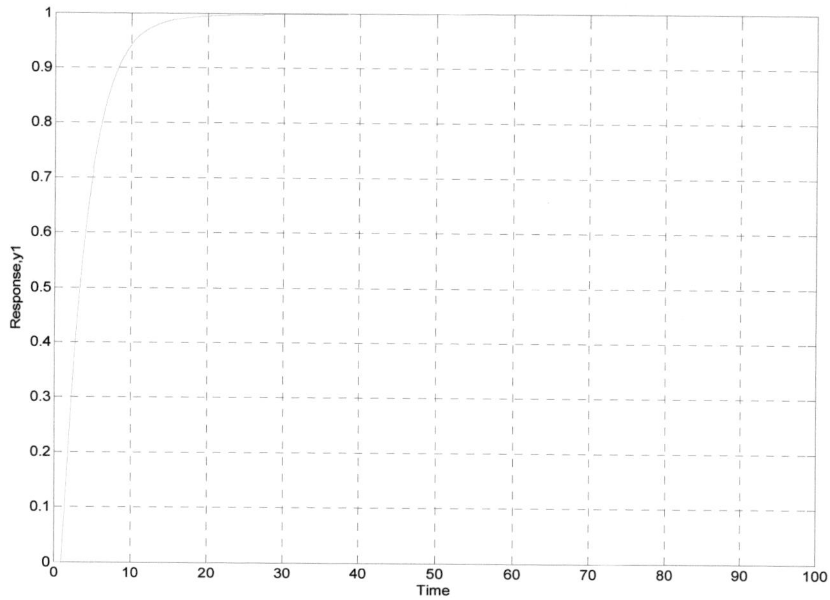

(a)

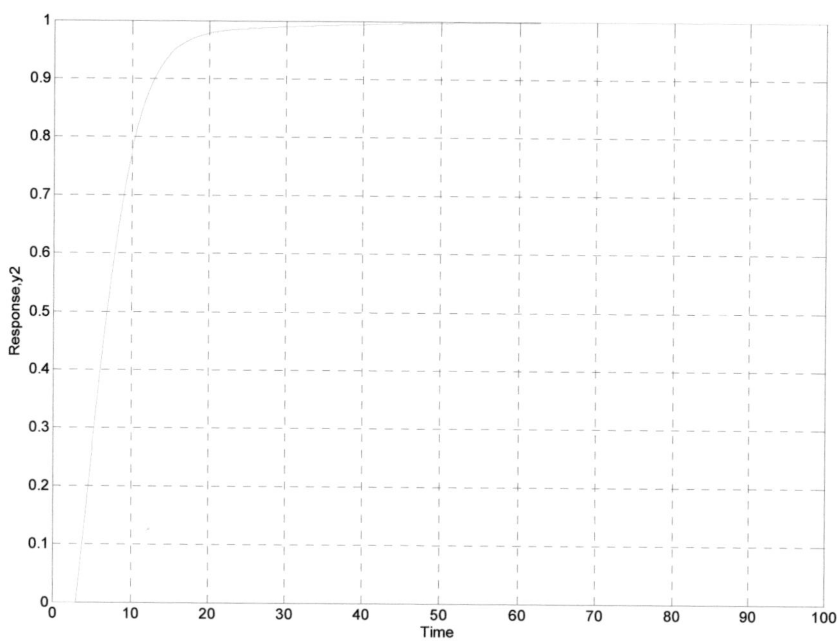

(b)

Fig. 5.1. Closed-loop responses (a: Loop-1 and b: Loop-2) to setpoint changes of example (5.1) -processes using PID controller

6.2 Nyquist Stability Theorem

Suppose that $G(s)$ is an n x n system with a decentralized control system $C(s) = diag\{c_1(s),......,c_n(s)\}$ and that the matrix, $1+G(s)C(s)$, is column diagonally dominant on the nyquist contour, i.e.

$$|1+g_{ll}(s)c_l(s)| > R_l(s)|c_l(s)| \tag{6.1}$$

where

$$R_l(s) = \sum_{k=1,k\neq 1}^{n} |g_{kl}(s)| \tag{6.2}$$

for $l = 1,2,........,n$ and for all s on the Nyquist contour

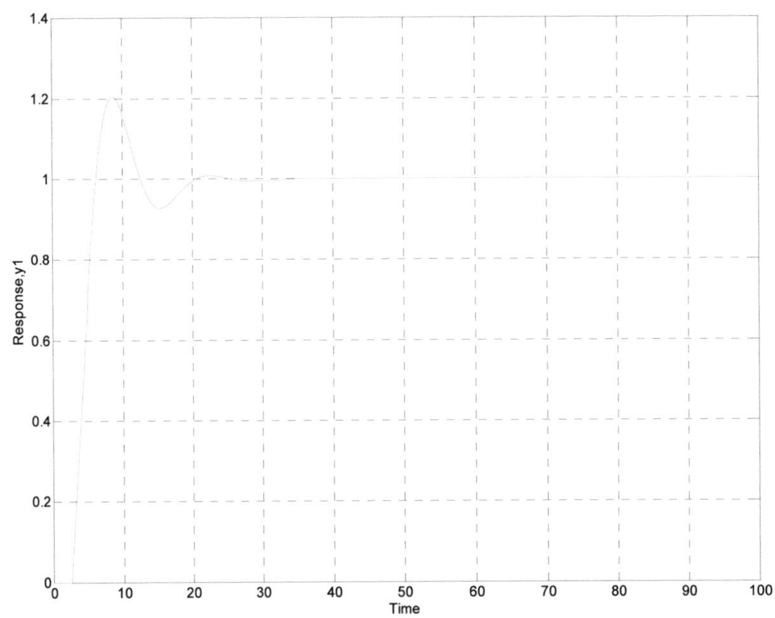

(a)

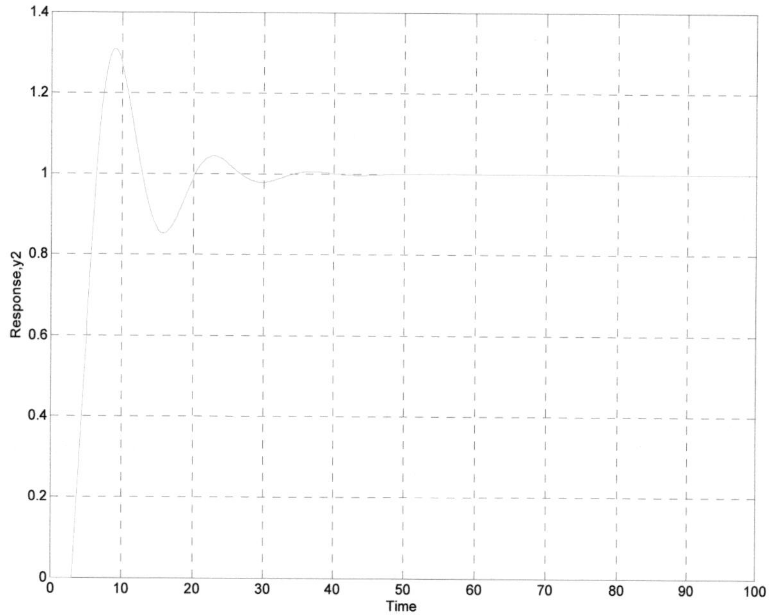

(b)

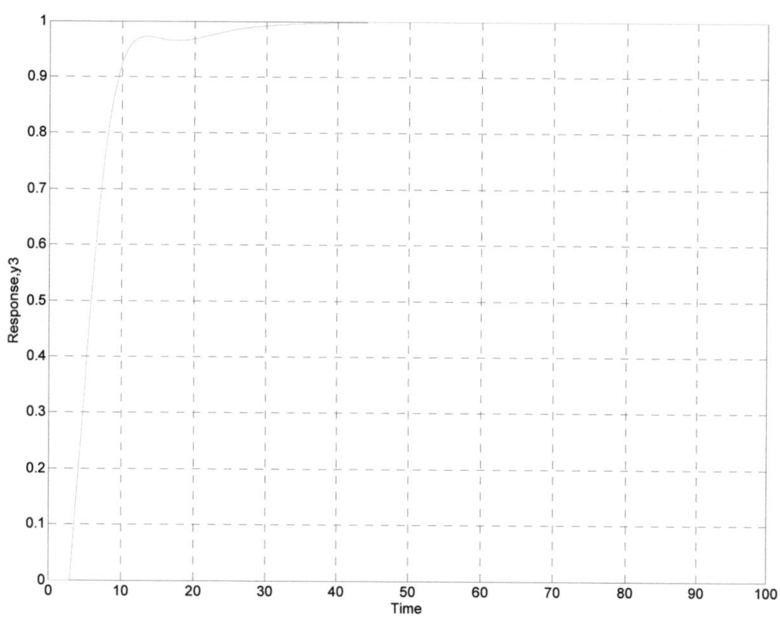

(c)

Fig. 5.2. Closed-loop responses (a: loop-1; b: loop-2 and c: loop-3) to setpoint changes of example 5.2 -processes using PID controller

6.3 INA design methodology

The following is the design procedure for the INA technique:

1. Obtain $G(s)$ and calculate its inverse, $\hat{G}(s)$.
2. Select an appropriate frequency range; usually $0 \le \omega \le \omega_c$, where ω_c is the frequency above which the response is certain to become and remain negligible.
3. Obtain the inverse nyquist array, which is the m^2 nyquist diagrams of the elements of $\hat{G}(s)$.
4. Design compensators, which transform the non dominant $\hat{G}(s)$ to a diagonally dominant.
5. To verify dominance, calculate the appropriate gershgorin circles for the diagonal elements of the INA at various frequencies. The size of the gershgorin circles measures the importance of off-diagonal (interacting) elements relative to diagonal elements.
6. The INA and gershgorin bands provide the amount of gain that may be applied to each of the loops without violating the stability requirement.

6.4 Example

Johansson and Koivo designed a multivariable controller for a boiler subsystem where the boiler was a 1.6MW water boiler using solid fuel. Significant interaction was present between the loops in the subsystem, which consisted of the boiler underpressure and flue gas oxygen content as outputs with damper position and motor speed of the secondary blower as associated inputs. The output vector is $y = [y_1 \quad y_2]^T$ where y_1 is the normalized boiler underpressure and y_2 is the percentage flue gas oxygen content. The input vector is $u = [u_1 \quad u_2]^T$ where u_1 is the damper position (%) and u_2 is secondary blower speed (rpm). The dynamics of the subsystem were determined from step response experiments. First order plus dead time responses were obtained, which produced the transfer function matrix:

$$G(s) = \begin{bmatrix} \dfrac{e^{-2s}}{(10s+1)} & \dfrac{-1}{(10s+1)} \\ 0 & \dfrac{e^{-10s}}{(60s+1)} \end{bmatrix} \tag{6.3}$$

The response of the flue gas oxygen content to step change in damper position was very slow and small in amplitude; therefore $g_{21}(s)$ was taken as zero. However, the secondary blower speed, u_2, affects both outputs.

The inverse of G can be written immediately as:

$$G^{-1}(s) = \begin{bmatrix} -e^{2s}(10s+1) & e^{12s}(60s+1) \\ 0 & e^{10s}(60s+1) \end{bmatrix} \tag{6.4}$$

Consider the g11 element, first replace s with jw which produces:

$$-e^{2s}(10s+1) = -e^{2j\omega}(10j\omega+1) \tag{6.5}$$

Using Euler's relation,

$$-e^{2j\omega}(10j\omega+1) = (10\omega\sin 2\omega - \cos 2\omega) + j(-10\omega\cos 2\omega - \sin 2\omega) \tag{6.6}$$

Consider w=0, $g_{11}(0)$=-1
To compute the radius, g12(w) is calculated as:

$$g_{12}(\omega) = -[(\cos 12\omega - 60\omega\sin 12\omega) - j(60\omega\cos\omega + \sin 12\omega)]$$

Recall that the magnitude of a complex number is the square root of the sum of real part squared and the imaginary part squared. Therefore, $g_{12}(0)$ =1
A constant pre-compensator was designed to obtain dominance. This was

$$k = \begin{bmatrix} 1 & -1 \\ 0 & 1 \end{bmatrix} \tag{6.7}$$

7. Conclusion

Thus in this chapter, it was found that least square and subspace methods have been used to identify process in open loop and sequential identification technique is used to estimate the process in closed loop. And the decentralized controllers are tuned using BLT method results in a stable controller. Finally, all the interaction tools are discussed as well the stability of the MIMO processes. The IMC-PID tuning rule suggested in this article yields fast and robust responses.
The following step-by-step procedure may be employed to solve a multi-variable control problem:

1. Choose an appropriate pairings of controlled and manipulated variables, by interaction analysis.
2. If interaction is modest, one may consider SISO controllers for the multi-variable system.
3. If interaction is significant, it may be possible to use decouplers to reduce interaction in conjunction with PID-type controllers.
4. An alternative to steps 2 and 3 is to use a full multi-variable control technique that inherently compensates for interactions.

Based on the concept of sequential identification-design, an approach for the automatic tuning of multivariable systems is discussed. Several system identification methods like subspace identification, least squares, relay feedback methods are used to determine dynamic parameters of a specific model structure from plant data (real time).

8. Acknowledgement

Authors wish to acknowledge the financial support of DST / SR-S3-CE-90-2009 in carrying out this research work

9. References

[1] Multivariable Process control, Pradeep B. Deshpande, ISA, Research Triangle Park, NC, U.S.A, 1989

[2] Autotuning of PID controllers, C.C.Yu, Springer Verlag, London, 2nd Edition, 2006

[3] Control configuration selection for multivariable plants, Ali Khaki-Sedigh and Bijan Moaveni, Springer, Berlin, Heidelberg, 2009.

[4] Multivariable control systems: An Engineering approach, P. Albertos and Sala, Springer Verlag, London, 2004

[5] Identification and control of multivariable systems-role of relay feedback, A Ph.D Thesis, Control Engg Lab, Electical Engg Dept., Anna University, Chennai, India, C. Selvakumar, 2009.

[6] Rames C. Panda, 2008, Synthesis of PID controller using desired closed loop criteria, Ind. & Engg. Chem Res., 47(22),1684-92

[7] Rames C. Panda, 2009, Synthesis of PID controller for unstable and integrating processes, Chem. Eng. Sci, 64 (12), 2807-16

6

Robust Decentralized PID Controller Design

Danica Rosinová and Alena Kozáková
Slovak University of Technology
Slovak Republic

1. Introduction

Robust stability of uncertain dynamic systems has major importance when real world system models are considered. A realistic approach has to consider uncertainties of various kinds in the system model. Uncertainties due to inherent modelling/identification inaccuracies in any physical plant model specify a certain uncertainty domain, e.g. as a set of linearized models obtained in different working points of the plant considered. Thus, a basic required property of the system is its stability within the whole uncertainty domain denoted as robust stability. Robust control theory provides analysis and synthesis approaches and tools applicable for various kinds of processes, including multi input - multi output (MIMO) dynamic systems. To reduce multivariable control problem complexity, MIMO systems are often considered as interconnection of a finite number of subsystems. This approach enables to employ decentralized control structure with subsystems having their local control loops. Compared with centralized MIMO controller systems, decentralized control structure brings about certain performance deterioration, however weighted against by important benefits, such as design simplicity, hardware, operation and reliability improvement. Robustness is one of attractive qualities of a decentralized control scheme, since such control structure can be inherently resistant to a wide range of uncertainties both in subsystems and interconnections. Considerable effort has been made to enhance robustness in decentralized control structure and decentralized control design schemes and various approaches have been developed in this field both in time and frequency domains (Gyurkovics & Takacs, 2000; Zečevič & Šiljak, 2004; Stankovič et al., 2007).

Recently, the algebraic approach has gained considerable interest in robust control, (Boyd et al., 1994; Crusius & Trofino, 1999; de Oliveira et al., 1999; Ming Ge et al., 2002; Grman et al., 2005; Henrion et al., 2002). Algebraic approach is based on the fact that many different problems in control reduce to an equivalent linear algebra problem (Skelton et al., 1998). By algebraic approach, robust control problem is formulated in algebraic framework and solved as an optimization problem, preferably in the form of Linear Matrix Inequalities (LMI). LMI techniques enable to solve a large set of convex problems in polynomial time (see Boyd et al., 1994). This approach is directly applicable when control problems for linear uncertain systems with a convex uncertainty domain are solved. Still, many important control problems even for linear systems have been proven as NP hard, including structured linear control problems such as decentralized control and simultaneous static output feedback (SOF) designs. In these cases the prescribed structure of control feedback matrix (block diagonal for decentralized control) results in nonconvex problem formulation. There are basically two approaches to solve the respective nonconvex control problem: 1) to reformulate the problem as LMI using certain convex relaxations (e.g. deOliveira et al., 2000; Rosinová & Veselý, 2003) or, alternatively, adopt an iterative procedure; 2) to formulate and solve the bilinear matrix inequalities (BMI) respective to robust control design problem. A nice review and basic characteristics of LMI and BMI in various control problems can be found in (Van Antwerp & Braatz, 2000).

To reduce the problem size in decentralized control design for large scale systems, the diagonal dominance or block diagonal dominance concept can be adopted. Recently, the so called Equivalent Subsystems Method has been developed for decentralized control in frequency domain, (Kozáková & Veselý, 2009). The main concept of the Equivalent Subsystems Method, originally developed as a Nyquist based frequency domain decentralized controller design technique, is the so called equivalent subsystem; equivalent subsystems are generated by shaping Nyquist plot of each decoupled subsystem using any selected characteristic locus of the matrix of interactions. The point of this approach consists in that local controllers of equivalent subsystems can be independently tuned for stability and required performance specified in terms of a suitable (preferably frequency domain) performance measure (e.g. degree of stability, phase margin, bandwidth), so that the resulting decentralized controller guarantees equivalent performance of the full system.

When designing decentralized control, besides robust stability, performance requirements have to be considered. Performance objectives can be of two basic types: a) achieving required performance in different subsystems; or b) achieving plant-wide desired performance. In this chapter two alternative approaches belonging to the latter group are presented, based on recent research results on robust decentralized PID controller design in the frequency and time domains.

The present chapter further extends the robust decentralized PID controller design techniques from (Kozáková et al., 2009; 2010; 2011; Rosinová et al., 2003; Rosinová & Veselý, 2007; 2011), bringing novel robust control design approaches. The results are illustrated on the case study dealing with robust decentralized controller design for the quadruple tank process. This laboratory process recently presented in (Johansson, 2000; Johansson et al., 1999) is an illustrative two input - two output laboratory plant for studying multivariable dynamic systems for both minimum and nonminimum-phase configurations.

The first presented approach is based on formulation and solution of BMI or LMI for uncertain linear polytopic system to design robust controller in the state space. In the time domain, we introduce the augmented model for closed-loop linear uncertain system with PID controller; this model is in general form, comprising both continuous- and discrete-time cases. For both cases, a general robust stability condition is formulated; the particular design procedures differ only in parameterization of augmented model matrices. A decentralized control design strategy is adopted, where robust PID control design approach is applied for structured - block diagonal controller matrices respective to decentralized controller.

The second approach is based on the Nyquist-type decentralized control design technique for uncertain MIMO systems described by a transfer function matrix. The decentralized controller is designed on subsystem level using the recently developed Equivalent Subsystem Method (Kozáková et al., 2009). Application of this method in the design for robust stability and nominal performance can be found e.g. in (Kozáková & Veselý, 2009) within a two-stage design scheme: 1. design of decentralized controller for nominal performance; 2. controller redesign with modified performance requirements to meet the robust stability conditions. A direct "one-shot" robust DC design methodology based on integration of robust stability conditions in the Equivalent Subsystems Method enables to design local controllers of equivalent subsystems with regard to robust stability of the full system. The frequency domain approach is applicable for both continuous- and discrete-time PID controller designs.

2. Motivation: Case study - Quadruple tank process

This section aims at description, and analysis of two input - two output process from literature, which will be later used to demonstrate our proposed methods for decentralized PID controller design. The quadruple-tank process shown in Fig.1 has been introduced in (Johansson et al., 1999; Johansson, 2000) to provide a case study to analyze both minimum and nonminimum phase MIMO systems on the same plant. The aim is to control the level in the lower two tanks using two pumps. The inputs v_1 and v_2 are pump 1 and 2 flows respectively, the controlled outputs y_1 and y_2 are levels in lower tanks 1 and 2 respectively.

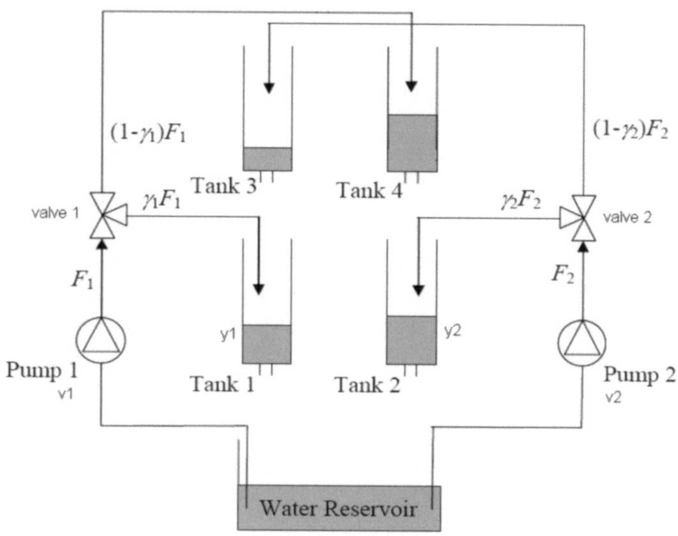

Fig. 1. Quadruple tank process scheme.

The nonlinear model of the four tanks can be described by state equations

$$\begin{aligned}
\frac{dh_1}{dt} &= -\frac{a_1}{A_1}\sqrt{2gh_1} + \frac{a_3}{A_1}\sqrt{2gh_3} + \frac{\gamma_1 k_1}{A_1} v_1 \\
\frac{dh_2}{dt} &= -\frac{a_2}{A_2}\sqrt{2gh_2} + \frac{a_4}{A_2}\sqrt{2gh_4} + \frac{\gamma_2 k_2}{A_2} v_2 \\
\frac{dh_3}{dt} &= -\frac{a_3}{A_3}\sqrt{2gh_3} + \frac{(1-\gamma_2)k_2}{A_3} v_2 \\
\frac{dh_4}{dt} &= -\frac{a_4}{A_4}\sqrt{2gh_4} + \frac{(1-\gamma_1)k_1}{A_4} v_1
\end{aligned} \tag{1}$$

where A_i is cross-section of tank i, a_i is cross-section of the outlet hole of tank i, h_i is water level in tank i, g is acceleration of gravity, the flow corresponding to pump i is $k_i v_i$. Parameter γ_1 denotes position of the valve dividing the pump 1 flow into the lower tank 1 and related upper tank 4 and similarly γ_2 divides flow from pump 2 to the tanks 2 and 3. The flow to tank 1 is $\gamma_1 k_1 v_1$ and to tank 4 it is $(1-\gamma_1)k_1 v_1$, analogically for the tanks 2 and 3. The nonlinear model (1) can be linearized around the working point given by the water levels in tanks $h_{10}, h_{20}, h_{30}, h_{40}$. The deviation state space model was considered with $x_i = h_i - h_{i0}$ and the respective control variables $u_i = v_i - v_{i0}$. The linearized state space model for quadruple tank (1) is then

$$\begin{bmatrix} \dot{x}_1 \\ \dot{x}_3 \\ \dot{x}_2 \\ \dot{x}_4 \end{bmatrix} = \begin{bmatrix} \frac{-1}{T_1} & \frac{A_3}{T_3 A_1} & 0 & 0 \\ 0 & \frac{-1}{T_3} & 0 & 0 \\ 0 & 0 & \frac{-1}{T_2} & \frac{A_4}{T_4 A_2} \\ 0 & 0 & 0 & \frac{-1}{T_4} \end{bmatrix} \cdot \begin{bmatrix} x_1 \\ x_3 \\ x_2 \\ x_4 \end{bmatrix} + \begin{bmatrix} \frac{\gamma_1 k_1}{A_1} & 0 \\ 0 & \frac{(1-\gamma_2)k_2}{A_3} \\ 0 & \frac{\gamma_2 k_2}{A_1} \\ \frac{(1-\gamma_1)k_1}{A_4} & 0 \end{bmatrix} \cdot \begin{bmatrix} u_1 \\ u_2 \end{bmatrix} \tag{2}$$

where $T_i = \frac{A_i}{a_i}\sqrt{\frac{2h_{i0}}{g}}, \; i = 1,\ldots,4$.

The argument t has been omitted; the state variables corresponding to levels in tanks 2 and 3 have been interchanged in state vector so that subsystems respective to input u_1 from pump 1 (tanks 1 and 3) and u_2 from pump 2 (tanks 2 and 4) are more apparent. This decomposition into two subsystems is used for decentralized control design.
The respective transfer function matrix having inputs v_1 and v_2 and outputs y_1 and y_2 is

$$G(s) = \begin{bmatrix} \frac{c_1 \gamma_1}{T_1 s + 1} & \frac{c_1(1-\gamma_2)}{(T_3 s + 1)(T_1 s + 1)} \\ \frac{c_2(1-\gamma_1)}{(T_4 s + 1)(T_2 s + 1)} & \frac{c_2 \gamma_2}{T_2 s + 1} \end{bmatrix} \tag{3}$$

where $c_i = \frac{T_i k_i}{A_i}\sqrt{\frac{2h_{i0}}{g}}, \; i = 1,2$.

The plant can be shifted from minimum to nonminimum phase configuration and vice versa simply by changing a valve controlling the flow ratios γ_1 and γ_2 between lower and upper tanks. The minimum-phase configuration corresponds to $1 < \gamma_1 + \gamma_2 < 2$ and the nonminimum-phase one to $0 < \gamma_1 + \gamma_2 < 1$.

2.1 Decentralized control of quadruple tank – problem formulation and pairing selection

The basic control aim for quadruple tank is to reach the given level in the lower two tanks, i.e. prescribed values of y_1 and y_2 by controlling input flows v_1 and v_2 delivered by two pumps. To achieve this aim, the decentralized control structure is employed, with two control loops respective to output values y_1 and y_2.

Decentralized control design consists of several steps, the crucial ones for controller design are

- choice of appropriate pairing of inputs to outputs;
- structural stability test respective to chosen pairing;
- robust decentralized controller design.

We consider the standard approach for the former two steps presented below; in Sections 3 and 4 we concentrate on the last step - robust decentralized control design.

Pairing and structural stability

Frequently used index to assess input-output pairing is the Relative Gain Array (RGA) index, see e.g. (Ogunnaike & Ray, 1994), (Skogestad & Postletwhaite, 2009), computed as

$$RGA(s) = G(s).*[G(s)^T]^{-1} \tag{4}$$

where G(s) is a square transfer function matrix of the linearized system.

Individual subsystems are then specified by the chosen pairing and their transfer functions are placed in the diagonal of the transfer function matrix. To check structural stabilizability using the chosen control configuration, the Niederlinski index is applied:

$$NI = \frac{\det(G(0))}{\Pi(diag(G(0))} \tag{5}$$

If $NI < 0$, the system cannot be stabilized using the chosen pairing and the pairing must be modified.

In our case study, the steady state RGA(0) is considered to choose appropriate pairing with the respective RGA elements positive and closest possible to 1.

$$RGA(0) = G(0).*\left[G(0)^{-1}\right]^T = \begin{bmatrix} \lambda & 1-\lambda \\ 1-\lambda & \lambda \end{bmatrix} \tag{6}$$

where $\lambda = \dfrac{\gamma_1\gamma_2}{\gamma_1+\gamma_2-1}$ depends on valve parameters γ_1 and γ_2 exclusively. The diagonal elements λ are positive for $1<\gamma_1+\gamma_2<2$ (minimum phase system) and the respective pairing is v_1-y_1, v_2-y_2. For $0<\gamma_1+\gamma_2<1$ (nonminimm phase system), the opposite pairing v_1-y_2, v_2-y_1 is indicated. This result is approved by Niederlinski index.

2.2 Quadruple tank process – uncertainty domain

For quadruple tank system (1), we consider the uncertainty to be a change of valve position, i.e. change of γ_1 and γ_2, uncertainty domain is specified by three working points.

In minimum phase region:

WP1: $\gamma_1 = 0.4,\ \gamma_2 = 0.8$; WP2: $\gamma_1 = 0.8,\ \gamma_2 = 0.4$

WP3: $\gamma_1 = 0.8,\ \gamma_2 = 0.8$ (7)

In nonminimum phase region:

WP1: $\gamma_1 = 0.1,\ \gamma_2 = 0.3$; WP2: $\gamma_1 = 0.3,\ \gamma_2 = 0.1$

WP3: $\gamma_1 = 0.1,\ \gamma_2 = 0.1$ (8)

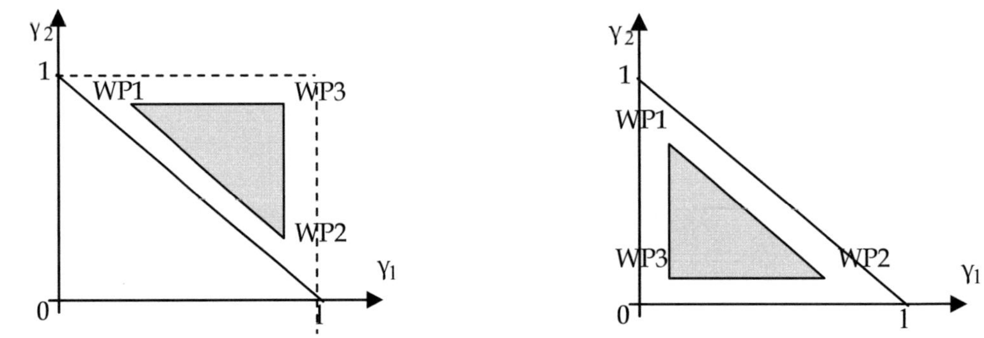

a) minimum phase configuration b) nonminimum phase configuration

Fig. 2. Uncertainty domain specified by working points

3. Robust decentralized PID controller design in the time domain

In this section, robust decentralized controller in time domain is designed based on robust stability conditions formulated and solved as linear (or bilinear) matrix inequalities. To include performance evaluation, the quadratic performance index is used. Decentralized robust control problem is formulated in general framework for augmented system, including the model of controlled system as well as controller dynamics. The robust stability conditions from literature are recalled, using D-stability concept which enables unified formulation for continuous-time and discrete-time cases. Our modification of these results includes derivative term of PID controller as well as a term for guaranteed cost. Thus, the decentralized control design procedure is presented in the general form comprising both continuous and discrete-time system models.

Notation: for a symmetric square matrix X, $X > 0$ denotes positive definiteness; * in matrices denotes the respective transposed term to make the matrix symmetric, 0 in matrices denotes zero block of the corresponding dimensions, I_n denotes identity matrix of dimensions nxn; dimension index is often omitted, when the dimension is clear from the context. Argument t denotes either continuous time for continuous-time, or sampled time for discrete-time system models; we intentionally use the same symbol t for both cases to underline that the formulation of developed results is general, applicable for both cases.

3.1 Preliminaries and problem formulation

3.1.1 Decentralized control of uncertain system, guaranteed cost control

Consider a linearized model of interconnected system, where subsystems with polytopic uncertainty are assumed, described by

$$S_i:\ \delta x_i(t) = A_i(\alpha)x_i(t) + B_i(\alpha)u_i(t) + \sum_{\substack{j=1 \\ j\neq i}}^{N} (A_{ij}(\alpha)x_j(t) + B_{ij}(\alpha)u_j(t))$$

$$y_i(t) = C_i x_i(t);\ \text{i=1,...,N} \tag{9}$$

where $\delta x(t) = \dot{x}(t)$ for continuous-time system model; $\delta x(t) = x(t+1)$ for discrete-time system model; $x_i(t) \in R^{n_i}$, $u_i(t) \in R^{m_i}$, $y_i(t) \in R^{p_i}$ are the subsystem state, control and output vectors respectively, $\sum_{i=1}^{N} n_i = n$, $\sum_{i=1}^{N} m_i = m$, $\sum_{i=1}^{N} p_i = p$; C_i are matrices with corresponding dimensions. Uncertain model matrices $A_i(\alpha)$, $B_i(\alpha)$, $A_{ij}(\alpha)$, $B_{ij}(\alpha)$ are from polytopic uncertainty domains

$$A_i(\alpha) \in \left\{ \sum_{k=1}^{K} \alpha_k \tilde{A}_{ik}, \sum_{k=1}^{K} \alpha_k = 1, \alpha_k \geq 0 \right\}, B_i(\alpha) \in \left\{ \sum_{k=1}^{K} \alpha_k \tilde{B}_{ik}, \sum_{k=1}^{K} \alpha_k = 1, \alpha_k \geq 0 \right\},$$

$$A_{ij}(\alpha) \in \left\{ \sum_{k=1}^{K} \alpha_k \tilde{A}_{ijk}, \sum_{k=1}^{K} \alpha_k = 1, \alpha_k \geq 0 \right\} B_{ij}(\alpha) \in \left\{ \sum_{k=1}^{K} \alpha_k \tilde{B}_{ijk}, \sum_{k=1}^{K} \alpha_k = 1, \alpha_k \geq 0 \right\}. \quad (10)$$

The whole interconnected system model in the compact form is

$$\text{S: } \delta x(t) = A(\alpha)x(t) + B(\alpha)u(t)$$

$$y(t) = C_d x(t) \quad (11)$$

uncertain system matrix $A(\alpha) = \left(A_d(\alpha) + A_m(\alpha)\right)$, $B(\alpha) = \left(B_d(\alpha) + B_m(\alpha)\right)$ and

$$A(\alpha) \in \left\{ \sum_{k=1}^{K} \alpha_k \tilde{A}_{(k)}, \sum_{k=1}^{K} \alpha_k = 1, \alpha_k \geq 0 \right\}, \; B(\alpha) \in \left\{ \sum_{k=1}^{K} \alpha_k \tilde{B}_{(k)}, \sum_{k=1}^{K} \alpha_k = 1, \alpha_k \geq 0 \right\} \quad (12)$$

where $\tilde{A}_{(k)}$ has diagonal blocks $\tilde{A}_{ik}$ and off-diagonal blocks $\tilde{A}_{ijk}$, $\tilde{B}_{(k)}$ has diagonal blocks $\tilde{B}_{ik}$ and off-diagonal blocks $\tilde{B}_{ijk}$ respective to (10); and

$x(t) = (x_1^T \, x_2^T \ldots x_N^T)$, $u(t) = (u_1^T \, u_2^T \ldots u_N^T)$, $y(t) = (y_1^T \, y_2^T \ldots y_N^T)$ are state, control and output vectors of the overall system S;

$A_d(\alpha) = diag\{A_1(\alpha), \ldots, A_N(\alpha)\}$, $B_d(\alpha) = diag\{B_1(\alpha), \ldots, B_N(\alpha)\}$, $C_d = diag\{C_1, \ldots, C_N\}$ are overall system matrices of corresponding dimensions respective to the subsystems, matrices $A_m(\alpha)$, $B_m(\alpha)$ correspond to interconnections.

A closed loop system performance is assessed considering the guaranteed cost notion; the quadratic cost function known from LQ theory is used.

$$J_c = \int_0^{\infty} [x(t)^T Q x(t) + u(t)^T R u(t)] dt \qquad \text{for a continuous-time and}$$

$$J_d = \sum_{k=0}^{\infty} [x(t)^T Q x(t) + u(t)^T R u(t)] \text{ for a discrete-time systems} \quad (13)$$

where $Q \in R^{n \times n}, R \in R^{m \times m}$ are symmetric positive semidefinite and positive definite block diagonal matrices respectively, with block dimensions respective to the subsystems. The concept of guaranteed cost control is used in a standard way: let there exist a control law $u(t)$ and a constant J_0 such that

$$J \leq J_0 \tag{14}$$

holds for the closed loop system (9). Then the respective control $u(t)$ is called the *guaranteed cost control* and the value J_0 is the *guaranteed cost.*

Decentralized Control Problem

The control design aim is to find decentralized control law $u_i(x_i(t))$, or $u_i(y_i(t))$, i=1,...,N, i.e. the overall system is controlled using local control loops for subsystems, such that uncertain dynamic system (11) is robustly stable in uncertainty domain (12) with guaranteed cost.

Basically, control design problem will be transformed into the output feedback form: $u_i(t) = F_i y_i(t)$, employing augmented system model to include controller dynamics, as it is using PID controller.

3.1.2 Augmented system model for continuous and discrete-time PID controller

The augmented system model including PID controller dynamics is developed in this section in general form appropriate both for continuous and discrete-time PID controllers. Firstly, recall PID control algorithms for both cases.

Control algorithm for continuous-time PID is

$$u(t) = K_P e(t) + K_I \int_0^t e(t)dt + K_D \dot{e}(t) \tag{15}$$

where $e(t) = y(t) - w(t)$ is control error, $w(k)$ is reference value (negative feedback sign is included in matrices K_P, K_I, K_D); K_P, K_I, K_D are controller parameter matrices (for SISO system they are scalars) to be designed.

Generally, different output variables can be considered for proportional, integral and derivative controller terms, for better readability we assume that all outputs enter all three controller terms. We further assume that the reference value is constant, $w(k) = w$ and that the system states in model (11) correspond to the deviations from working point (these assumptions correspond to step change of reference value). Then the control law (15) can be rewritten as

$$u(t) = K_P y(t) + K_I \int_0^t y(t)dt + K_D \dot{y}(t) \,. \tag{16}$$

Integral term can be included into the state vector in the common way defining the auxiliary state $z = \int_0^t y(t)$, i.e. $\dot{z}(t) = y(t) = C_d x(t)$ and PID controller algorithm is

$$u(t) = K_P C_d x(t) + K_I C_d z(t) + K_D C_d \dot{x}(t)\,. \tag{17}$$

Then the closed-loop system (11) with PID controller (17) can be described by augmented model

$$\dot{x}_n = \begin{bmatrix} \dot{x} \\ \dot{z} \end{bmatrix} = \begin{bmatrix} A(\alpha) & 0 \\ C_d & 0 \end{bmatrix}\begin{bmatrix} x \\ z \end{bmatrix} + \begin{bmatrix} B(\alpha) \\ 0 \end{bmatrix} u =$$

$$= \begin{bmatrix} A(\alpha) & 0 \\ C_d & 0 \end{bmatrix}\begin{bmatrix} x \\ z \end{bmatrix} + \begin{bmatrix} B(\alpha) \\ 0 \end{bmatrix}\begin{bmatrix} K_P C_d & K_I C_d \end{bmatrix}\begin{bmatrix} x \\ z \end{bmatrix} + \begin{bmatrix} B(\alpha) \\ 0 \end{bmatrix}\begin{bmatrix} K_D C_d & 0 \end{bmatrix}\begin{bmatrix} \dot{x} \\ \dot{z} \end{bmatrix}$$

or

$$\left(\begin{bmatrix} I & 0 \\ 0 & I \end{bmatrix} - \begin{bmatrix} B(\alpha)K_D C_d & 0 \\ 0 & 0 \end{bmatrix}\right)\begin{bmatrix} \dot{x} \\ \dot{z} \end{bmatrix} = \begin{bmatrix} A(\alpha) & 0 \\ C_d & 0 \end{bmatrix}\begin{bmatrix} x \\ z \end{bmatrix} + \begin{bmatrix} B(\alpha) \\ 0 \end{bmatrix}\begin{bmatrix} K_P C_d & K_I C_d \end{bmatrix}\begin{bmatrix} x \\ z \end{bmatrix}$$

which in a compact form yields

$$M_d(\alpha)\dot{x}_n = A_C(\alpha)x_n \tag{18}$$

where

$$\begin{aligned} M_d(\alpha) &= \left(\begin{bmatrix} I & 0 \\ 0 & I \end{bmatrix} - \begin{bmatrix} B(\alpha) \\ 0 \end{bmatrix}\begin{bmatrix} K_D & 0 \end{bmatrix}\begin{bmatrix} C_d & 0 \\ 0 & C_d \end{bmatrix}\right), \\ A_C(\alpha) &= \begin{bmatrix} A(\alpha) & 0 \\ C_d & 0 \end{bmatrix} + \begin{bmatrix} B(\alpha) \\ 0 \end{bmatrix}\begin{bmatrix} K_P & K_I \end{bmatrix}\begin{bmatrix} C_d & 0 \\ 0 & C_d \end{bmatrix} \end{aligned} \tag{19}$$

argument t is omitted for brevity.
A discrete-time PID (often denoted as PSD) controller is described by control algorithm

$$u(t) = k_P e(t) + k_I \sum_{i=0}^{t} e(i) + k_D[e(t) - e(t-1)] \tag{20}$$

where $u(t)$, $e(t) = y(t) - w(t)$, $w(t)$ are discrete time counterparts to the continuous time signals; k_P, k_I, k_D are controller parameter matrices to be designed. By analogy with continuous time case, for constant $w(t)$ we write

$$u(t) = k_P y(t) + k_I \sum_{i=0}^{t} y(i) + k_D[y(t) - y(t-1)] \tag{21}$$

State space description of PID controller can be derived in the following way. The dynamics of PID controller (21) requires two state variables, since besides $\sum_{i=0}^{t} y(i)$, also y(t-1) is needed. One possible choice of controller state variables is: $z(t) = [z_1^T(t)\;\; z_2^T(t)]^T$, $z_1(t) = \sum_{i=0}^{t-2} y(i),\; z_2(t) = \sum_{i=0}^{t-1} y(i)$, then $y(t-1) = z_2(t) - z_1(t)$. Rewriting (21) as

$$\begin{aligned} u(t) &= k_P y(t) + k_I \sum_{i=0}^{t-1} y(i) + k_I y(t) + k_D[y(t) - y(t-1)] \\ &= (k_P + k_I + k_D)y(t) + k_I z_2(t) - k_D(z_2(t) - z_1(t)) \end{aligned} \tag{21}$$

we obtain the respective description of the discrete-time PID controller in state space as

$$\begin{aligned} z(t+1) &= \begin{bmatrix} 0 & I \\ 0 & I \end{bmatrix} z(t) + \begin{bmatrix} 0 \\ I \end{bmatrix} y(t) = A_R z(t) + B_R y(t) \\ u(t) &= \begin{bmatrix} k_D & k_I - k_D \end{bmatrix} z(t) + (k_P + k_I + k_D)y(t) = \\ &= C_R z(t) + D_R y(t) \end{aligned} \tag{22}$$

where z(t) is controller dynamics state vector, $z(t) \in R^{2p}$.

The respective augmented model for discrete-time version of system (11) with PID controller is

$$x_n(t+1) = \begin{bmatrix} x(t+1) \\ z(t+1) \end{bmatrix} = \begin{bmatrix} A(\alpha) & 0 \\ B_R C_d & A_R \end{bmatrix} \begin{bmatrix} x(t) \\ z(t) \end{bmatrix} + \begin{bmatrix} B(\alpha) \\ 0 \end{bmatrix} [(D_R C_d \quad C_R] \begin{bmatrix} x(t) \\ z(t) \end{bmatrix} \tag{23}$$

where $A_R \in R^{2p\times 2p}, A_R = \begin{bmatrix} 0 & I \\ 0 & I \end{bmatrix}$, $B_R \in R^{2p\times p}, B_R = \begin{bmatrix} 0 \\ I \end{bmatrix}$, $C_R \in R^{m\times 2p}, C_R = \begin{bmatrix} k_D & k_I - k_D \end{bmatrix}$,

$D_R = k_P + k_I + k_D$.

Analogically as in continuous time case, the augmented system (23) can be rewritten in a compact form as

$$x_n(t+1) = A_C(\alpha) x_n(t) \tag{24}$$

where

$$A_C(\alpha) = \begin{bmatrix} A(\alpha) & 0 \\ B_R C_d & A_R \end{bmatrix} + \begin{bmatrix} B(\alpha) \\ 0 \end{bmatrix} [D_R \quad C_R] \begin{bmatrix} C_d & 0 \\ 0 & I_{2p} \end{bmatrix}. \tag{25}$$

Summarizing the augmented closed loop system models (18), (19) and (24), (25) for continuous and discrete-time PID controllers respectively, we can finally, using denotation $\delta x(t)$, introduced in (9), rewrite both of them in general form

$$M_d(\alpha)\delta x_n(t) = A_C(\alpha) x_n(t) \tag{26}$$

where $M_d(\alpha)$ is assumed to be invertible,

$A_C(\alpha) = A_{aug}(\alpha) + \begin{bmatrix} B(\alpha) \\ 0 \end{bmatrix} [F_1 \quad F_2] C_{aug} = A_{aug}(\alpha) + B_{aug}(\alpha) F C_{aug}$ and

for a continuous PID: $A_{aug}(\alpha) = \begin{bmatrix} A(\alpha) & 0 \\ C_d & 0 \end{bmatrix}$, $C_{aug} = \begin{bmatrix} C_d & 0 \\ 0 & C_d \end{bmatrix}$ and

$$M_d(\alpha) = \left(\begin{bmatrix} I & 0 \\ 0 & I \end{bmatrix} - \begin{bmatrix} B(\alpha) K_D C_d & 0 \\ 0 & 0 \end{bmatrix} \right); \tag{27a}$$

for a discrete-time PID: $A_{aug}(\alpha)=\begin{bmatrix} A(\alpha) & 0 \\ B_R C_d & A_R \end{bmatrix}$, $C_{aug}=\begin{bmatrix} C_d & 0 \\ 0 & I_{2p} \end{bmatrix}$ and $M_d(\alpha)=I$. (27b)

PID controller parameters are:

$$F=[F_1 \quad F_2]=[K_P \quad K_I] \text{ and } K_D \text{ included in } M_d(\alpha) \text{ for a continuous-time case;} \tag{28a}$$

$$F=[F_1 \quad F_2];\, F_1=k_P+k_I+k_D,\, F_2=[k_D \quad k_I-k_D] \text{ for a discrete-time case.} \tag{28b}$$

In a decentralized PID controller design, controller gain matrices are restricted to block diagonal structure respective to subsystem dimensions.
The presented general closed loop augmented system polytopic model (26) is advantageously used in next developments.

3.1.3 Robust stability

In this section we recall several recent results on robust stability for linear uncertain systems with polytopic model. These results are formulated as robust stability conditions in LMI form. Let us start with basic notions concerning Lyapunov stability and D-stability concept (Peaucelle et al., 2000; Henrion et al., 2002), used to receive the robust stability conditions in more general form.
Definition 3.1 (D-stability)
Consider the D-domain in the complex plain defined as

$$D=\{s \text{ is complex number}: \begin{bmatrix} 1 \\ s \end{bmatrix}^* \begin{bmatrix} r_{11} & r_{12} \\ r_{12}^* & r_{22} \end{bmatrix} \begin{bmatrix} 1 \\ s \end{bmatrix} < 0\} \tag{29}$$

Linear system is D-stable if and only if all its poles lie in the D-domain.
(For simplicity, we use in Def. 3.1 scalar values of parameters r_{ij}, in general, the stability domain can be defined using matrix values of parameters r_{ij} with the respective dimensions.) The standard choice of r_{ij} is $r_{11}=0$, $r_{12}=1$, $r_{22}=0$ for a continuous-time system; $r_{11}=-1$, $r_{12}=0$, $r_{22}=1$ for a discrete-time system, corresponding to open left half plane and unit circle respectively.
The D-stability concept enables to formulate robust stability condition for uncertain polytopic system in general way, (deOliveira et al., 1999; Peaucelle et al., 2000). The following robust stability condition is based on the existence of Lyapunov function $V(t)=x(t)P(\alpha)x(t)$ for linear uncertain polytopic system

$$\delta x(t)=A(\alpha)x(t) \tag{30}$$

where $A(\alpha)$ is from uncertainty domain (12).

Definition 3.2 (Robust stability)

Uncertain system (30) is *robustly D-stable* in the convex uncertainty domain (12) if and only if there exists a matrix $P(\alpha)=P(\alpha)^T>0$ such that

$$r_{12}P(\alpha)A(\alpha)+r_{12}^*A^T(\alpha)P(\alpha)+r_{11}P(\alpha)+r_{22}A^T(\alpha)P(\alpha)A(\alpha)<0 \tag{31}$$

For one Lyapunov function for the whole uncertainty domain, i.e. $P(\alpha) = P > 0$, the *quadratic D-stability* is guaranteed by (31). Generally, robust stability condition (31) with parameter dependent matrix $P(\alpha)$ is less conservative (provides bigger stability domain for $A(\alpha)$ than quadratic stability one), however stability is guaranteed only for relatively slow changes of system parameters within uncertainty domain (12) (in comparison with system dynamics). On the other hand, quadratic stability guards against arbitrary quick changes of system parameters within uncertainty domain (12) at the expense of sufficient, relatively strong, stability condition; which can be overly conservative for the case of slow parameter changes.

We consider the parameter dependent Lyapunov function (PDLF) defined as

$$V(t) = x(t)P(\alpha)x(t) \tag{32}$$

$$P(\alpha) = \sum_{k=1}^{K} \alpha_k P_k \text{ where } P_k = P_k^T > 0 \tag{33}$$

PDLF given by (32), (33) enables to transform robust stability condition (31) for uncertain linear polytopic system (9), (10) into the set of N Linear Matrix Inequalities (LMIs). Several respective sufficient robust stability conditions have been developed in the literature, e.g. (deOliveira et al., 1999; Peaucelle et al., 2000; Henrion et al., 2002). Recall the sufficient robust *D*-stability condition proposed in (Peaucelle et al., 2000), which to the authors best knowledge belongs to the least conservative (Grman et al., 2005).

Lemma 3.1

If there exist matrices $H \in R^{nxn}, G \in R^{nxn}$ and K symmetric positive definite matrices $P_k \in R^{nxn}$ such that for all k = 1,..., K:

$$\begin{bmatrix} r_{11}P_k + \tilde{A}_{(k)}^T H^T + H\tilde{A}_{(k)} & r_{12}P_k - H + \tilde{A}_{(k)}^T G \\ r_{12}^* P_k - H^T + G^T \tilde{A}_{(k)} & r_{22}P_k - (G + G^T) \end{bmatrix} < 0 \tag{34}$$

then uncertain system (30) is robustly D-stable in uncertainty domain (12).

Note that matrices H and G are not restricted to any special form; they were included to relax the conservatism of the sufficient condition. Robust stability condition for more general dynamic system model (26), including also the term for guaranteed cost will be presented in the next section.

3.2 Robust decentralized PID controller design

In this section, the robust decentralized PID controller is designed, based on robust stability condition developed in our recent papers, (Rosinová & Veselý, 2007; Veselý & Rosinová, 2011). Robust stability condition with guaranteed cost for closed loop uncertain system (26) is provided in the next theorem.

Theorem 3.1

Consider uncertain linear system (26) with cost function (13). If there exist symmetric matrix $P(\alpha) > 0$ and matrices H, G and F of the respective dimensions such that

$$\begin{bmatrix} r_{11}P(\alpha)+A_C(\alpha)^T H^T + HA_C(\alpha)+Q+C^TF^TRFC & r_{12}P(\alpha)-HM_d(\alpha)+A_C(\alpha)^T G \\ r_{12}^*P(\alpha)-M_d(\alpha)^T H^T + G^T A_C(\alpha) & r_{22}P_k - M_d(\alpha)G - G^T M_d(\alpha)^T \end{bmatrix} < 0 \quad (35)$$

then the system (26) is robustly *D*-stable with guaranteed cost: $J \le J_0 = x^T(0)P(\alpha)x(0)$.

Proof. The proof is analogical to the one presented in (Rosinová & Veselý, 2007) for the continuous-time PID. Firstly, we formulate the sufficient stability condition for uncertain system (26) using the respective Lyapunov function. The assumption that $M_d(\alpha)$ is invertible, enables us to rewrite (26) as $\delta x(t) = M_d^{-1}(\alpha)A(\alpha)x(t)$ and use parameter dependent Lyapunov function (32) to write robust stability condition.

Denote $\delta V(t) = \dot{V}(t)$ for a continuous-time system, $\delta V(t) = V(t+1) - V(t)$ for a discrete-time system. Then the sufficient D-stability condition (31) can be rewritten in the following form (known from LQ theory, for details see e.g. Rosinová et al., 2003)

$$\begin{aligned} & r_{12}P(\alpha)M_d^{-1}(\alpha)A(\alpha) + r_{12}^* A^T(\alpha)\left(M_d^{-1}(\alpha)\right)^T P(\alpha) + r_{11}P(\alpha) + \\ & r_{22}A^T(\alpha)\left(M_d^{-1}(\alpha)\right)^T P(\alpha)M_d^{-1}(\alpha)A(\alpha) + Q + C_d^T F^T RFC_d < 0 \end{aligned} \quad (36)$$

where the term $Q + C_d^T F^T RFC_d$ has been appended to $\delta V(t)$ to consider the guaranteed cost. To prove Theorem 3.1, it is sufficient to prove that (35) implies (36). This can be shown applying congruence transformation on (35):

$$\begin{bmatrix} I & A_C^T(\alpha)\left(M_d^{-1}(\alpha)\right)^T \end{bmatrix} \{\textit{left hand side of}\ (35)\} \begin{bmatrix} I \\ \left(M_d^{-1}(\alpha)\right)A_C(\alpha) \end{bmatrix} < 0 \quad (37)$$

which immediately yields (36).

It is important to note that robust stability condition (35) is linear with respect to parameter α. Therefore, for convex polytopic uncertainty domain (12) and PDLF (33), matrix inequality (35) is equivalent to the set of matrix inequalities respective to the polytope vertices, as summarized in Corollary 3.1.

Corollary 3.1

Uncertain linear system (26) with cost function (13) is robustly *D*-stable with parameter dependent Lyapunov function (32), (33) and guaranteed cost $J \le J_0 = x^T(0)P(\alpha)x(0)$ if the following matrix inequalities hold

$$\begin{bmatrix} r_{11}P_k + A_{Ck}^T H^T + HA_{Ck} + Q + C^T F^T RFC & r_{12}P_k - HM_{dk} + A_{Ck}^T G \\ r_{12}^* P_k - M_{dk}^T H^T + G^T A_{Ck} & r_{22}P_k - M_{dk}G - G^T M_{dk}^T \end{bmatrix} < 0, \ \text{k=1,...,K} \quad (38)$$

where $A_C(\alpha) = A_{aug}(\alpha) + B_{aug}(\alpha)FC_{aug} \in \left\{ \sum_{k=1}^{K} \alpha_k A_{Ck}, \sum_{k=1}^{K} \alpha_k = 1, \alpha_k \ge 0 \right\}$,

$A_{Ck} = A_{augk} + B_{augk}FC_{aug}$, and A_{augk}, B_{augk} correspond to the k-th vertex of uncertainty domain of the overall system (10), (12);

$M_d(\alpha) \in \left\{ \sum_{k=1}^{K} \alpha_k M_{dk}, \sum_{k=1}^{K} \alpha_k = 1, \alpha_k \geq 0 \right\}$, M_{dk} is for PID controller given by (27a) or (27b), and $B(\alpha)$ is given by (12).

Robust stability condition (38) is LMI for stability analysis, for controller synthesis it is in the BMI form. Therefore, (38) can be used for robust controller design either directly - using appropriate BMI solver (Henrion et al., 2005) or using some convexifying approach, (for discrete-time case see e.g. (Crusius & Trofino, 1999; deOliveira et al., 1999)). We have relatively good experience with the following simple convexified LMI procedure for static output feedback discrete-time controller design, which is directly applicable for discrete-time PID controller design problem formulated by (26), (27b), (28b).

The controller gain block diagonal matrix F is obtained by solving LMIs (39) for unknown matrices F, M, G and P_k of appropriate dimensions, the P_k being block diagonal symmetric, and M, G block diagonal with block dimensions conforming to subsystem dimensions. This convexifying approach does not allow including a term corresponding to performance index, therefore the resulting control guarantees only robust stability within considered uncertainty domain.

$$\begin{bmatrix} -P_k & A_{aug\,k}G + B_{aug\,k}KC_{aug} \\ G^T A_{aug\,k}^T + C^T K^T B_{aug\,k}^T & -G - G^T + P_k \end{bmatrix} < 0, \quad , \text{k=1,...,K}$$

$$MC_{aug} = C_{aug}G \tag{39}$$

$$F = KM^{-1}$$

F is the corresponding output feedback gain matrix.

The main advantage of the use of LMI (39) for controller design is its simplicity. The major drawbacks are, that the performance index cannot be considered, and that due to convexifying constraint ($MC_{aug} = C_{aug}G$), it need not provide a solution even in a case when feasible solution is received through BMI (38). (This is the case in our example in Section 3.3, in nonminimum phase configuration.)

To conclude this section we summarize the described decentralized PID controller design procedure, assuming that the state space model is in the form of (9) with polytopic uncertainty domain given by (10), where columns of control input matrix B are arranged respectively to chosen pairing.

Design procedure for decentralized PID design in time domain

Step 1. Formulate the augmented state space model (26) for given system and chosen type of PID controller.

Step 2. Compute decentralized PID controller parameters using one of design alternatives:

- LMI alternative for discrete-time case - guarantees robust stability: solve LMI (39) for unknown block diagonal matrices F, M, G and $P_k>0$, of appropriate dimensions; PID controller parameters are given by F respectively to (28b).
- BMI alternative - guarantees robust stability and guaranteed cost for quadratic performance index (13): solve BMI (38) for unknown block diagonal matrices F, $P_k>0$ and matrices G, H, of appropriate dimensions, PID controller parameters are given by F and M_{dk} respectively to (28) and (27), M_{dk} is for PID controller given by (27).

3.3 Decentralized PID controller design for the Quadruple tank process

We consider quadruple tank linearized model (2) with parameters:

$$A_1 = A_3 = 30[cm^2]; \quad A_2 = A_4 = 35\,[cm^2];$$

$$a_1 = a_3 = 0.0977\,[cm^2]; \quad a_2 = a_4 = 0.0785\,[cm^2];$$

$$h_{10} = h_{20} = 20\,[cm]; \; h_{30} = 2.75\,[cm]; \; h_{40} = 2.22\,[cm]\,;$$

$$g = 981\,[cm/s^2]; \quad k_1 = 1.790; \; k_2 = 1.827\,.$$

$$\begin{bmatrix} \dot{x}_1 \\ \dot{x}_3 \\ \dot{x}_2 \\ \dot{x}_4 \end{bmatrix} = \left[\begin{array}{cc:cc} -0.0161 & 0.0435 & 0 & 0 \\ 0 & -0.0435 & 0 & 0 \\ \hdashline 0 & 0 & -0.0111 & 0.0333 \\ 0 & 0 & 0 & -0.0333 \end{array}\right] \cdot \begin{bmatrix} x_1 \\ x_3 \\ x_2 \\ x_4 \end{bmatrix} + \left[\begin{array}{c:c} 0.0596\gamma_1 & 0 \\ 0 & 0.0595(1-\gamma_2) \\ \hdashline 0 & 0.0522\gamma_2 \\ 0.052(1-\gamma_1) & 0 \end{array}\right] \cdot \begin{bmatrix} u_1 \\ u_2 \end{bmatrix}$$

$$\begin{bmatrix} y_1 \\ y_2 \end{bmatrix} = \left[\begin{array}{cc:cc} 1 & 0 & 0 & 0 \\ \hdashline 0 & 0 & 1 & 0 \end{array}\right] \begin{bmatrix} x_1 \\ x_2 \\ \hdashline x_3 \\ x_4 \end{bmatrix}$$

Subsystems are indicated via the splitting dashed lines. Polytope vertices respective to working points (7) or (8) for minimum phase or nonminimum phase configurations respectively determine the corresponding uncertainty domains indicated in Fig.2. State space model has been discretized with sampling period $T_s = 5[s]$ (sampling period was chosen with respect to the process dynamics).

Minimum phase configuration

In the minimum phase case, robust decentralized controller is designed for chosen pairing $v_1 - y_1$, $v_2 - y_2$ (see Section 2.1) using alternatively solution of LMI (39) or BMI (38) for decentralized discrete-time PI controller design. The resulting controller parameters are in Tab.1, the respective simulation results are illustrated and compared on step responses in one tested point from uncertainty domain, in Fig. 3.

Design approach	1st subsyst. controller	2nd subsyst. controller
LMI (39)	$\dfrac{5.862 - 2.602z^{-1}}{1 - z^{-1}}$	$\dfrac{6.45 - 2.578z^{-1}}{1 - z^{-1}}$
BMI (38) Q=0.01*I, R=5*I	$\dfrac{1.3002 - 1.0351z^{-1}}{1 - z^{-1}}$	$\dfrac{1.3833 - 1.1361z^{-1}}{1 - z^{-1}}$

Table 1. Decentralized PID controller parameters – minimum phase case

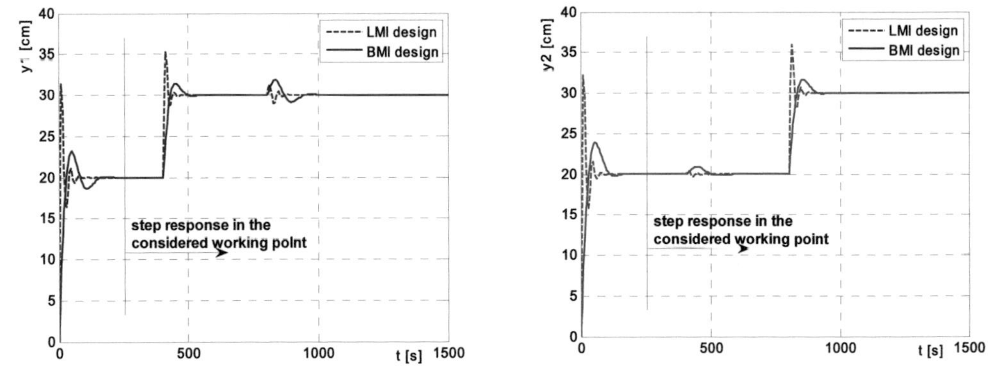

Fig. 3. Step response of y_1 and y_2 to setpoint step changes: w_1 in 400s and w_2 in 800s; comparison of LMI and BMI design results from Tab.1

Obviously, the results for the BMI solution including performance index outperform the ones obtained using simpler LMI approach.

Nonminimum phase configuration

In the nonminimum phase case, robust decentralized controller is designed for chosen pairing $v_1 - y_2$, $v_2 - y_1$ (see Section 2.1) using a solution of BMI (38) for decentralized discrete-time PI controller design, (in this case LMI procedure (39) does not provide a feasible solution). The resulting controller parameters are in Tab.2, the respective simulation results are illustrated on step responses in one tested point from uncertainty domain, in Fig. 4.

Design approach	1st subsyst. controller	2nd subsyst. controller
BMI (38) Q=0.01*I, R=5*I	$\frac{0.5371 - 0.5099z^{-1}}{1 - z^{-1}}$	$\frac{0.7221 - 0.6941z^{-1}}{1 - z^{-1}}$

Table 2. Decentralized PID controller parameters - nonminimum phase case

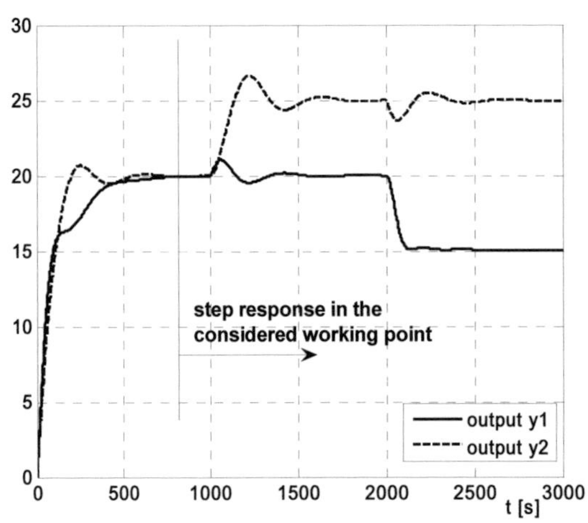

Fig. 4. Step response of y_1 and y_2 to setpoint step changes: w_1 (for y_2) in 1000s and w_2 (for y_1) in 2000s

Comparison of simulation results for minimum and nonminimum phase cases shows the deteriorating influence of nonminimum phase on settling time.

4. Robust decentralized PID controller design in the frequency domain

This section deals with an original frequency domain robust decentralized controller design methodology applicable for uncertain systems described by a set of transfer function matrices. The design methodology is based on the Equivalent Subsystems Method (ESM) - a frequency domain decentralized controller design technique to guarantee stability and specified performance of multivariable systems and is applicable for both continuous- and discrete-time controller designs (Kozáková et al., 2009). In contrast to the two stage robust decentralized controller design method based on the M-Δ structure stability conditions (Kozáková & Veselý, 2009), the recent innovation (Kozáková et al., 2011) consists in that robust stability conditions are directly integrated into the ESM, thus providing a one-step (direct) robust decentralized controller design for robust stability and plant-wide performance.

4.1 Preliminaries and problem formulation

Consider a MIMO system described by a transfer function matrix $G(s) \in R^{m\times m}$ and a controller $R(s) \in R^{m\times m}$ in the standard feedback configuration according to Fig. 5,

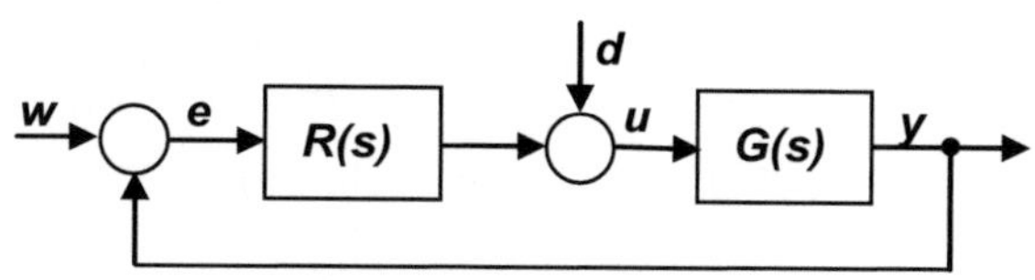

Fig. 5. Standard feedback configuration

where w, u, y, e, d are respectively vectors of reference, control, output, control error and disturbance of compatible dimensions. Necessary and sufficient conditions for closed-loop stability are given by the Generalized Nyquist Stability Theorem applied to the closed-loop characteristic polynomial

$$\det F(s) = \det[I + Q(s)] \tag{40}$$

where $Q(s) = G(s)R(s) \in R^{m\times m}$ is the open-loop transfer function matrix.

Characteristic functions of $Q(s)$ are the set of m algebraic functions $q_i(s), i = 1,...,m$ defined as follows:

$$\det[q_i(s)I_m - Q(s)] = 0 \quad i = 1,...,m \tag{41}$$

Characteristic loci (CL) are the set of loci in the complex plane traced out by the characteristic functions of Q(s), $s = j\omega$.

Theorem 4.1 (Generalized Nyquist Stability Theorem)

The closed-loop system in Fig. 1 is stable if and only if

$$\text{a. } \det F(s) \neq 0 \qquad \forall s$$

$$\text{b. } N[0, \det F(s)] = \sum_{i=1}^{m} N\{0, [1 + q_i(s)]\} = n_q \tag{42}$$

where $F(s) = (I + Q(s))$ and n_q is the number of unstable poles of Q(s).
Let the uncertain plant be given as a set Π of N transfer function matrices

$$\Pi = \{G^k(s)\},\ k = 1,2,\ldots,N \quad \text{where} \quad G^k(s) = \left\{G_{ij}^k(s)\right\}_{m\times m} \tag{43}$$

The simplest uncertainty model is the unstructured perturbation. A set of unstructured perturbations D_U is defined as

$$D_U := \{E(j\omega) : \sigma_{\max}[E(j\omega)] \leq \ell(\omega),\ \ \ell(\omega) = \max_k\ \sigma_{\max}[E(j\omega)]\} \tag{44}$$

where $\ell(\omega)$ is a scalar weight on the norm-bounded perturbation $\Delta(s) \in R^{m\times m}$, $\sigma_{\max}[\Delta(j\omega)] \leq 1$ over given frequency range, $\sigma_{\max}(\cdot)$ is the maximum singular value of (.), hence

$$E(j\omega) = \ell(\omega)\Delta(j\omega) \tag{45}$$

Using unstructured perturbation, the set Π can be generated by either additive (E_a), multiplicative input (E_i) or multiplicative output (E_o) uncertainties, or their inverse counterparts (Skogestad & Postlethwaite, 2009) thus specifying pertinent uncertainty regions. In the sequel, just additive (a) and multiplicative output (o) perturbations will be considered; results for other uncertainty types can be obtained by analogy.
Denote $G(s)$ any member of a set of possible plants $\Pi_k, k = a,i$; $G_0(s)$ the nominal model used to design the controller, and $\ell_k(\omega)$ the scalar weight on a normalized perturbation. The sets Π_k generated by the two considered uncertainty forms are:
Additive uncertainty:

$$\begin{aligned} &\Pi_a := \{G(s) : G(s) = G_0(s) + E_a(s),\ E_a(j\omega) \leq \ell_a(\omega)\Delta(j\omega)\} \\ &\ell_a(\omega) = \max_k\ \sigma_{\max}[G^k(j\omega) - G_0(j\omega)],\ k = 1,2,\ldots,N \end{aligned} \tag{46}$$

Multiplicative output uncertainty:

$$\begin{aligned} &\Pi_o := \{G(s) : G(s) = [I + E_o(s)]G_0(s),\ \Delta(j\omega) \leq \ell_o(j\omega)\Delta(j\omega)\} \\ &\ell_o(\omega) = \max_k\ \sigma_{\max}\{[G^k(j\omega) - G_0(j\omega)]G_0^{-1}(j\omega)\},\ k = 1,2,\ldots,N \end{aligned} \tag{47}$$

Standard feedback configuration with uncertain plant modelled using any unstructured uncertainty form can be recast into the $M-\Delta$ structure (for additive perturbation see Fig. 6) where M(s) is the nominal model and $\Delta(s)$ is the norm-bounded complex perturbation.

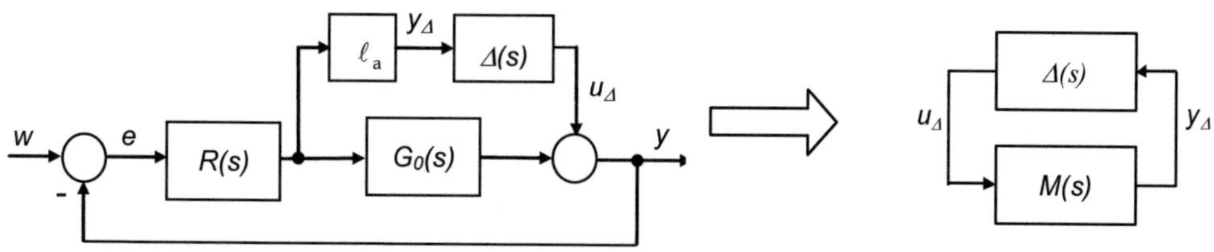

Fig. 6. Standard feedback configuration with additive perturbation (left) recast into the $M-\Delta$ structure (right)

According to the general robust stability condition (Skogestad & Postlethwaite, 2009), if both the nominal closed-loop system M(s) and the perturbations $\Delta(s)$ are stable, the $M-\Delta$ system in Fig. 2 is stable for all perturbations $\Delta(s): \sigma_{\max}(\Delta) \leq 1$ if and only if

$$\sigma_{\max}[M(j\omega)] < 1\,, \ \forall\omega \tag{48}$$

For individual uncertainty forms $M(s) = \ell_k M_k(s),\ k = a,o$ the corresponding matrices $M_k(s)$ are given by (49) and (50), respectively (disregarding negative signs which do not affect resulting robustness condition). The nominal model $G_0(s)$ is usually obtained as a model of mean parameter values.

$$M(s) = \ell_a(s)R(s)[I + G_0(s)R(s)]^{-1} = \ell_a(s)M_a(s) \tag{49}$$

$$M(s) = \ell_o(s)G_0(s)R(s)[I + G_0(s)R(s)]^{-1} = \ell_o(s)M_o(s) \tag{50}$$

4.1.1 Problem formulation

Consider an uncertain system that consists of m subsystems and is given as a set of N transfer function matrices obtained in N working points of plant operation. Let the uncertain system be described by a nominal model $G_0(s)$ and any unstructured uncertainty form (46), (47). Consider the following splitting of $G_0(s)$:

$$G_0(s) = G_d(s) + G_m(s) \tag{51}$$

where

$$G_d(s) = diag\{G_i(s)\}_{m\times m}, \quad \det G_d(s) \neq 0 \tag{52}$$

$$G_m(s) = G_0(s) - G_d(s) \tag{53}$$

A decentralized controller

$$R(s) = diag\{R_i(s)\}_{m\times m} \quad \det R(s) \neq 0 \tag{54}$$

is to be designed to guarantee stability over the whole operating range of the plant specified by (46) or (47) (robust stability) and a specified plant-wide performance (nominal performance).

To solve this problem, a frequency domain robust decentralized controller design technique has been developed (Kozáková and Veselý, 2009; Kozáková et. al., 2011); the core of it is the Equivalent Subsystems Method (ESM).

4.2 Decentralized controller design for performance: Equivalent Subsystems Method

The Equivalent Subsystems Method (ESM) is a Nyquist-based technique to design decentralized controller for stability and specified plant-wide performance. According to it, local controllers $R_i(s), i=1,...,m$ are designed independently for so-called equivalent subsystems obtained from frequency responses of decoupled subsystems by shaping each of them using one of m characteristic loci of the interactions matrix $G_m(s)$. If local controllers are independently tuned for specified degree-of-stability of equivalent subsystems, the resulting decentralized controller guarantees the same degree-of-stability plant-wide (Kozáková et al., 2009). Unlike standard robust approaches, the proposed technique considers full nominal model of mean parameter values, thus reducing conservatism of resulting robust stability conditions. In the context of robust decentralized controller design, the Equivalent Subsystems Method is directly applicable to design DC for the nominal model (Fig. 3).

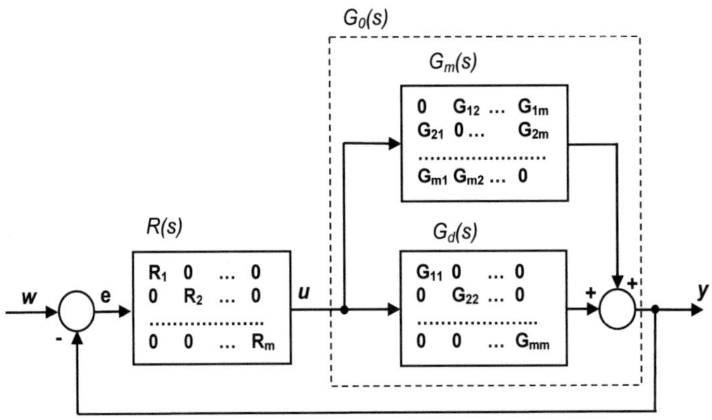

Fig. 7. Standard feedback loop under decentralized controller

The key idea behind the method is factorization of the closed-loop characteristic polynomial (40) in terms of the nominal system (51) under the decentralized controller (54). Then

$$\det F(s) = \det[R^{-1}(s) + G_d(s) + G_m(s)] \det R(s) \tag{55}$$

Denote the sum of the diagonal matrices in (55) as

$$R^{-1}(s) + G_d(s) = P(s) \tag{56}$$

where $P(s) = diag\{p_i(s)\}_{m \times m}$.

In order to "counterbalance" interactions $G_m(s)$, consider the closed-loop being at the limit of instability and choose the diagonal matrix $P(s) = p_k(s)I$ to have identical entries $p_k(s)$; then by similarity with (41) the bracketed term in (55) defines the k-th of the m characteristic functions of $[-G_m(s)]$ (the set of characteristic functions are denoted $g_i(s), i = 1,2,...,m$); thus

$$\det[P(s) + G_m(s)] = \det[p_k I + G_m] = \prod_{i=1}^{m}[-g_k(s) + g_i(s)] = 0,\ \ k = 1,2,...m \tag{57}$$

With respect to stability, the interactions matrix $G_m(s)$ can thus be replaced by [-P(s)] yielding the important relationship

$$\begin{aligned}&\det[I + G(s)R(s)] = \det\{I + [G_d(s) + G_m(s)]R(s)\} = \\ &= \det[R^{-1}(s) + G_d(s) - P(s)]\det R(s) = \det[I + G^{eq}(s)R(s)]\end{aligned} \tag{58}$$

where

$$G^{eq}(s) = diag\{G_i^{eq}(s)\}_{m\times m} \tag{59}$$

is a diagonal matrix of m equivalent subsystems generated as follows

$$G_{ik}^{eq}(s) = G_i(s) + g_k(s), \quad i = 1,2,\ldots,m \tag{60}$$

As all matrices are diagonal, on subsystems level (58) breaks down into m equivalent closed-loop characteristic polynomials (CLCP)

$$CLCP_i^{eq}(s) = 1 + R_i(s)G_i^{eq}(s) \quad i = 1,2,...,m \tag{61}$$

Considering (58)-(61), stability conditions stated in the Generalized Nyquist Stability Theorem modify as follows:

Corollary 4.1

The closed-loop in Fig. 3 comprising the system (51) and the decentralized controller (54) is stable if and only if there exists a diagonal matrix $P(s) = p_k(s)I(s)$ such that

$$\begin{aligned}&1.\ \det[p_k(s)I + G_m] = 0, \quad \text{for a fixed } k \in \{1,...,m\} \\ &2.\ \text{all equivalent characteristic polynomials (61) have roots with } \operatorname{Re}\{s\} < 0; \\ &3.\ N[0, \det F(s)] = n_q\end{aligned} \tag{62}$$

where N[0,g(s)] is number of anticlockwise encirclements of the complex plane origin by the Nyquist plot of g(s); n_q is number of open loop poles with $\operatorname{Re}\{s\} > 0$.

The decentralized controller design technique for nominal stability resulting from *Corollary 4.1* enables to independently design stabilizing local controllers for individual single input-single output equivalent subsystems using any standard frequency-domain design method, e.g. (Bucz et al., 2010; Drahos, 2000). In the originally developed ESM version (Kozáková et al., 2009) it was proved that local controllers tuned for a specified feasible degree-of-stability of equivalent subsystems constitute the decentralized controller guaranteeing the same degree-of-stability plant-wide. To design local controllers of equivalent subsystems, the general conditions in *Corollary 4.1* allow using any frequency domain performance measure that can appropriately be interpreted for the full system. In the next subsection, the plant wide performance is specified in terms of maximum overshoot which is closely related to phase margins of equivalent subsystems.

4.2.1 Decentralized controller design for guaranteed maximum overshoot and specified settling time

The ESM can be applied to design decentralized controller to guarantee specified maximum overshoot of output variables of the multivariable system. The design procedure evolves from the known relationship between the phase margin (PM) and the maximum peak of the complementary sensitivity (Skogestad & Postlethwaite, 2009)

$$PM \geq 2\arcsin\left(\frac{1}{2M_T}\right) \geq \frac{1}{M_T}[rad] \tag{63}$$

where

$$M_T = \sigma_{\max}[T(j\omega)] \tag{64}$$

is the maximum peak of the complementary sensitivity T(s) defined as

$$T(s) = G(s)R(s)[I + G(s)R(s)]^{-1} \tag{65}$$

Relation between the maximum overshoot $\eta_{\max}$ and M_T is given by (Bucz et al., 2010)

$$\eta_{\max} \leq \frac{1.18M_T - |T(0)|}{|T(0)|}100[\%] \tag{66}$$

According to the ESM philosophy, local controllers are designed using frequency domain methods; if PID controller is considered, the most appropriate ones are e.g. the Bode diagram design or the Neymark D-partition method. If using the Bode diagram design, in addition to $\eta_{\max}$ it is also possible to specify the required settling time t_s related with the closed-loop bandwidth frequency ω_0 defined as the gain crossover frequency. The following relations between t_s and ω_0 are useful (Reinisch, 1974).

$$t_s \approx \frac{3}{\omega_0} \text{ for } M_T \in (1.3;\, 1.5)$$

$$\frac{\pi}{t_s} < \omega_0 < \frac{4\pi}{t_s} \tag{67}$$

In general, a larger bandwidth corresponds to a smaller rise time, since high frequency signals are more easily passed on to the outputs. If the bandwidth is small, the time response will generally be slow and the system will usually be more robust.

Design procedure:

1. Generating frequency responses of equivalent subsystems.
2. Specification of performance requirements in terms of $\eta_{\max}$, t_s and M_T using (66), (67).
3. Specification of a minimum phase margin PM for equivalent subsystems using (63).
4. Local controller design for specified PM in equivalent subsystems using appropriate frequency domain method.
5. Verification of achieved performance by evaluating frequency domain performance measure and via simulation.

4.3 Decentralized controller design for robust stability using the Equivalent Subsystems Method

In the context of robust control approach, the ESM method in its original version is inherently appropriate to design decentralized controller guaranteeing stability and specified performance of the nominal model (nominal stability, nominal performance). If, in addition, the decentralized controller has to guarantee closed-loop stability over the whole operating range of the plant specified by the chosen uncertainty description (robust stability), the ESM can be used either within a two-stage design procedure or a direct design procedure for robust stability and nominal performance.

1. Two stage robust decentralized controller design for robust stability and nominal performance

In the first stage, the decentralized controller for the nominal system is designed using ESM, afterwards, fulfilment of the M-Δ stability condition (48) is examined; if satisfied, the design procedure stops, otherwise in the second stage the controller parameters are modified to satisfy robust stability conditions in the tightest possible way, or local controllers are redesigned using modified performance requirements (Kozáková & Veselý, 2009).

2. Direct decentralized controller design for robust stability and nominal performance

By direct integration of robust stability condition (48) in the ESM, a "one-shot" design of local controllers for both nominal performance and robust stability can be carried out. In case of decentralized controller design for guaranteed maximum overshoot and specified settling time, the upper bound for the maximum peak of the nominal complementary sensitivity over the given frequency range

$$M_T = \max_{\omega}\{\sigma_{max}[T_0(j\omega)]\} \qquad T_0(s) = G_0(s)R(s)[I + G_0(s)R(s)]^{-1} \tag{68}$$

can be obtained using the singular value properties in manipulations of the M-Δ condition (48) considering (49) or (50). The following bounds for the nominal complementary sensitivity have been derived:

$$\sigma_{\max}[T_0(j\omega)] < \frac{\sigma_{\min}[G_0(j\omega)]}{|\ell_a(\omega)|} = L_A(\omega) \qquad \forall\omega \ \textit{additive uncertainty} \tag{69}$$

$$\sigma_{\max}[T_0(j\omega)] < \frac{1}{|\ell_o(\omega)|} = L_O(\omega) \qquad \forall\omega \ \text{multiplicative output uncertainty} \tag{70}$$

Expressions on the r.h.s. of (69) and (70) do not depend on a particular controller and can be evaluated prior to designing the controller. In this way, if

$$M_T = \max_{\omega}\{\sigma_{\max}[T_0(j\omega)]\} \tag{71}$$

is used in the Design procedure, the resulting decentralized controller will simultaneously guarantee achieving the required maximum overshoot of all output variables (nominal performance) and stability over the whole operating range of the plant specified by selected working points (robust stability).

4.4 Discrete-time robust decentralized controller design using the Equivalent Subsystems Method

Controllers for continuous-time plants are mostly implemented as discrete-time controllers. A common approach to discrete-time controller design is the continuous controller redesign i.e. conversion of the already designed continuous controller into its discrete counterpart. This approach, however, is only an approximate scheme; performance under these controllers deteriorates with increasing sampling period. This drawback may be improved by modifying the continuous controller design before it is discretized which can often allow significantly larger sampling periods (Lewis, 1992). Then, the ESM design methodology can be applied in a similar way as in the continuous-time case using discrete characteristic loci, discrete Nyquist plots and discrete Bode diagrams of equivalent subsystems. Local controllers designed as continuous-time ones are subsequently converted into their discrete-time counterparts. Closed-loop performance under a discrete-time controller is verified using simulations and the discrete-time maximum singular value of the sensitivity $\sigma_M[S(z)]$ where

$$S(z) = [I + G(z)R(z)]^{-1},\ z = e^{j\omega T_s} \tag{72}$$

The maximum singular value $\max_{\omega} \sigma_{max}[S(e^{j\omega T})]$ plotted as function of frequency ω should be small at low frequencies where feedback is effective, and approach 1 at high frequencies, as the system is strictly proper, having a peak larger than 1 around the crossover frequency. The peak is unavoidable for real systems. Bandwidth frequency is defined as frequency where $\sigma_M[S(e^{j\omega T_s})]$ crosses 0.7 from below (Skogestad & Postlethwaite, 2009). Similarly, a discretized version of robust stability conditions (69), (70) based on (46) and (47) is applied.

4.4.1 Design of continuous controllers for discretization

The crucial step for the discrete controller design is proper choice of the sampling time T. Then, frequency response of the discretized system matches the one of the continuous time system up to a certain frequency $\omega < \omega_S / 2$, and the discrete controller can be obtained by converting the continuous–time controller designed from the discrete frequency responses to its discrete-time counterpart.

The sampling period T is to be selected according to the Shannon-Kotelnikov sampling theorem, or using common rules of thumb, e.g. as ~ 1/10 of the settling time of the plant step response, or from control system bandwidth according to the relation

$$20 < \frac{\omega_s}{\omega_0} < 40 \tag{72}$$

where ω_s is sampling frequency, and ω_0 is control system bandwidth, i.e. the maximum frequency at which the system output still tracks and input sinusoid in a satisfactory manner (Lian et al., 2002). A proper choice of sampling period is crucial for achievable bandwidth and feasibility of the required phase margin. Given a discrete-time transfer function $G(z)$, the frequency response can be studied by plotting Nyquist or Bode plots of $G(z)|_{z=e^{j\omega T}}$. The discrete-time robust controller design for maximum overshoot and settling time is described in the next Section.

4.5 Decentralized discrete-time PID Controller design for the Quadruple tank process

In the frequency domain, the direct robust decentralized PID design procedure has been applied for the transfer function matrix (3) identified in three working points within the minimum and nonminimum phase regions (7) and (8), respectively. In both cases the nominal model is a mean value parameter model.

Minimum phase configuration

From three plant models (3) evaluated in working points taken from the minimum phase uncertainty region as specified in (7), the resulting continuous-time nominal model is

$$G_0(s) = \begin{bmatrix} \dfrac{2.4667}{62s+1} & \dfrac{1.2333}{(23s+1)(62s+1)} \\ \dfrac{1.5667}{(30s+1)(90s+1)} & \dfrac{3.1333}{90s+1} \end{bmatrix} \tag{73}$$

All three transfer function matrices were discretized using the sampling period $T_S = 30s$ chosen as approx. $1/10$ of the settling time of plant step responses in Fig. 8.

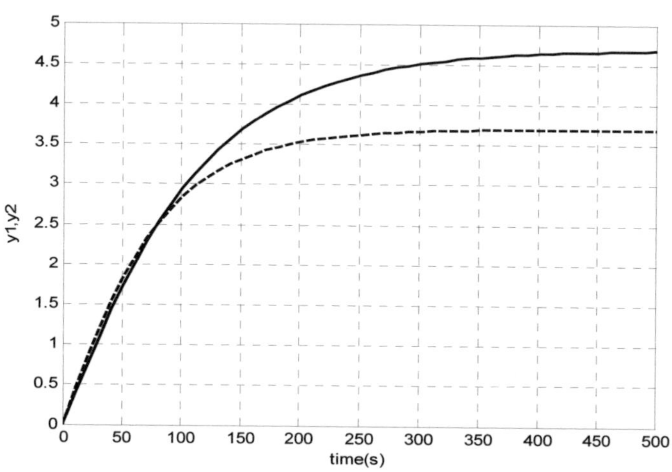

Fig. 8. Step response of the quadruple tank process

Discrete-time transfer function matrix of the nominal plant is

$$G(z)=\begin{bmatrix} \dfrac{0.9462z^{-1}}{1-0.6164z^{-1}} & \dfrac{0.2221z^{-1}+0.1226z^{-2}}{1-0.8877z^{-1}+0.1673z^{-2}} \\ \dfrac{0.1710z^{-1}+0.1097z^{-2}}{1-1.0840z^{-1}+0.2636z^{-2}} & \dfrac{0.8882z^{-1}}{1-0.7165z^{-1}} \end{bmatrix} \quad (74)$$

From the discretized transfer function matrices and the nominal model (74), upper bounds for $\sigma_{\max}[T_0(j\omega)]$ were evaluated according to (69) and (70).

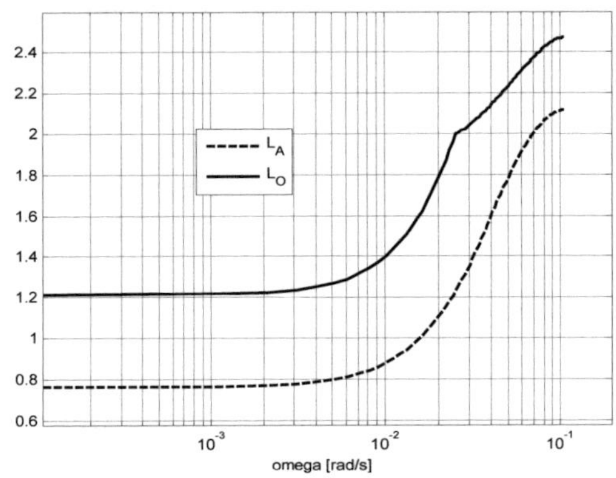

Fig. 9. Upper bounds for $\sigma_{\max}[T_0(j\omega)]$ evaluated according to (69) and (70)

Inspection of Fig. 9 reveals, that $M_{T\_A}=\min_{\omega} L_A = 0.77 < 1$ is not feasible for the local controller design (closed-loop design magnitude less than 1 does not guarantee proper setpoint tracking, even at ω=0); hence $M_T < M_{T\_O} = \min_{\omega} L_O = 1.22$ has been considered in the sequel.

Characteristic loci $g_1(z)$, $g_2(z)$ of $G_m(z)$ were calculated; $g_2(z)$ was selected to generate the equivalent subsystems according to (60). Bode plots of resulting equivalent subsystems are shown in Fig. 10.

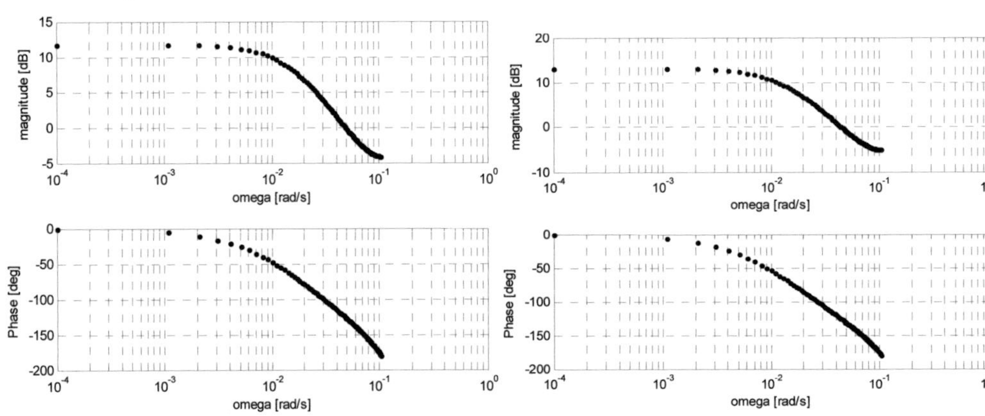

Fig. 10. Discrete Bode plots of equivalent subsystems generated by $g_2(z)$: $G_{12}^{eq}(z)$ (left), $G_{22}^{eq}(z)$ (right) (min. phase case)

Relevant parameters read form discrete Bode plots of uncompensated equivalent subsystems in Fig. 10 are summarized in Tab. 3.

Equivalent subsystem	PM	Crossover frequency
$G_{12}^{eq}(z)$	53.9^0	0.048 rad/ s^{-1}
$G_{22}^{eq}(z)$	58.35^0	0.0448 rad/s^{-1}

Table 3. Relevant parameters of equivalent subsystems generated by $g_2(z)$

For both equivalent subsystems the required settling time and maximum overshoot were chosen with respect to plant dynamics: $t_s = 600s$, $M_T = 1.05$ corresponding to $\eta_{\max} = 5\%$. Related values of other design parameters obtained from (63) and (67) respectively are: $PM_{\min} = 56.87^0$ and required crossover frequency $\omega_0 = 0.0131$. The required phase margin $PM_{req} > PM_{\min}$ was chosen $PM_{req} = 65^0$. To design local controllers, Bode design procedure (Kuo, 2003) has been applied independently for each equivalent subsystem to achieve the required phase margin: $PM(\omega_0)$ is found on the magnitude Bode plot; if $PM(\omega_0) > PM_{req}$, a PI controller $G_{PI}(s) = K_P + \frac{K_I}{s}$ is designed. If $PM(\omega_0) < PM_{req}$, a PD controller $G_{PD}(s) = 1 + K_D s$ is designed first, to provide $PM_{req}(\omega_0)$, and subsequently a PI controller is designed. The resulting PID controller is obtained in the series form $G_{PID}(s) = (K_P + \frac{K_I}{s})(1 + K_D s)$. Achieved design results are summarized in Tab. 4.

Eq. subsyst.	$R_i(s)$	$R_i(z)$	$PM_{achieved}$	$\omega_{0\_achieved}$
$G_{12}^{eq}(z)$	$R_1(s) = 0.1988 + \frac{0.0039}{s}$	$R_1(z) = \frac{0.199 - 0.082z^{-1}}{1 - z^{-1}}$	58.35^0	0.0122 rad/s^{-1}
$G_{22}^{eq}(z)$	$R_2(s) = 0.2212 + \frac{0.0034}{s}$	$R_2(z) = \frac{0.221 - 0.119z^{-1}}{1 - z^{-1}}$	65.70	0.0121 rad/s^{-1}

Table 4. Design results and achieved frequency domain performance measures (minimum phase configuration)

Design results in Tab. 4 along with Bode plots of compensated equivalent subsystems in Fig.11 prove achieving required design parameters. Closed-loop step responses are in Fig. 12.

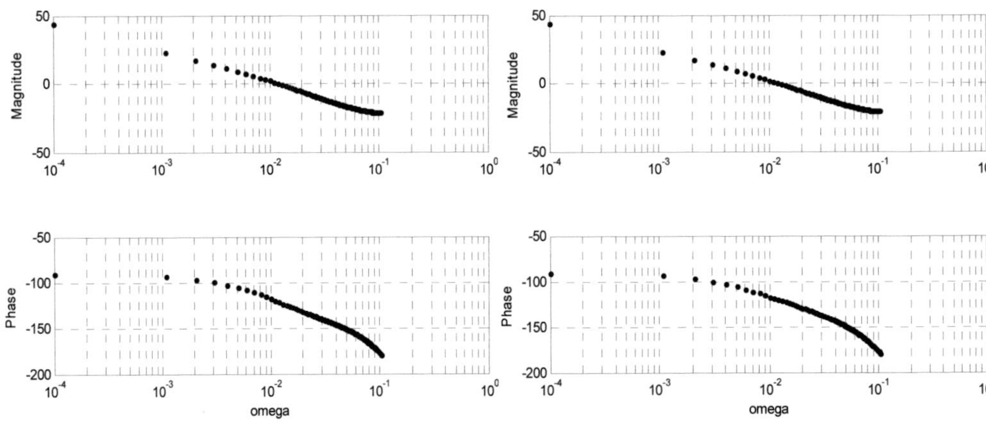

Fig. 11. Discrete Bode plots of equivalent subsystems under designed PI controllers: $G_{12}^{eq}(z)$ (left), $G_{22}^{eq}(z)$ (right)

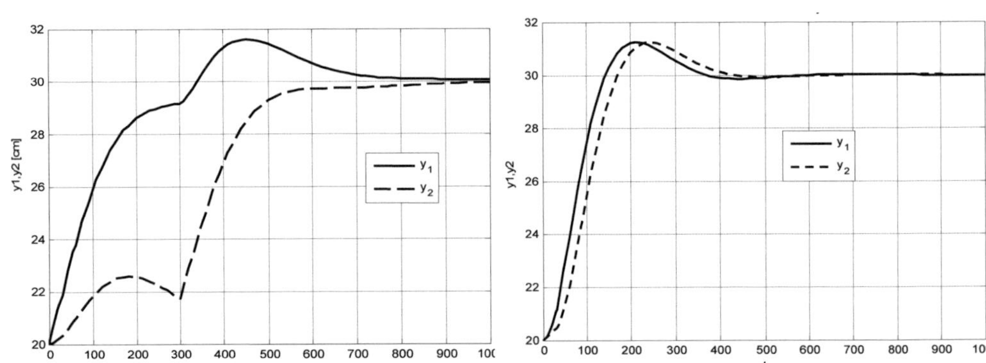

Fig. 12. Nominal closed-loop step responses of the quadruple tank process (reference steps 0.1m occurred at t=0s at the input of the 1st subsystem, and at t=300s and t=10s, respectively, at the input of the 2nd subsystem). Maximum overshoot and settling time (600s) were kept in both cases.

Nominal closed-loop stability was verified both by calculating closed-loop poles and using the Generalized Nyquist encirclement criterion (Fig. 13).

$$\text{Roots\_of\_CLCP} = \{\, 0.7019 \pm 0.2572i, 0.8313, 0.7167, 0.7165, 0.6164, 0.3720, 0.2637 \,\}$$

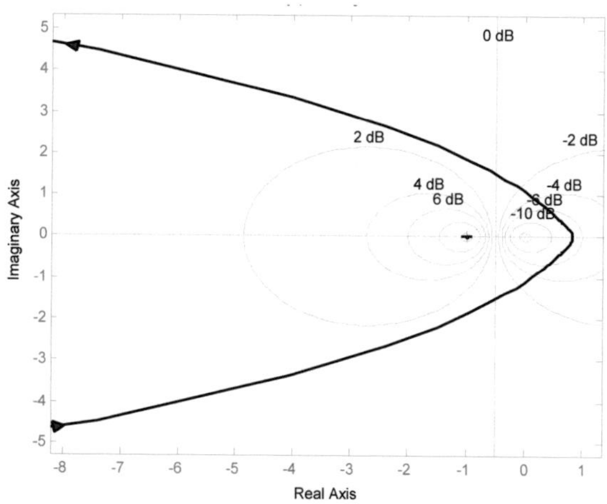

Fig. 13. Stability test using the Nyquist plot of $\det[I + G(z)R(z)]$

Achieved nominal performance was verified via plotting sensitivity magnitude plot in Fig. 14. Sensitivity peak $\max_{\omega}\{\sigma_M[S(j\omega)]\} < 2$ around the crossover frequency proves good closed-loop performance.

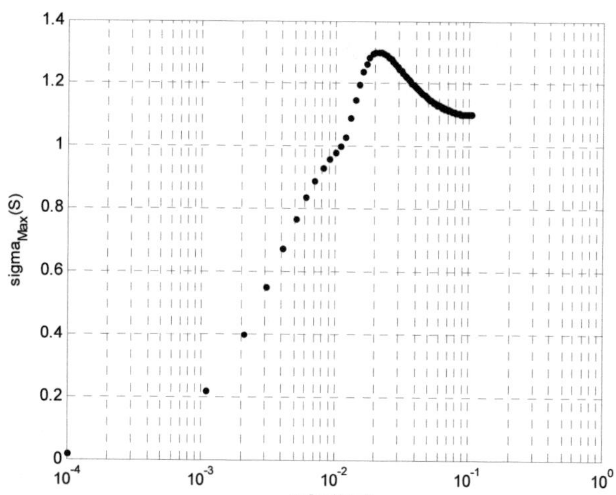

Fig. 14. $\sigma_M[S(z)]_{z=e^{j\omega T}}$ - versus –frequency plot

Fulfilment of robust stability condition (70) is examined in Fig. 15. The closed-loop system is stable over the whole minimum phase region (7).

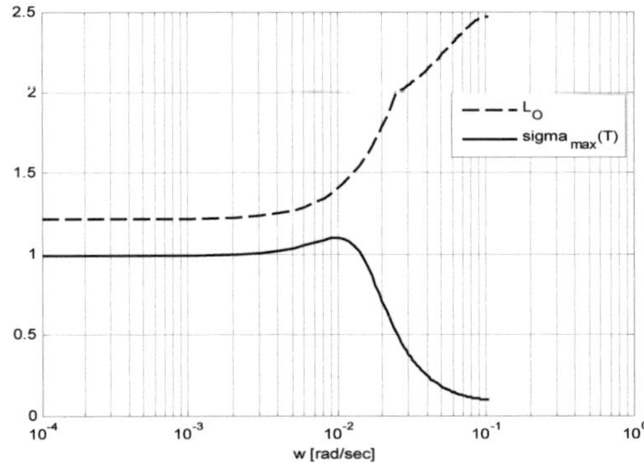

Fig. 15. Verification of the robust stability condition $\sigma_{\max}[T_0(j\omega)] < \dfrac{1}{|\ell_o(\omega)|} = L_O(\omega)$

Non-minimum phase configuration

To design robust decentralized PI controller for the non-minimum phase configuration, the continuous-time nominal model was evaluated for γ_1, γ_2 taken from the non-minimum phase uncertainty region (8) and interchanged columns of the transfer function matrix (due to opposite pairing as suggested in Section 2):

$$G_0(s) = \begin{bmatrix} \dfrac{3.0830}{(23s+1)(62s+1)} & \dfrac{0.6167}{62s+1} \\ \dfrac{0.7833}{90s+1} & \dfrac{3.9170}{(30s+1)(90s+1)} \end{bmatrix} \tag{73}$$

Discrete-time transfer function matrix of the nominal plant obtained for $T_S = 30s$ is

$$G(z) = \begin{bmatrix} \dfrac{0.5554z^{-1} + 0.3065z^{-2}}{1 - 0.8877z^{-1} + 0.1673z^{-2}} & \dfrac{0.2366z^{-1}}{1 - 0.6164z^{-1}} \\ \dfrac{0.2220z^{-1}}{1 - 0.7165z^{-1}} & \dfrac{0.4275z^{-1} + 0.2743z^{-2}}{1 - 1.0840z^{-1} + 0.2636z^{-2}} \end{bmatrix} \tag{74}$$

Upper bounds for $\sigma_{\max}[T_0(j\omega)]$ evaluated according to (69) and (70) are in Fig. 16.

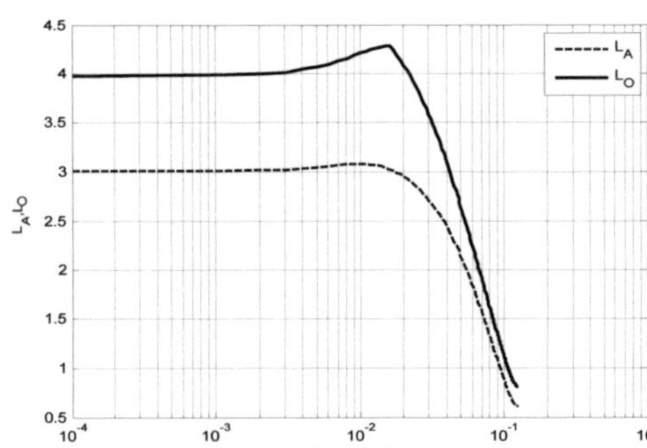

Fig. 16. Upper bounds for $\sigma_{\max}[T_0(j\omega)]$ evaluated according to (69) and (70)

Obviously, proper setpoint tracking can be guaranteed for both uncertainty types, just on a limited frequency range. Hence, $M_T = 1.05$ and multiplicative output uncertainty will be considered in the sequel.
Bode plots of equivalent subsystems generated using $g_2(z)$ are shown in Fig. 17, and their relevant parameters are summarized in Tab. 5.

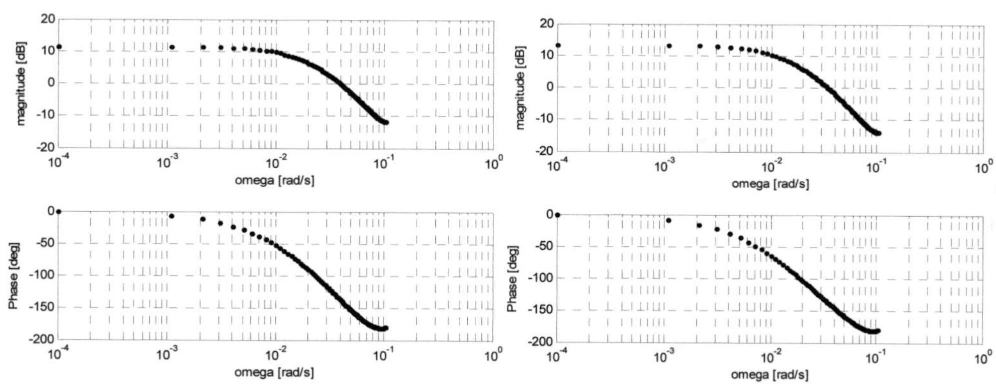

Fig. 17. Discrete Bode plots of equivalent subsystems generated by g2(z): $G_{12}^{eq}(z)$ (left), $G_{22}^{eq}(z)$ (right) (non-minimum phase case)

Equivalent subsystem	PM	Crossover frequency
$G_{12}^{eq}(z)$	43.81^0	0.040rad/s^{-1}
$G_{22}^{eq}(z)$	44.04^0	0.0344 rad/s^{-1}

Table 5. Relevant parameters of equivalent subsystems generated by g2(z).

For both equivalent subsystems the required settling time and maximum overshoot were chosen the same as in the minimum phase case: $t_s = 600s$, $M_T = 1.05$ corresponding to $\eta_{\max} = 5\%$. Related values of other design parameters are: $PM_{\min} = 56.87^0$ and required crossover frequency $\omega_0 = 0.0131$. The required phase margin $PM_{req} > PM_{\min}$ was chosen $PM_{req} = 60^0$. Achieved design results are summarized in Tab. 6 and Bode plots of compensated equivalent subsystems in Fig.18 prove achieving required design parameters.

Eq. subsyst.	$R_i(s)$	$R_i(z)$	$PM_{achieved}$	$\omega_{0\_achieved}$
$G_{12}^{eq}(z)$	$R_1(s) = 0.2083 + \dfrac{0.0039}{s}$	$R_1(z) = \dfrac{0.2083 - 0.0923z^{-1}}{1 - z^{-1}}$	54.17^0	0.0122 rad/s^{-1}
$G_{22}^{eq}(z)$	$R_2(s) = 0.2376 + \dfrac{0.0030}{s}$	$R_2(z) = \dfrac{0.2376 - 0.1832z^{-1}}{1 - z^{-1}}$	56.89	0.0120 rad/s^{-1}

Table 6. Design results and achieved frequency domain performance measures for the non-minimum phase case

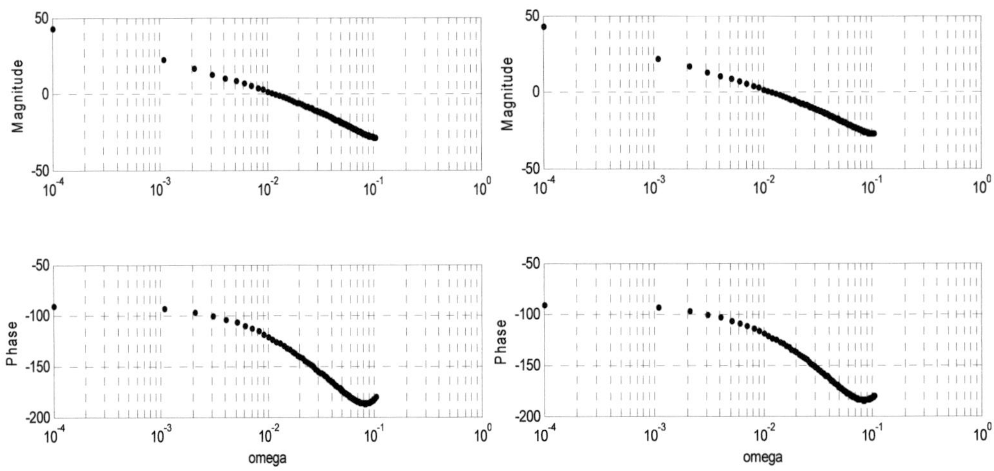

Fig. 18. Discrete Bode plots of equivalent subsystems under designed PI controllers: $G_{12}^{eq}(z)$ (left), $G_{22}^{eq}(z)$ (right)

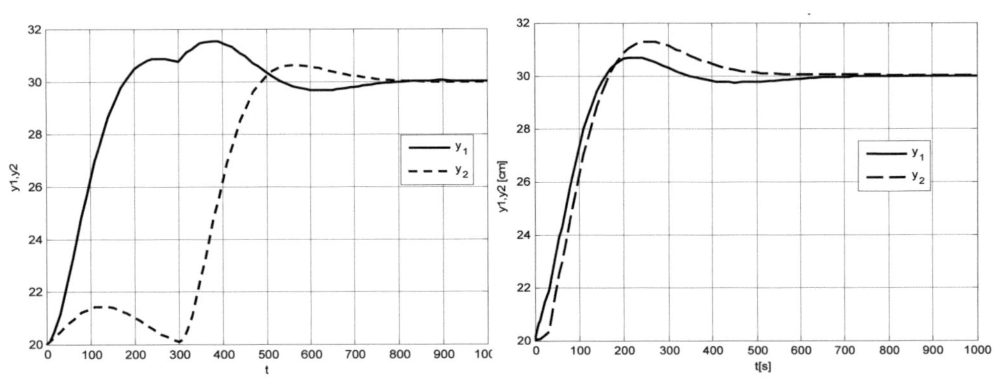

Fig. 19. Nominal closed-loop step responses of the quadruple tank system in non-minimum phase configuration (reference steps 0.1m occurred at t=0s at the input of the 1st subsystem, and at t=300s and t=10s, respectively, at the input of the 2nd subsystem). Maximum overshoot and settling time (600s) were kept in both cases.

Nominal closed-loop poles verify nominal stability.

$$\text{Roots\_of\_CLCP} = \{0.6768 \pm 0.2761i, 0.7335 \pm 0.2262i, 0.7165, 0.6164, 0.5876, 0.3313\}$$

The sensitivity magnitude plot in Fig. 20 with the peak $\max_{\omega}\{\sigma_M[S(j\omega)]\} < 2$ around the crossover frequency proves good closed-loop nominal performance.

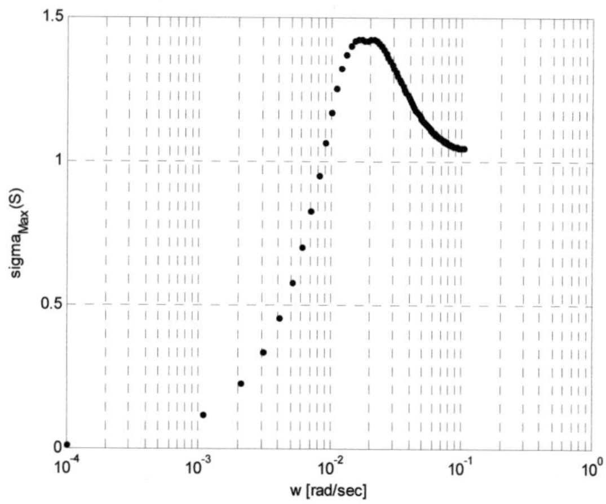

Fig. 20. $\sigma_M[S(z)]_{z=e^{j\omega T}}$ - versus –frequency plot

Fulfilment of robust stability condition (70) is examined in Fig. 21. The closed-loop system is stable over the whole non-minimum phase region (8).

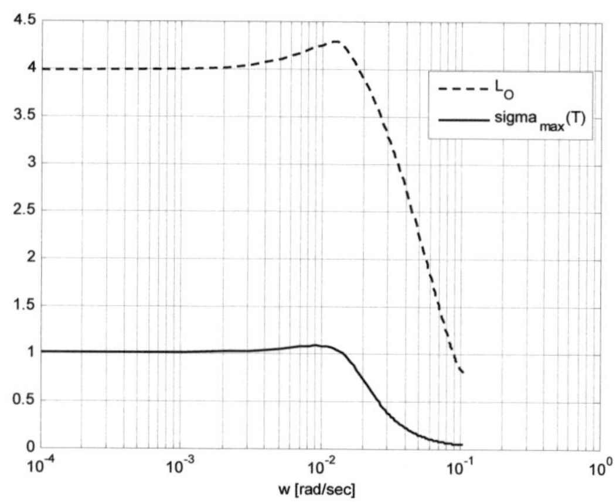

Fig. 21. Verification of the robust stability condition $\sigma_{\max}[T_0(j\omega)] < \dfrac{1}{|\ell_o(\omega)|} = L_O(\omega)$

5. Conclusion

The robust decentralized PID controller design procedures have been developed both in frequency and time domains. The proposed controller design schemes are based on different principles, with the same control aim: to achieve robust stability and specified performance. The comparative study of both approaches is presented on robust decentralized discrete-time PID controller design for quadruple-tank process model, for minimum and nonminimum phase configurations. Both proposed approaches provide promising results verified by simulation on nonlinear process model.

6. Acknowledgment

This research work has been supported by the Scientific Grant Agency of the Ministry of Education of the Slovak Republic, Grants No. 1/0544/09 and 1/0592/10, and by Slovak Research and Development Agency, Grant APVV-0211-10.

7. References

Boyd, S., El Ghaoui, L., Feron, E. & Balakrishnan, V. (1994). *Linear matrix inequalities in system and control theory*, SIAM Studies in Applied Mathematics, ISBN 0-89871-334-X, Philadelphia

Bucz, Š., Marič, L., Harsányi, L. & Veselý, V. (2010). A simple robust PID controller design method based on sine-wave identification of the uncertain plant. *Journal of Electrical Engineering*, Vol. 61, No.3, pp.164-170, ISSN 1335-3632

Drahoš, P. (2000). Position control of SMA drive. *Int. Carpathian Control Conference*, pp. 189-192, Podbanské, Slovak Republic, 2000.

Crusius, C.A.R. & Trofino, A. (1999). LMI Conditions for Output Feedback Control Problems. *IEEE Trans. on Automatic Control*, Vol. 44, pp. 1053-1057, ISSN 0018-9286

de Oliveira, M.C., Bernussou, J. & Geromel, J.C. (1999). A new discrete-time robust stability condition. *Systems and Control Letters*, Vol. 37, pp. 261-265, ISSN 0167-6911

de Oliveira, M.C., Camino, J.F. & Skelton, R.E. (2000). A convexifying algorithm for the design of structured linear controllers. *Proc. 39nd IEEE CDC*, pp. 2781-2786, Sydney, Australia, 2000.

Ming Ge, Min-Sen Chiu & Qing-Guo Wang (2002). Robust PID controller design via LMI approach. *Journal of Process Control*, Vol.12, pp. 3-13, ISSN 0959-1524

Grman, Ľ. , Rosinová, D. , Kozáková, A. & Veselý, V. (2005). Robust stability conditions for polytopic systems. *International Journal of Systems Science*, Vol. 36, No. 15, pp. 961-973, ISSN 1464-5319 (electronic) 0020-7721 (paper)

Gyurkovics, E. & Takacs, T. (2000). Stabilisation of discrete-time interconnected systems under control constraints. *IEE Proceedings - Control Theory and Applications*, Vol. 147, No. 2, pp. 137-144, ISSN 1350-2379

Henrion, D., Arzelier, D. & Peaucelle, D. (2002). Positive polynomial matrices and improved LMI robustness conditions. 15th *IFAC World Congress*, CD-ROM, Barcelona, Spain, 2002

Henrion, D., Löfberg, J., Kocvara, M. & Stingl, M. (2005). Solving polynomial static output feedback problems with PENBMI. LAAS-CNRS Research Report No. 05165, March 2005. *Proceedings of the joint IEEE Conference on Decision and Control and European Control Conference*, Sevilla, Spain, December 2005

Johansson, K.H. (2000). The Quadruple-Tank Process: A Multivariable Laboratory Process with an Adjustable Zero, *IEEE Transactions on Control Systems Technology*, Vol. 8, No. 3, pp.456-465, ISSN 1063-6536

Johansson, K. H., Horch, A., Wijk, O. &Hansson, A. (1999). Teaching Multivariable Control Using the Quadruple-Tank Process. *Proc. 38nd IEEE CDC*, Phoenix, AZ, 1999

Kozáková, A. & Veselý, V. (2009). Design of robust decentralized controllers using the M-□ structure robust stability conditions. *Int. Journal of Systems Science,* Vol. 40, No.5, pp.497-505, ISSN 1464-5319 (electronic) 0020-7721 (paper)

Kozáková, A., Veselý, V. & Osuský, J. (2009). A new Nyquist-based technique for tuning robust decentralized controllers. *Kybernetika,* Vol. 45, No.1, pp. 63-83, ISSN 0023-5954

Kozáková, A., Veselý, V. & Osuský, J. (2010). Decentralized digital PID controller design for performance. *12th IFAC Symposium on Large Scale Systems: Theory and Applications,* CD ROM, Villeneuve d'Ascq, France, 2010.

Kozáková, A., Veselý, V., Osuský, J. (2011). Direct Design of Robust Decentralized Controllers. 18th *IFAC World Congress.* CD ROM, Milan, Italy, 2011

Kuo, B.C. (2003). *Automatic Control Systems.* Prentice Hall International, Inc. ISBN-13: 9788122418095.

Lewis, F.L. (1992). Applied optimal control & estimation: digital design & implementation. Prentice-Hall and Texas Instruments, Englewood Cliffs, NJ, ISBN 978-0130403612

Lian, F-L., Moyne, J, Tilbury, D. (2002). Network design consideration for distributed control systems. *IEEE Trans. Control Systems Technology,* Vol. 10, No. 2, pp. 297 - 304, ISSN 1063-6536

Ogunnaike, A. & Ray W. H. (1994). *Process Dynamics Modeling and Control,* Oxford University Press, Inc., ISBN 0-19-509119-1, New York

Peaucelle, D., Arzelier, D., Bachelier, O. & Bernussou, J. (2000). A new robust *D*-stability condition for real convex polytopic uncertainty. *Systems and Control Letters,* Vol. 40, pp. 21-30, ISSN 0167-6911.

Reinisch, K. (1974). *Kybernetische Grundlagen und Beschreibung kontinuierlicher Systeme,* VEB Verlag Technik, Berlin, 1974 (in German)

Rosinová, D., Veselý, V. & Kučera, V. (2003). A necessary and sufficient condition for static output feedback stabilizability of linear discrete-time systems. *Kybernetika,* Vol. 39, pp. 447-459, ISSN 0023-5954

Rosinová, D. & Veselý, V. (2003). Robust output feedback design of discrete-time systems – linear matrix inequality methods. *Proceedings* 2nd *IFAC Conf. CSD'03* (CD-ROM), Bratislava, Slovakia, 2003

Rosinová, D. & Veselý, V. (2007). Robust PID decentralized controller design using LMI. *International Journal of Computers, Communication & Control,* Vol II, No.2, pp. 195-204, ISSN 1841 - 9836

Rosinová, D. & Veselý, V. (2011). Decentralized stabilization of discrete-time systems: subsystem robustness approach. 18th *IFAC World Congress,* CD-ROM, Milan, Italy, 2011

Skelton, R.E.; Iwasaki, T. & Grigoriadis, K. (1998). *A Unified Algebraic Approach to Linear Control Design,* Taylor and Francis, Ltd, ISBN 0-7484-0592-5, London, UK

Skogestad, S. & Postlethwaite, I. (2009). *Multivariable feedback control: analysis and design,* John Wiley & Sons Ltd., ISBN 13 978-0-470-01167-6 Chichester, West Sussex, UK

Stankovič, S.S., Stipanovič, D.M. & Šiljak, D.D. (2007). Decentralized dynamic output feedback for robust stabilization of a class of nonlinear interconnected systems. *Automatica,* Vol.43, pp. 861-867, ISSN 0005-1098

Van Antwerp, J. G. & Braatz, R. D. (2000). A tutorial on linear and bilinear matrix inequalities. *Journal of Process Control,* Vol. 10, pp. 363–385, ISSN 0959-1524

Veselý, V. (2003). Robust output feedback synthesis: LMI Approach, *Proceedings 2th IFAC Conference CSD'03* (CD-ROM), Bratislava, Slovakia, 2003

Veselý, V. & Rosinová, D. Robust PID-PSD controller design: BMI approach. *submitted to Asian Journal of Control*

Zečevič, A.I. & Šiljak, D.D. (2004). Design of robust static output feedback for large-scale systems. *IEEE Trans. on Automatic Control,* Vol.49, No.11, pp. 2040-2044, ISSN 0018-9286

Zheng Feng, Qing-Guo Wang & Tong Heng Lee (2002). On the design of multivariable PID controllers via LMI approach. *Automatica*, Vol. 38, pp. 517-526, ISSN 0005-1098

Part 4

Intelligent PID Control

7

Tuning Fuzzy PID Controllers

Constantin Volosencu
"Politehnica" University of Timisoara
Romania

1. Introduction

After the development of fuzzy logic, an important application of it was developed in control systems and it is known as fuzzy PID controllers. They represent interest in order to be applied in practical applications instead of the linear PID controllers, in the feedback control of a variety of processes, due to their advantages imposed by the non-linear behavior. The design of fuzzy PID controllers remains a challenging area that requires approaches in solving non-linear tuning problems while capturing the effects of noise and process variations. In the literature there are many papers treating this domain, some of them being presented as references in this chapter.

Fuzzy PID controllers may be used as controllers instead of linear PID controller in all classical or modern control system applications. They are converting the error between the measured or controlled variable and the reference variable, into a command, which is applied to the actuator of a process. In practical design it is important to have information about their equivalent input-output transfer characteristics. The main purpose of research is to develop control systems for all kind of processes with a higher efficiency of the energy conversion and better values of the control quality criteria.

What has been accomplished by other researchers is reviewed in some of these references, related to the chapter theme, making a short review of the related work form the last years and other papers. The applications suddenly met in practice of fuzzy logic, as PID fuzzy controllers, are resulted after the introduction of a fuzzy block into the structure of a linear PID controller (Buhler, 1994, Jantzen, 2007). A related tuning method is presented in (Buhler, 1994). That method makes the equivalence between the fuzzy PID controller and a linear control structure with state feedback. Relations for equivalence are derived. In the paper (Moon, 1995) the author proves that a fuzzy logic controller may be designed to have an identical output to a given PI controller. Also, the reciprocal case is proven that a PI controller may be obtained with identical output to a given fuzzy logic controller with specified fuzzy logic operations. A methodology for analytical and optimal design of fuzzy PID controllers based on evaluation approach is given in (Bao-Gang et all, 1999, 2001). The book (Jantzen, 2007) and other papers of the same author present a theory of fuzzy control, in which the fuzzy PID controllers are analyzed. Tuning fuzzy PID controller is starting from a tuned linear PID controller, replacing it with a linear fuzzy controller, making the fuzzy controller nonlinear and then, in the end, making a fine tuning. In the papers (Mohan & Sinha, 2006, 2008), there are presented some mathematical models for the simplest fuzzy PID controllers and an approach to design fuzzy PID controllers. The paper (Santos & all, 1096) shows that it is possible to apply the empirical tools to predict the achievable performance of the conventional PID controllers to evaluate the performance of a fuzzy logic controller based on the equivalence between a fuzzy controller and a PI controller. The paper (Yame, 2006) analyses the analytical structure of a simple class of Takagi-Sugeno PI controller with respect to conventional control theory. An example shows an approach to Takagi-Sugeno fuzzy PI controllers tuning. In the paper (Xu & all, 1998) a tuning method based on gain and phase margins has been proposed to determine the weighting coefficients of the fuzzy PI controllers in the frame of a linear plant control. There are presented numerical simulations. Mamdani fuzzy PID controllers are studied in (Ying, 2000). The author has published his theory on tuning fuzzy PID controllers at international conferences and on journals (Volosencu, 2009).

This chapter presents some techniques, under unitary vision, to solve the problem of tuning fuzzy PID controllers, developed based on the most general structure of Mamdani type of fuzzy systems, giving some tuning guidelines and recommendations for increasing the quality of the control systems, based on the practical experience of the author. There is given a method in order to make a pseudo-equivalence between the linear PID controllers and the fuzzy PID controllers. Some considerations related to the stability analysis of the control systems based on fuzzy controllers are made. Some methods to design fuzzy PID controllers are there presented. The tuning is made using a graphical-analytical analysis based on the input-output transfer characteristics of the fuzzy block, the linear characteristic of the fuzzy block around the origin and the usage of the gain in origin obtained as an origin limit of the variable gain of the fuzzy block. Transfer functions and equivalence relations between controller's parameters are obtained for the common structures of the PID fuzzy controllers. Some algorithms of equivalence are there presented. The linear PID controllers may be designed based on different methods, for example the modulus or symmetrical criterion, in Kessler's variant. The linear controller may be used for an initial design. Refining calculus and simulations must follow the equivalence algorithm. The author used this equivalence theory in fuzzy control applications as the speed control of electrical drives, with good results. The unitary theory presented in this chapter may be applied to the most general fuzzy PID controllers, based on the general Mamdani structure, which may be developed using all kind of membership functions, rule bases, inference methods and defuzzification methods. A case study of a control system using linear and fuzzy controllers is there also presented. Some advantages of this method are emphasized. Better control quality criteria are demonstrated for control systems using fuzzy controllers tuned, by using the presented approach.

In the second paragraph there are presented some considerations related to the fuzzy controllers with dynamics, the structures of the fuzzy PI, PD and PID controllers. In the third paragraph there are presented: the transfer characteristics of the fuzzy blocks, the principle of linearization, with the main relations for pseudo-equivalence of the PI, PD and PID controllers. A circuit of correction for the fuzzy PI controller, to assure stability, is also presented. In the fourth paragraph there are presented some considerations for internal and external stability assurance. There is also presented a speed fuzzy control system for electrical drives based on a fuzzy PI controller, emphasizing the better control quality criteria obtained using the fuzzy PI controller.

2. Fuzzy controllers

2.1 Fuzzy controllers with dynamics

The basic structure of the fuzzy controllers with dynamics is presented in Fig. 1.

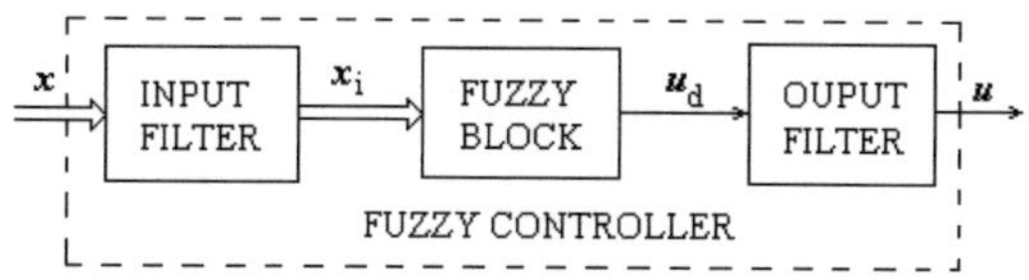

Fig. 1. The block diagram of a fuzzy controller with dynamics

So, the following fuzzy controllers, with dynamics, have, as a central part a fuzzy block FB, an input filter and an output filter. The two filters give the dynamic character of the fuzzy controller. The fuzzy block has the well-known structure, from Fig. 2.

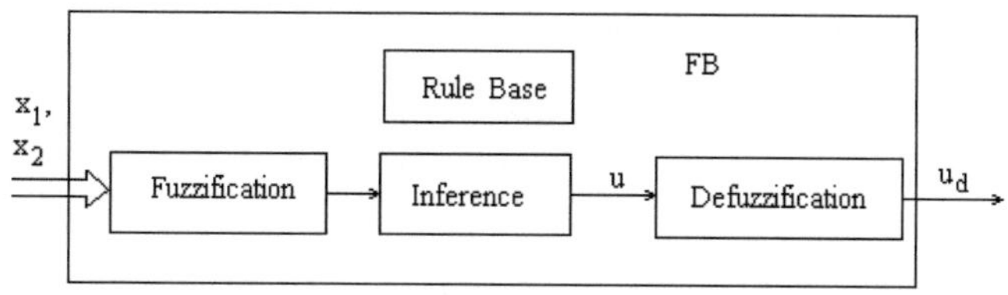

Fig. 2. The structure of fuzzy block

The fuzzy block does not treat a well-defined mathematical relation (a control algorithm), as a linear controller does, but it is using the inference with many rules, based on linguistic variables. The inference is treated with the operators of the fuzzy logic. The fuzzy block from Fig. 2 has three distinctive parts, in Mamdani type: fuzzyfication, inference and defuzzification. The fuzzy controller is an inertial system, but the fuzzy block is a non-inertial system. The fuzzy controller has in the most common case two input variables x_1 and x_2 and one output variable u. The input variables are taken from the control system. The inference interface of the fuzzy block releases a treatment by linguistic variables of the input variables, obtained by the filtration of the controller input variables. For the linguistic treatment, a definition with membership functions of the input variable is needed. In the interior of the fuzzy block the linguistic variables are linked by rules that are taking account of the static and dynamic behavior of the control system and also they are taking account of the limitations imposed to the controlled process. In particular, the control system must be stable and it must assure a good amortization. After the inference we obtain fuzzy information for the output variable. The defuzzification is used because, generally, the actuator that follows the controller must be commanded with a crisp value u_{d},. The command variable u, furnished by the fuzzy controller, from Fig. 1, is obtained by filtering the defuzzified variable u_d. The output variable of the controller is the command input for the process. The fuzzification, the inference and the defuzzification bring a nonlinear behavior of the fuzzy block. The nonlinear behavior of the fuzzy block is transmitted also to the fuzzy PID controllers. By an adequate choosing of the input and output filters we may realize different structures of the fuzzy controllers with imposed dynamics, as are the general PI, PD and PID dynamics.

2.2 Fuzzy PI controller

The structure of a PI fuzzy controller with integration at its output (FC-PI-OI) is presented in Fig. 3.

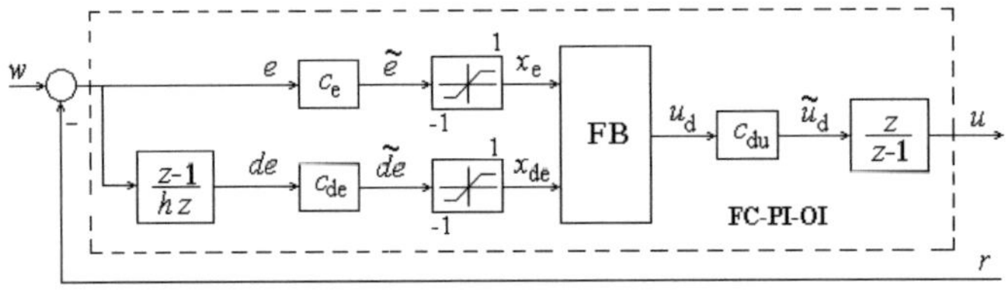

Fig. 3. The block diagram of the fuzzy PI controller

The controller is working after the error *e* between the input variable reference and the feedback variable *r*. In this structure we may notice that two filter were used. One of them is placed at the input of the fuzzy block FB and the other at the output of the fuzzy block. In the approach of the PID fuzzy controllers the concepts of integration and derivation are used for describing that these filters have mathematical models obtained by discretization of a continuous time mathematical models for integrator and derivative filters.
The structure of the linear PI controller may be presented in a modified block diagram from Fig. 4.

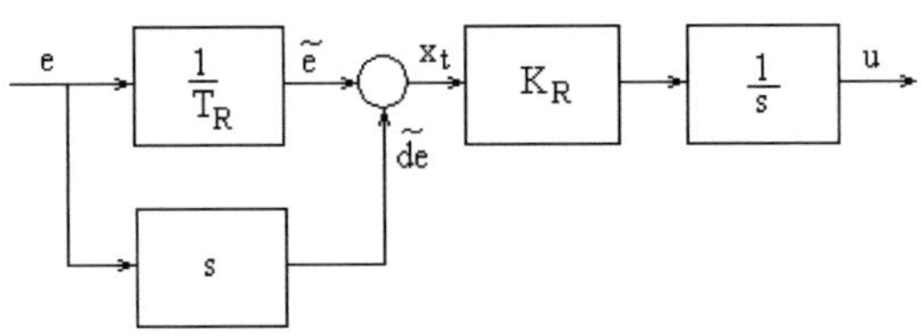

Fig. 4. The modified block diagram of the linear PI controller

For this structure the following modified form of the transfer function may be written:

$$u(s) = K_R \frac{1}{s}(s + \frac{1}{T_R})e(s) = K_R \frac{1}{s} x_t(s) \tag{1}$$

where

$$\begin{aligned} x_t &= \tilde{e} + \tilde{de} \\ \tilde{e} &= \frac{1}{T_R} e \\ \tilde{de} &= s.e \end{aligned} \tag{2}$$

In the next paragraph we shall show that the fuzzy block BF may be described using its input-output transfer characteristics, its variable gain and its gain in origin, as a linear function around the origin ($\tilde{e} = 0, \tilde{de} = 0, u_d = 0$).
The block diagram of the linear PI controller may be put similar as the block diagram of the fuzzy PI controller as in Fig. 5.

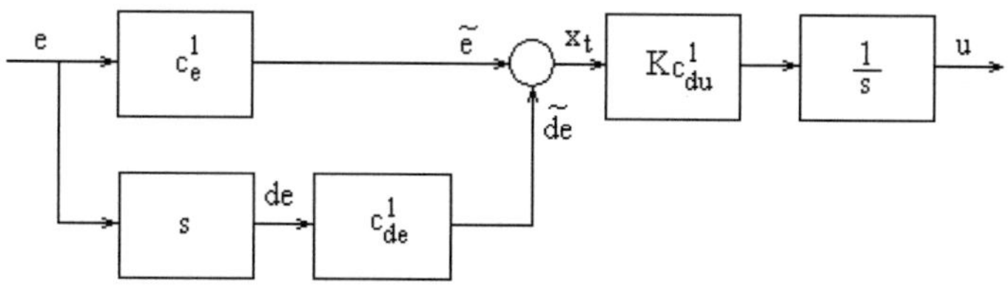

Fig. 5. The block diagram of the linear PI controller with scaling coefficients

For the transfer function of the linear PI controller with scaling coefficients the following relation may be written:

$$H_R(s) = K_R.\frac{1}{s}(s+\frac{1}{T_R}) = K.c^l_{du}.\frac{1}{s}.(c^l_e + c^l_{de}s) \tag{3}$$

In the place of the summation block from Fig. 4 the fuzzy block BF from Fig. 2 is inserted. The derivation and integration are made in discrete time and specific scaling coefficients are there introduced. The saturation elements are introduced because the fuzzy block is working on scaled universes of discourse [-1, 1].

The filter from the controller input, placed on the low channel, takes the operation of digital derivation; at its output we obtain the derivative *de* of the error *e*:

$$de(t) = \frac{d}{dt}e(t) \circ\!-\!\bullet\ de(z) = \frac{z-1}{hz}e(z) \tag{4}$$

where h is the sampling period. In the domain of discrete time the derivative block has the input-output model:

$$de(t+h) = \frac{1}{h}e(t+h) - \frac{1}{h}e(t) \tag{5}$$

That shows us that the digital derivation is there accomplished based on the information of error at the time moments $t=t_k=k.h$ and $t_{k+1}=t_k+h$:

$$\begin{aligned} e_k &= e(kh) \\ e_{k+1} &= e((k+1)h) \end{aligned} \tag{6}$$

So, the digital equipment is making in fact the substraction of the two values.

The error e and its derivative de are scaled with two scaling coefficients c_e and c_{de}, as it follows:

$$\tilde{e}(t) = c_e e(t) \tag{7}$$

$$\tilde{de}(t) = c_{de} de(t) \tag{8}$$

The variables x_e and x_{de} from the inputs of the fuzzy block FB are obtained by a superior limitation to 1 and an inferior limitation to –1, of the scaled variables e and de. This limitation is introduced because in general case the numerical calculus of the inference is made only on the scaled universe of discourse [-1, 1].

The fuzzy block offers the defuzzified value of the output variable u_d. This value is scaled with an output scaling coefficient c_{du}:

$$\tilde{u}_d = c_{du} u_d \tag{9}$$

In the case of the PI fuzzy controller with integration at the output the scaled variable $\tilde{u}_d$ is the derivative of the output variable u of the controller. The output variable is obtained at the output of the second filter, which has an integrator character and it is placed at the output of the controller:

$$u(t) = \int_0^t \tilde{u}_d(\tau)d\tau \circ\!-\!\bullet\, u(z) = \frac{z}{z-1}\tilde{u}_d(z) \tag{10}$$

The input-output model in the discrete time of the output filter is:

$$u(t+1) = u(t) + \tilde{u}_d(t+1) \tag{11}$$

The above relation shows that the output variable is computed based on the information from the time moments t and $t+h$:

$$\begin{aligned} u_{k+1} &= u((k+1)h) \\ u_k &= u(kh) \\ \tilde{u}_{dk+1} &= \tilde{u}_d((k+1)h) \end{aligned} \tag{12}$$

From the above relations we may notice that the "integration" is reduced in fact at a summation:

$$u_{k+1} = u_k + \tilde{u}_{dk+1} \tag{13}$$

This equation could be easily implemented in digital equipments.
Due to this operation of summation, the output scaling coefficient c_{du} is called also the increment coefficient.
Observation: The controller presented above could be called "fuzzy controller with summation at the output" and not with "integration at the output".

2.3 Fuzzy PD controller

The structure of the fuzzy PD controller (RF-PD) is presented in Fig. 6.

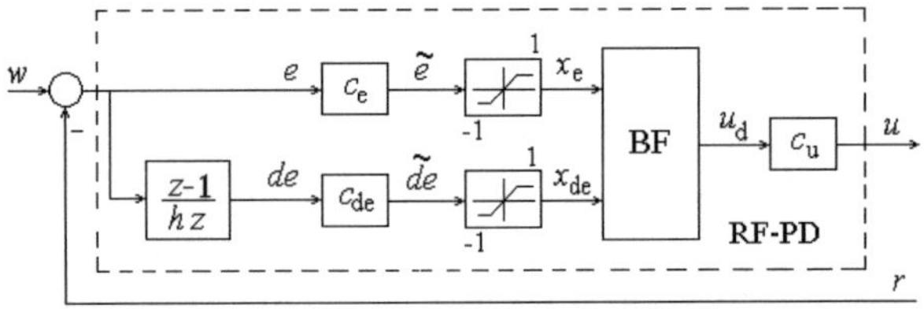

Fig. 6. The block diagram of the fuzzy PD controller with scaling coefficients

In this case the derivation is made at the input of the fuzzy bock, on the error *e*.
For the fuzzy controller FC-PD there is obtained the following relation in the z-domain:

$$u(z) = \tilde{c}_u[x_e(z) + x_{de}(z)] = \tilde{c}_u\left[c_e + c_{de}\frac{z-1}{hz}\right]e(z) \tag{14}$$

With this relation the transfer function results:

$$H_{RF}(z) = \frac{u(z)}{e(z)} = \tilde{c}_u\left(c_e + c_{de}\frac{z-1}{hz}\right) \tag{15}$$

For the PD linear controller we take the transfer function:

$$H_{RG}(s) = K_{RG}(1 + T_D s) \tag{16}$$

2.4 Fuzzy PID controller

The structure of the fuzzy PID controller is presented in Fig. 7.
In this case the derivation and integration is made at the input of the fuzzy bock, on the error *e*. The fuzzy block has three input variables x_e, x_{ie} and x_{de}.

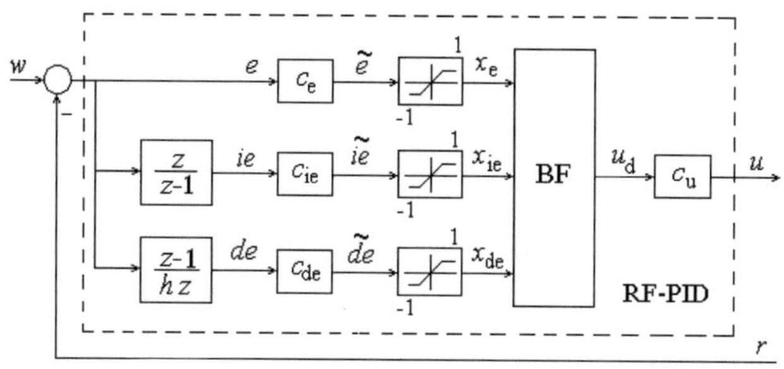

Fig. 7. The block diagram of the fuzzy PID controller

The transfer function of the PID controller is obtained considering a linearization of the fuzzy block BF around the origin, for x_e=0, x_{ie}=0, x_{de}=0 şi u_d=0 with a relation of the following form:

$$u_d = K_0(x_e + x_{ie} + x_{de}) \tag{17}$$

A relation, as the fuzzy block from the PID controller - which has 3 input variables - may describe, is:

$$K_{BF}(x_t; x_{de}, x_{ie} = 0) = \frac{u_d}{x_t},\ x_t \neq 0 \tag{18}$$

where:

$$x_t = x_e + x_{ie} + x_{de} \tag{19}$$

The value K_0 is the limit value in origin of the characteristics of the function:

$$K_0 = \lim_{x_t \to 0} K_{BF}(x_t; x_{de}, x_{ie} = 0) \tag{20}$$

Taking account of the correction made on the fuzzy block with the incremental coefficient c_u, the characteristic of the fuzzy block corrected and linearized around the origin is given by the relation:

$$u = c_u K_0 (x_e + x_{ie} + x_{de}) \tag{21}$$

We are denoting:

$$\tilde{c}_u = c_u K_0 \tag{22}$$

For the fuzzy controller RF-PID, with the fuzzy block BF linearized, the following input-output relation in the z domain may be written:

$$u(z) = \tilde{c}_u [x_e(z) + x_{ie}(z) + x_{de}(z)] = \tilde{c}_u \left[c_e + c_{ie} \frac{z}{z-1} + c_{de} \frac{z-1}{hz} \right] e(z) \tag{23}$$

With these observations the transfer function of the fuzzy ID controller becomes:

$$H_{RF}(z) = \frac{u(z)}{e(z)} = \tilde{c}_u \left(c_e + c_{ie} \frac{z}{z-1} + c_{de} \frac{z-1}{hz} \right) \tag{24}$$

For the linear PID controller, the following relation for the transfer function is considered:

$$H_{RG}(s) = K_{RG} \left(1 + T_D s + \frac{1}{T_I s} \right) \tag{25}$$

3. Pseudo-equivalence

3.1 Fuzzy block description using I/O transfer characteristics. Linearization

The fuzzy block has a MISO transfer characteristic:

$$u_d = f_{FB}(x_e, x_{de}),\ x_e, x_{de} \in [-a, a] \tag{26}$$

From this transfer characteristic, a SISO transfer characteristic may be obtained:

$$u_d = f_e(x_e; x_{de}),\ x_e \in [-a, a] \tag{27}$$

where x_{de} is a parameter.
We introduce a composed variable:

$$x_t = x_e + x_{de} \tag{28}$$

Using this new, composed variable, a family of translated characteristics may be obtained:

$$u_d = f_t(x_t; x_{de}),\ x \in [-2a, 2a] \tag{29}$$

with x_{de} as a parameter. The passing from a frequency model to the parameter model is reduced to the determination of the parameters of the transfer impedance. The steps in such identification procedure are: organization and obtaining of experimental data on the transducer, interpretation of measured data, model deduction with its structure definition and model validation. Using the above translated characteristics we may obtain the characteristic of the variable gain of the fuzzy block:

$$K_{FB}(x_t;x_{de}) = f_t(x_t;x_{de}) / x_t,\ x_t \neq 0 \tag{30}$$

The MISO transfer characteristic of the fuzzy block may be written as follows:

$$\begin{aligned} u_d &= f_{FB}(x_e,x_{de}) = K_{FB}(x_e,x_{de}). \\ &.(x_e + x_{de}) = K_{FB}(x_t;x_{de}).x_t \end{aligned} \tag{31}$$

If the fuzzy bloc is linearized around the point of the origin, in the permanent regime: x_e=0, x_{de}=0 and u_d=0, the following relation will be obtained:

$$u_d = K_0(x_e + x_{de}) \tag{32}$$

The value K_0 is the value at the limit, in origin of the characteristic $K_{BF}(x_t;\ x_{de})$:

u		x_e		
		NB	ZE	PB
x_{de}	NB	NB	NB	ZE
	ZE	NB	ZE	PB
	PB	ZE	PB	PB

Table 1. The 3x3 (primary) rule base

$$K_0 = \lim_{x_e \to 0} K_{FB}(x_t;x_{de}), x_{de} = 0 \tag{33}$$

This value may be determined with a good approximation, at the limit, from the gain characteristics.

We show here an example of the above characteristics for the fuzzy block with max-min inference, defuzzification with center of gravity, were the variables have the 3x3 primary rule base from Tab. 1 and three membership values from Fig. 8.

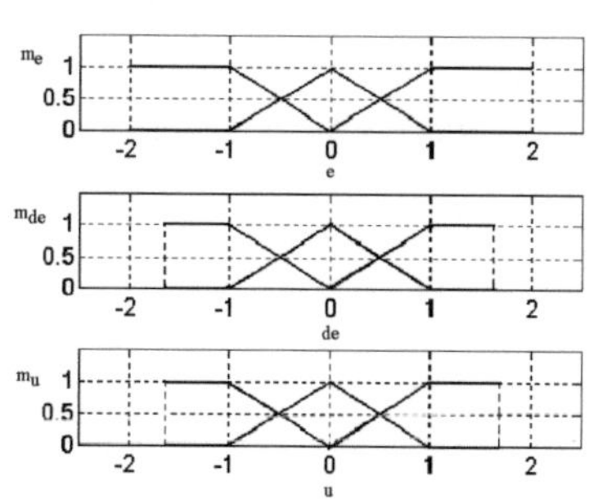

Fig. 8. Membership functions

The MISO characteristic is presented in Fig. 9.a). The SISO characteristics are presented in Fig. 9.b). The translated characteristics are presented in Fig. 9.c). The characteristics of the variable gain are presented in Fig. 9.d).

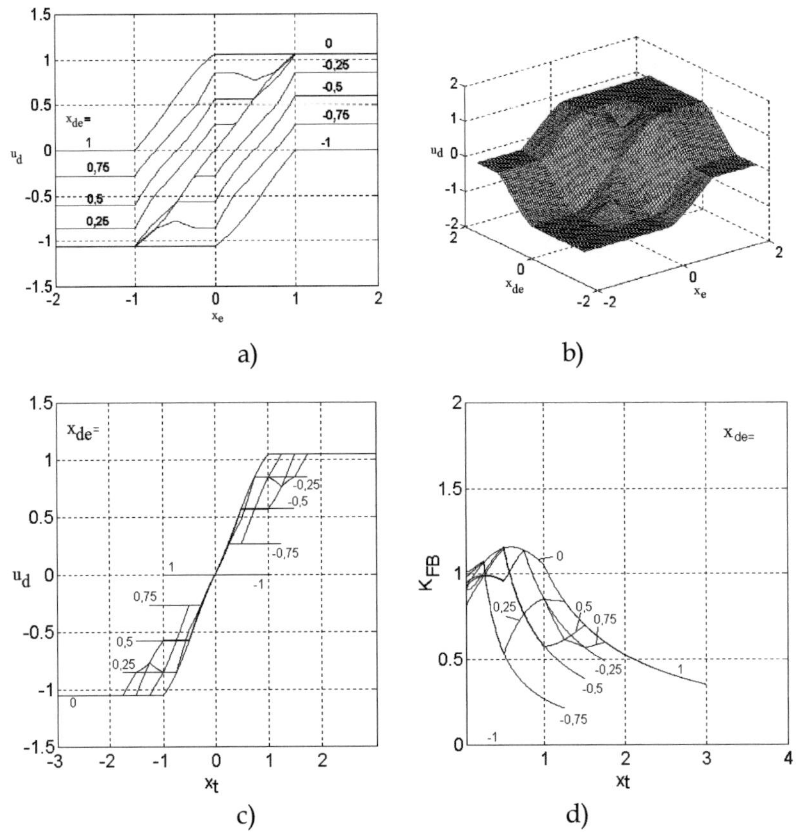

Fig. 9. Transfer characteristics: a) MISO transfer characteristic b) SISO transfer characteristic c) Translated transfer characteristic d) Gain characteristic

From the Fig. 9.d) we may notice that the value of the gain in origin is $K_0 \approx 1{,}2$.
Taking account of the correction made upon the fuzzy block with the scaling coefficient c_{du}, the characteristic of the fuzzy bloc around the origin is given by the relation:

$$\tilde{u}_d = c_{du} K_0 (x_e + x_{de}) \tag{34}$$

We use:

$$\tilde{c}_{du} = c_{du} K_0 \tag{35}$$

3.2 Pseudo-equivalence of the fuzzy PI controller

For the fuzzy controller with the fuzzy block BF linearized around the origin, we may write the following input-output relation in the *z*-domain:

$$u(z)=\frac{z}{z-1}\tilde{c}_{du}(e(z)+de(z))=\frac{z}{z-1}\tilde{c}_{du}\left[c_e+c_{de}\frac{z-1}{hz}\right]e(z) \tag{36}$$

The transfer function of the PI fuzzy controller with integration at the output becomes:

$$H_{RF}(z)=\frac{u(z)}{e(z)}=\frac{z}{z-1}\tilde{c}_{du}\left(c_e+c_{de}\frac{z-1}{hz}\right) \tag{37}$$

A pseudo-equivalence may be made for the fuzzy controller with a linear PI controller in the continuous time, used in common applications. The equivalence is a false one, because the fuzzy controller is not linear, so we use the word "pseudo".
The PI controller has the general transfer function:

$$H_{RG}(s)=\frac{u(s)}{e(s)}=K_{RG}\left(1+\frac{1}{sT_{RG}}\right) \tag{38}$$

We use the quasi-continual form of the transfer function, obtained by the conversion from the discrete time in the continuous time with the transformation:

$$z=\frac{1+sh/2}{1-sh/2} \tag{39}$$

where h is the sampling period for the conversion of the transfer function:

$$H_{RF}(s)=\frac{u(s)}{e(s)}=H_{RF}(z)\Big|_{z=\frac{1+sh/2}{1-sh/2}}=\frac{\tilde{c}_{du}}{h}\left(c_{de}+\frac{h}{2}c_e\right)\left[1+\frac{c_e}{(c_{de}+c_e h/2)s}\right] \tag{40}$$

We notice that the above transfer function matches the general transfer function of the linear PI controller.
From the identification of the coefficients of the two transfer functions, the following relations results:

$$K_{RG}=\frac{\tilde{c}_{du}}{h}\left(c_{de}+\frac{h}{2}c_e\right) \tag{41}$$

$$T_{RG}=\frac{c_{de}+\frac{h}{2}c_e}{c_e} \tag{42}$$

From relation (41) we may notice that the value of the gain coefficient K_{RG} of the PI fuzzy controller depends on the all three scaling coefficients, and what it is the most important, it depends on the gain in the origin of the fuzzy block.
And from the relation (42) we may notice that the time constant T_{RG} depends only on the scaling coefficients c_e and c_{de} from the inputs of the fuzzy block. At the limit, for $h\rightarrow 0$, the gain coefficient of the fuzzy controller has the value

$$K_{RG}=c_{de}K_0c_{du}/h \tag{43}$$

and the time constant of the fuzzy controller has the value

$$T_{RG} = c_{de} / c_e \tag{44}$$

Observations: A great value of c_e insures a small value of time constant of the fuzzy controller based on the relation (42). The value $c_e=1/e_M$, were e_M is the superior limit of the universe of discourse of the variable e and it insures a dispersion of the values from the input e of the fuzzy block on the entire universe of discourse, without limitation for large variations of the error e. A great value of c_{de} makes a great value for the time constant of the controller. A small value of c_{de} makes smalls values for the time constant and also for the gain. But, by increasing c_{du} , we may compensate the decreasing of the gain due to the decreasing of c_{de}.
Chosen of other fuzzy block with other membership functions and inference method is equivalent to the chosen of other K_0, greater or smaller.
From these relations we obtain the relation for designing the scaling coefficients based on the parameters of the linear PI controller:

$$c_e = \frac{hK_{RG}}{c_{du}K_0T_{RG}} \tag{45}$$

$$c_{de} = c_e(T_{RG} - h/2) \tag{46}$$

We may notice the influence of the gain in origin on c_e and also c_{de}.
The linear PI controller may be designed with different methods taken from the linear control theory.
Because the gain in origin is the main issue in this equivalence, we present the algorithm of computation of the gain in origin is:

1. Obtaining the MIMO transfer characteristic of the fuzzy block.
2. Obtaining the family of SISO transfer characteristics from the MIMO characteristic, using one of the input variables as a parameter.
3. Obtaining the family of translated characteristic from the SISO characteristic, using a compound variable as summation of the two input variables.
4. Obtaining the gain characteristic by dividing the translated characteristic to the compound variable.
5. Obtaining the gain in origin by computing the limit in origin of the families of gain characteristics.

3.3 Anti-wind-up circuit

As in the case of the analogue linear PI controllers for the digital fuzzy controllers with integration, there is needed an anti-wind-up circuit. For the PI controller with integration at the output, an equivalent anti-wind-up circuit may be implemented as it is shown in Fig. 10.

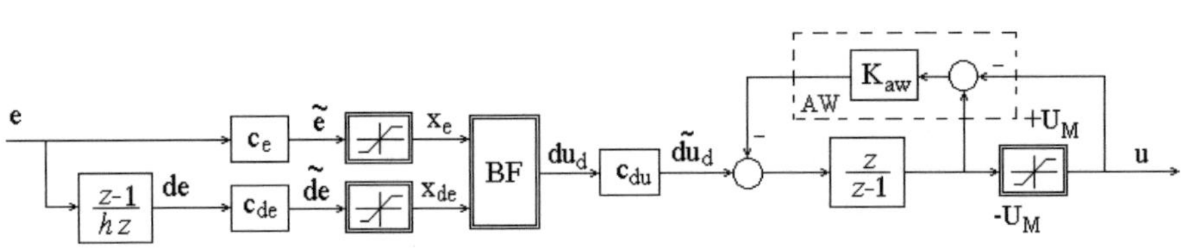

Fig. 10. The structure of the fuzzy PI controller with an anti-wind-up circuit

This structure is different from the first structure. Because of the integration block, a feedback is made with the anti-wind-up circuit AW. The circuit is needed because the output of the controller is limited at maximum and minimum values +/-U_M.
The limitations are imposed by the maximum value of the command u of the process.

3.4 Correction of the fuzzy block

To assure stability to control systems using fuzzy PI controllers, we need a correction in order to modify the input-output transfer characteristic and a quasi-fuzzy controller results, with the structure from Fig. 11.

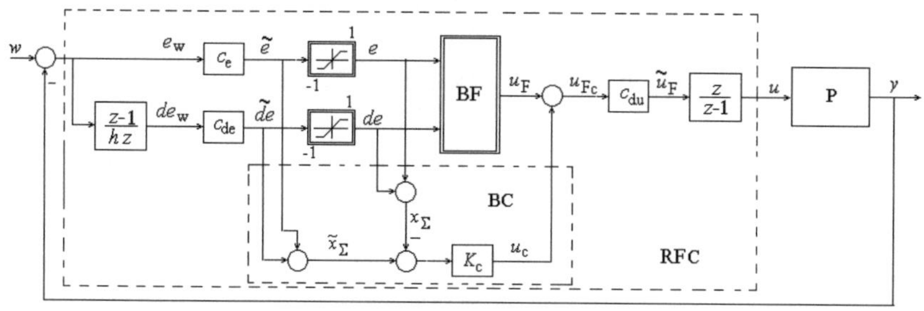

Fig. 11. The structure of the fuzzy PI controller (RFC) with an anti-wind-up circuit

The characteristic of the nonlinear part of the control system is placed only in the I-st and III-rd quadrants, like in Fig. 12.

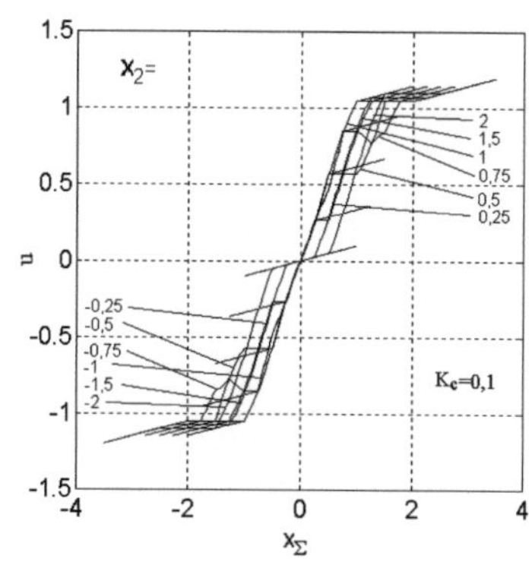

Fig. 12. The translated characteristics with a correction of $K_c = 0{,}1$

With the correction circuit from Fig. 11, the correction command is given by the relation:

$$u_c = K_c[(\tilde{e} + \tilde{de}) - (e + de)] \tag{47}$$

Even if the quasi-fuzzy structure in parallel with the fuzzy block BF a linear structure is introduced, the correction will be nonlinear.

3.5 Pseudo-equivalence of the fuzzy PD controller

As in the case of the fuzzy PI controller, a quasi-continual form is obtained:

$$H_{RF}(s)=\frac{u(s)}{e(s)}=\tilde{c}_u\left(c_e+c_{de}s\right)=\frac{\tilde{c}_u}{c_e}\left(1+\frac{c_{de}}{c_e}s\right) \tag{48}$$

From the identification of the coefficients, the following relations of tuning result:

$$K_{RG}=\frac{\tilde{c}_u}{c_e} \tag{49}$$

$$T_{RG}=\frac{c_{de}}{c_e} \tag{50}$$

From these equations, the expressions of the scaling coefficients results:

$$c_e=\frac{\tilde{c}_u}{K_{RG}} \tag{51}$$

$$c_{de}=\frac{T_{RG}\tilde{c}_u}{K_{RG}} \tag{52}$$

3.6 Pseudo-equivalence of the fuzzy PID controller

As in the case of the fuzzy PI controller, there is obtained a quasi-continual form:

$$H_{RF}(s)=\frac{u(s)}{e(s)}=H_{RF}(z)\Big|_{z=\frac{1+sh/2}{1-sh/2}}=\tilde{c}_u\left(c_e+c_{ie}/2\right)\left[1+\frac{c_{ie}}{h(c_e+c_{ie}/2)s}+\frac{c_{de}}{c_e+c_{ie}/2}s\right] \tag{53}$$

From the identification of the coefficients, the following relations of tuning are:

$$K_{RG}=\tilde{c}_u(c_e+c_{ie}/2) \tag{54}$$

$$T_I=\frac{h(c_e+c_{ie}/2)}{c_{ie}} \tag{55}$$

$$T_D=\frac{c_{de}}{c_e+c_{ie}/2} \tag{56}$$

From these equations, the expressions of the scaling coefficients are:

$$c_e=\left(\frac{T_I}{h}-\frac{1}{2}\right)\frac{hK_{RG}}{\tilde{c}_u T_I} \tag{57}$$

$$c_{ie}=\frac{hK_{RG}}{\tilde{c}_u T_I} \tag{58}$$

$$c_{de} = \frac{K_{RG}}{\tilde{c}_u} T_D \tag{59}$$

4. Stability assurance

4.1 Internal stability

For stability analysis, we are working with the structure from Fig. 13.

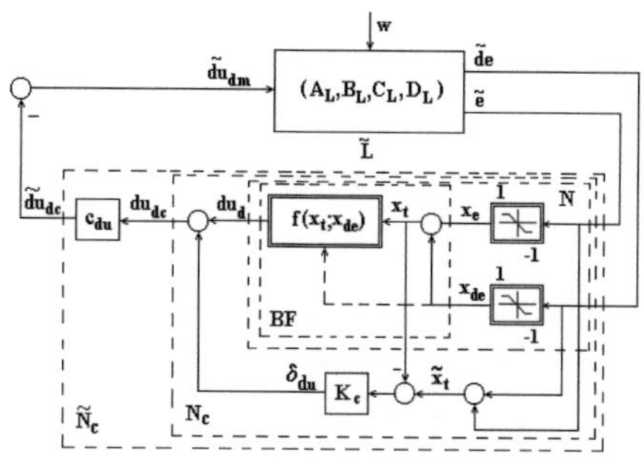

Fig. 13. The structure of the control system with the correction of the non-linear part *N*

The linear part L has the input-state-output model (60).

$$\dot{x}_{L1} = A_{L1}x_{L1} + b_{L1}K_{CNA}x_{a2} + \frac{1}{2}b_{L1}K_{CNA}\tilde{du}_d$$

$$\dot{x}_{a1} = -\frac{2c_{de}}{h}K_{CAN}c_{L1}^T x_{L1} - \frac{2}{h}x_{a1} + \frac{2c_{de}}{h}w$$

$$\dot{x}_{a2} = \frac{1}{h}\tilde{du}_d \tag{60}$$

$$\tilde{e} = -c_e K_{CAN}c_{L1}^T x_{L1} + c_e w$$

$$\tilde{de} = -\frac{2c_{de}}{h}K_{CAN}c_{L1}^T x_{L1} - \frac{2}{h}x_{a1} + \frac{2c_{de}}{h}w$$

With the new compound variable (61)

$$\tilde{x}_t = \begin{bmatrix}1 & 1\end{bmatrix}\tilde{y} = \tilde{e} + \tilde{de} \tag{61}$$

there may be introduced a new function of the compound variable $\tilde{x}_t$ and parameter $\tilde{de}$ (62).

$$K_N(\tilde{x}_t;\tilde{de}) = \frac{f_N(\tilde{e},\tilde{de})}{\tilde{x}_t},\ pt.\ \tilde{x}_t \neq 0 \tag{62}$$

The families of characteristics $du_d = \tilde{f}(\tilde{x}_t;\tilde{de})$ present the sector property to be placed only in the quadrants I and III and they are inducing the consideration of the relation (63).

$$0 \le K_N(\tilde{x}_t;\tilde{de}) \le K_M \tag{63}$$

The characteristic of the non-linear part has null intervention, due to the limitations placed at the inputs of the fuzzy block. To the fuzzy blocks we may attach a fuzzy relation of which characteristic is placed only in the quadrants I and III.
From the relation $f_{BF}(x_e, x_{de})$, which is describing the fuzzy block, a source of nonlinearity is there made by the membership functions. If the block will work on the universe of discourse [-1, 1], its characteristic will only be in the sector [K_1, K_2], $0<K_1<K_2$. By introducing the saturation elements with a role of limitation at the inputs of the fuzzy block, the non-linear part $\tilde{N}$ is placed in a sector [0, K]. To accomplish the sector condition, necessary for the stability insurance, a correction is used to the non-linear part. It consists in summation at the output du_d of the fuzzy block of the quantity δ_{du}:

$$\delta_{du} = K_c[(\tilde{e}-e)+(\tilde{de}-de)] = K_c(\tilde{x}_t - x_t) \tag{64}$$

The value $K_c>0$ will be chosen in a way that the nonlinearity $\tilde{N}_c$ characteristic is to be framed in an adequate sector [K_{min}, K_{max}].
The design method in order to obtain the value for the gain coefficient is presented as it follows:
The method recommended for stability insurance is as it follows:

1. For a certain fuzzy block type, the minimum value of K_m and the maximum value of K_M are chosen from the curve families $K_{Nc}=f(\tilde{x}_t)$, or $du_{dc}=f(\tilde{x}_t)$, with $\tilde{de}$ as a parameter.
2. The value of incremental coefficient of the command variable is limited by the capacity of control system to furnish the command variable to the process.
3. The incremental coefficient of the command variable may be determined with the relation that is describing the digital integration.
4. The maximum value of the command variable cannot overpass a maximum value.
5. At an incremental step, on a sampling period h, for the incremental of the command variable, a value is not recommended. For this, there may be chosen maximum a value of the incremental coefficient of $c_{duM}.K_M$.
6. The values of coefficients c_{du} and K_c may be chosen to insure sector stability.
7. In the choosing of c_{du} we must take account to the maximum values of K_M of the superior limit of the nonlinearity of the fuzzy block.
8. The chosen of K_c is done by taking account on the rapport $r_k=K_{min}/K_{max}$.

4.2 External stability

To assure external BIBO stability (Khalil, 1991) the following relation may be taken in consideration:

$$\dot{x}(t) = f_x(x(t), w(t)) \tag{65}$$

$$y(t) = f_y(x(t), w(t))$$

where the non-linear part $f(\tilde{e}, \tilde{de})$ is considered introduced in f_x.

According to [14], we may write the following conditions: x=0 is a stable point of equilibrium with w=0, and $f_x(0, 0)=0$, $\forall t \geq 0$; x=0 is a global equilibrium point of the system;

$$\dot{x} = f_x(x, 0) \tag{66}$$

Jacobian matrix $[\partial f_x / \partial x]$, evaluated for w=0, and $[\partial f_x / \partial w]$ are global limited; $f_y(t, x, w)$, satisfies:

$$\|f(x, w)\| \leq k_1 \|x\| + k_2 \|u\| + k_3 \tag{67}$$

global, for k_1, k_2, k_3>0. Then, for any $\|x(0)\| \leq \eta$, there are the constants γ>0 şi $\beta = \beta(\eta, k_3) \geq 0$ such as:

$$\sup_{t \geq 0} \|y(t)\| \leq \sup_{t \geq 0} \|w(t)\| + \beta \tag{68}$$

5. Control system example

A fuzzy control system, as it is in the example, has the block diagram from Fig. 14. A fuzzy PI controller RF-Ω is used in a speed control system of an electrical drive with the following elements: MCC - DC motor, CONV - power converter, RG-I - current controller, RF-Ω - speed controller, Ti - current sensor, TΩ - speed sensor, CAN, CNA - analogue to digital and digital to analogue converters.

The fuzzy controller has the structure from Fig. 15. It is a quasi-fuzzy PI controller with summation at the output, with an internal fuzzy block BF with the structure presented at the beginning, and a correction circuit to insure stability. The controller has also an anti wind-up circuit.

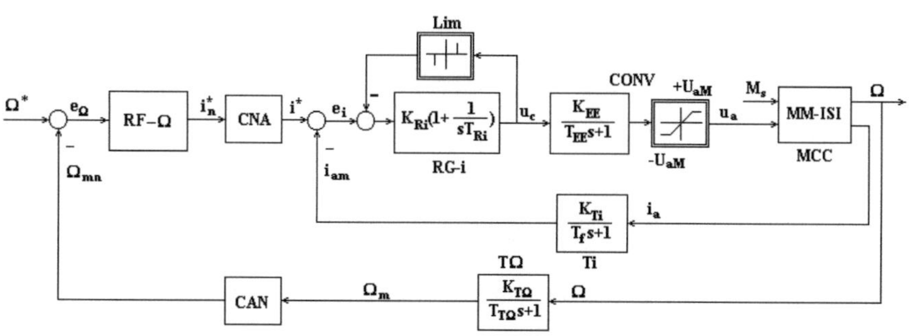

Fig. 14. The block diagram of the fuzzy control system

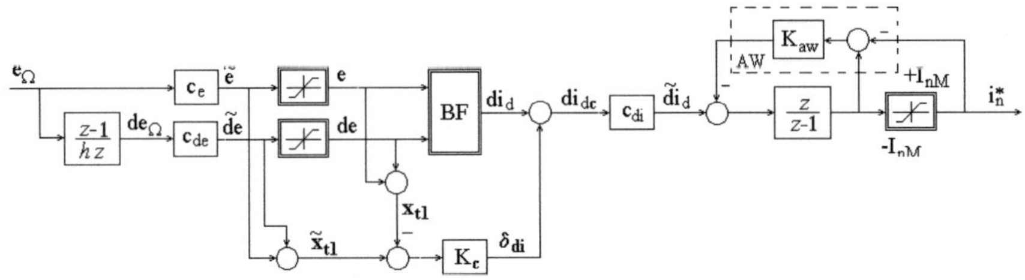

Fig. 15. The speed fuzzy PI controller, with anti-wind-up and correction circuit

A method to choose initial scaling coefficients based on the quality criteria of the control system is recommended, as it follows. The scaling coefficients were chosen after some iterative steps, using the quality criteria of the transient characteristics of the speed fuzzy control system at a step speed reference. The speed scaling coefficient c_e had the same value $c_e=1/e_M$. The first value of the derivative scaling coefficient was $c_{de}=1/de_M$.

1. Initial values are chosen c_{e1} and c_{de1}, based on operator knowledge.
2. An initial value for the output scaling coefficient is chosen c_{di1}, based on controller equivalence.
3. With the above values for c_e and c_{di} it is calculated a value for c_{de2}.
4. Maintaining the values of c_e and c_{de} and increasing the value of c_{di}.
5. Maintaining the values of c_e and c_{di} and decreasing c_{de}, and so on.

The adopted solution contains the values of the scaling coefficients from the sixth step. The transient characteristics obtained in the process of choosing the scaling coefficients are there presented in Fig. 16. The value of c_{de} was decreased to the final value from the sixth step. Decreasing more this scaling coefficient, the fuzzy control system becomes unstable.

Simulations are made for the control system with fuzzy PI controller and also for linear PI controller, for tuned and detuned system parameters. The transient characteristics for the current and speed are to be presented in Fig. 17. With continuous line, there are represented the characteristics for fuzzy control, and with dash-dot line, there are represented the characteristics for conventional control. The regime consists in starting the process unloaded, with a constant speed reference. A constant load torque, in the range of the rated process torque, is also introduced. Then, the motor is reversed, maintaining the constant load torque.

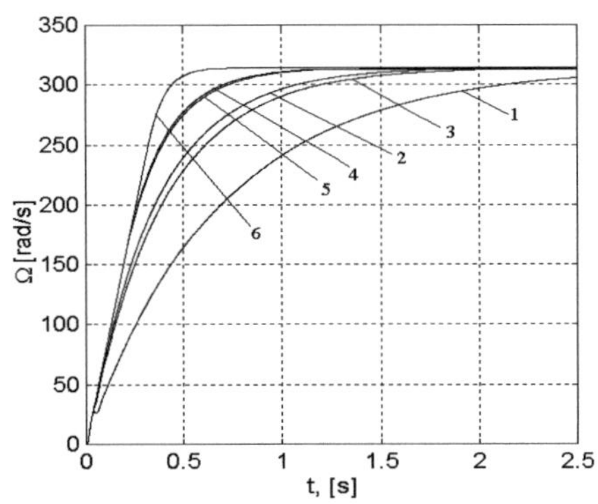

Fig. 16. The transient characteristics for scaling coefficients determination

The quality criteria of the control system, with linear (l) and fuzzy controller (f), for tuned (a) and detuned (d) parameter are there presented in Tab. 2.

Case	$\sigma_{1\Omega}$ [%]	$t_{r\Omega}$ [s]	σ_{1M} [%]	t_{rM} [s]	σ_{1r} [%]	t_{rr} [s]	$\Im$ 10^{-5}	$\Delta\sigma_{1\Omega}$ [%]	$\Delta\sigma_{1M}$ [%]	$\Delta t_{r\Omega}$ [s]	Δt_{rM} [s]
l-a	6,7	1	6,1	0,6	4,1	1,5	1,1	6,7	2,3	0,5	0,46
f-a	0	0,5	3,8	0,14	0	1,2	1,03				
l-d	8,3	1,5	6,1	0,65	4,1	3	2,0	8,3	2,3	0,7	0,51
f-d	0	0,8	3,8	0,14	0	2,2	1,89				

Table 2. The values of the quality criteria for the control system, for linear and fuzzy controllers, for tuned and detuned parameters of the electrical drive

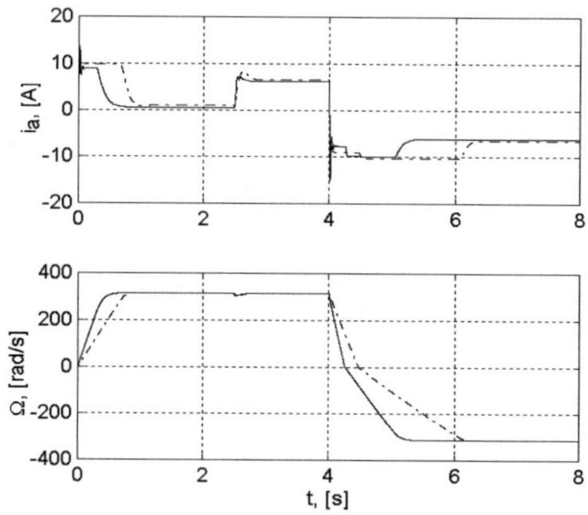

Fig. 17. Transient characteristics for the current and speed

Based on a comparative analysis of the speed performance criteria, better results were there obtained with the fuzzy PI controller designed, using the above methods as it follows:

- better quality criteria: zero overshot and shorter settling time;
- better performances for detuned parameters;
- the fuzzy control system is more robust at the identification errors and at the disturbance.

6. Conclusion

In this chapter, there were analyzed some digital controller, based on fuzzy blocks with Mamdani structure and PID dynamics.

A pseudo-equivalence of them with linear PID controllers was made, based on the input-output transfer characteristics of the fuzzy block, obtained by digital computer calculation.

The design of the fuzzy controller is based on the linearization of the fuzzy block around the origin, for the permanent regime. There is used the gain in the origin obtained as a limit in origin of the gain function, obtained from the translated SISO transfer characteristic.

For this type of controllers, the design relations were demonstrated. There was made an analysis of these design relations. There were also presented some observations related to the influences of the scaling coefficients.

The results presented in this chapter are important in the practice design of the control systems based on PID fuzzy controllers. This method for equivalence is valid for all kind of fuzzyfication and defuzzification methods, all types of membership functions, all inference methods, because it is based on analytic transfer characteristic, which may be obtained using computer calculations.

If there is a designed linear PID controller for a process control, we may use the equivalent fuzzy PID controller in its place in order to control the process with better control quality criteria. Based on the above notice, the method may be used also for tuning the fuzzy PID controller in a control system.
The term of "pseudo-equivalence" is used because there is no direct equivalence between the nonlinear digital fuzzy PI controller, with linearization only in the origin, and a linear analogue PI controller.
The theory presented in this paper is used and proved by the author in practical control applications, as speed control of electrical drives for dc motors, synchronous and induction motors.

7. References

Bao-Gang, H.; Mann, G.K.I. & Gosine, R.G. New methodology for analytical and optimal design of fuzzy PID controllers, *IEEE Trans. On Fuzzy Systems*, Vol. 7, Issue 5, Oct. 1999, p. 521.

Bao-Gang, H., Mann, G.K.I. & Gosine, R.G. A systematic study of fuzzy PID controllers function based evaluation approach, *IEEE Trans. On Fuzzy Systems*, Vol. 9, Issue 5, Oct. 2001, p. 699.

Buhler, H. *Reglage par logique floue*, Presses Polytechnique et Universitaires Romandes, Lausanne, 1994.

Jantzen, J. *Foundations of Fuzzy Control*, Wiley, 2007.

Khalil, H. K. *Nonlinear Systems*, Macmillan Pub. Co., N. Y., 1991.

Moon, B.S. Equivalence between fuzzy logic controllers and PI controllers for single input systems, *Fuzzy Sets and Systems*, Vol. 69, Issue 2, 1995, p. 105-113.

Mohan, B.M. & Sinha, A. The simplest fuzzy PID controllers: mathematical models and stability analysis, *Soft Computing - A Fusion of Foundations, Methodologies and Applications*, Springer Berlin / Heidelberg, Volume 10, Number 10 / August, 2006, p. 961-975.

Mohan, B.M. & Sinha, A. Analytical Structures for Fuzzy PID Controllers?, *IEEE Trans. On Fuzzy Systems*, Vol. 16, Issue 1, Feb., 2008.

Santos, M.; Dormido, S.; de Madrid, A.P.; Morilla F. & de la Cruz, J.M. Tuning fuzzy logic controllers by classical techniques, *Lecture Notes in Computer Science*, Volume 1105/1996, Springer Berlin/Heidelberg, p. 214-224.

Volosencu, C. Pseudo-Equivalence of Fuzzy PID Controllers, *WSEAS Transactions on Systems and Control*, Issue 4, Vol. 4, April 2009, p. 163-176.

Volosencu, C. Properties of Fuzzy Systems, *WSEAS Transactions On Systems*, Issue 2, Vol. 8, Feb. 2009, pp. 210-228.

Volosencu, C. Stabilization of Fuzzy Control; Systems, *WSEAS Transactions On Systems and Control*, Issue 10, Vol. 3, Oct. 2008, pp. 879-896.

Volosencu, C. Control of Electrical Drives Based on Fuzzy Logic, *WSEAS Transactions On Systems and Control*, Issue 9, Vol. 3, Sept. 2008, pp.809-822.

Yame, J.J. Takagi-Sugeno fuzzy PI controllers: Analytical equivalence and tuning, *Journal A*, Vol. 42, no. 3, p. 13-57, 2001.

Ying, H. Mamdani Fuzzy PID Controllers, *Fuzzy Control and Modeling: Analytical Foundations and Applications*, IEEE, 2000.

Xu; J.X.; Pok; Y.M.; Liu; C. & Hang, C.C. Tuning and analysis of a fuzzy PI controller based on gain and phase margins, *IEEE Transactions on Systems, Man and Cybernetics*, Part A, Volume 28, Issue 5, Sept. 1998, p. 685 - 691.

Part 5

Discrete Intelligent PID Controller

8

Discrete PID Controller Tuning Using Piecewise-Linear Neural Network

Petr Doležel, Ivan Taufer and Jan Mareš
University of Pardubice & Institute of Chemical Technology Prague
Czech Republic

1. Introduction

PID controller (which is an acronym to "proportional, integral and derivative") is a type of device used for process control. As first practical use of PID controller dates to 1890s (Bennett, 1993), PID controllers are spread widely in various control applications till these days. In process control today, more than 95% of the control loops are PID type (Astrom et al., 1995). PID controllers have experienced many changes in technology, from mechanics and pneumatics to microprocessors and computers.
Especially microprocessors have influenced PID controllers applying significantly. They have given possibilities to provide additional features like automatic tuning or continuous adaptation - and continuous adaptation of PID controller via neural model of controlled system (which is considered to be significantly nonlinear) is the aim of this contribution. Artificial Neural Networks have traditionally enjoyed considerable attention in process control applications, especially for their universal approximation abilities (Montague et al., 1994), (Dwarapudi, et al., 2007). In next sections, there is to be explained how to use artificial neural networks with piecewise-linear activation functions in hidden layer in controller design. To be more specific, there is described technique of controlled plant linearization using nonlinear neural model. Obtained linearized model is in a shape of linear difference equation and it can be used for PID controller parameters tuning.

2. Continuous-time and discrete PID controller

The basic structure of conventional feedback control using PID controller is shown in Fig. 1 (Astrom et al., 1995), (Doyle et al., 1990). In this figure, the SYSTEM is the object to be controlled. The aim of control is to make controlled system output variable $y_S(t)$ follow the set-point $r(t)$ using the manipulated variable $u(t)$ changes. Variable $e(t)$ is control error and is considered as PID controller input and t is continuous time.
Continuous-time PID controller itself is defined by several different algorithms (Astrom et al., 1995), (Doyle et al., 1990). Let us use the common version defined by (Eq. 1).

$$u(t) = K_p\left(e(t) + \frac{1}{T_i}\int_0^t e(\tau)d\tau + T_d\frac{de(t)}{dt}\right) \tag{1}$$

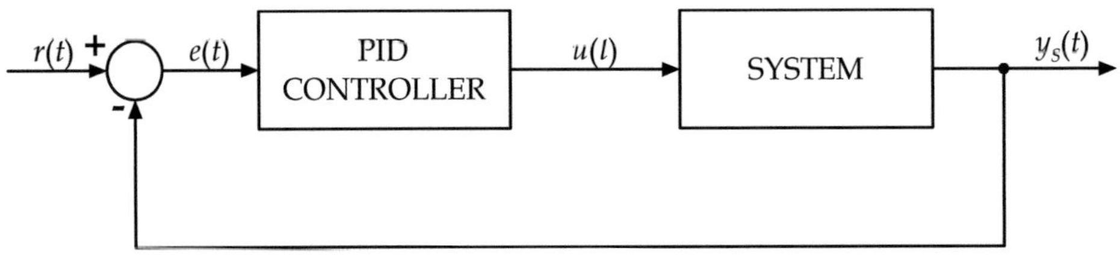

Fig. 1. Conventional feedback control loop

The control variable is a sum of three parts: proportional one, integral one and derivative one - see Fig. 2. The controller parameters are proportional gain K_p, integral time T_i and derivative time T_d.

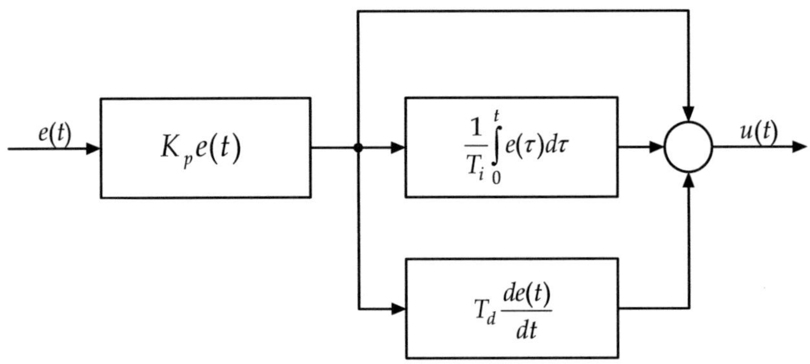

Fig. 2. Continuous-time PID controller

In applications, all three parameters have to be tuned to solve certain problem most appropriately while both stability and quality of control performance are satisfied. Many tuning techniques have been published in recent decades, some of them experimental, the others theoretically based.

As microprocessors started to set widely in all branches of industry, discrete form of PID controller was determined. Discrete PID controller computes output signal only at discrete time instants $k \cdot T$ (where T is sapling interval and k is an integer). Thus, conventional control loop (Fig. 1) has to be upgraded with zero order hold (ZOH), analogue-digital converter (A/D) and digital-analogue converter (D/A) - see Fig. 3 ($k \cdot T$ is replaced by k for formal simplification).

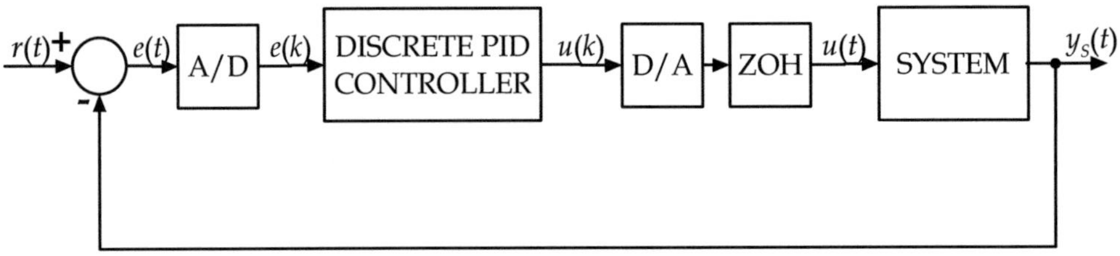

Fig. 3. Feedback control loop with discrete PID controller

Formula of discrete PID controller can be obtained by discretizing of (Eq. 1). From a purely numerical point of view, integral part of controller can be approximated by (Eq. 2) and derivative part by (Eq. 3).

$$\int_0^t e(\tau)d\tau \approx T\sum_{i=1}^{k}\frac{e(i)+e(i-1)}{2} \tag{2}$$

$$\frac{de(t)}{dt} \approx \frac{e(k)-e(k-1)}{T} \tag{3}$$

Then, discrete PID controller is defined by (Eq. 4).

$$u(k) = K_p\left(e(k)+\frac{T}{T_i}\sum_{i=1}^{k}\frac{e(i)+e(i-1)}{2}+\frac{T_d}{T}(e(k)-e(k-1))\right) \tag{4}$$

For practical application, incremental form of discrete controller is more suitable. Let us assume

$$\Delta u(k) = u(k)-u(k-1) \tag{5}$$

Then, with respect to (Eq. 4)

$$u(k)-u(k-1) = q_0 e(k)+q_1 e(k-1)+q_2 e(k-2) \tag{6}$$

where

$$q_0 = K_p\left(1+\frac{T}{2T_1}+\frac{T_d}{T}\right)$$

$$q_1 = -K_p\left(1-\frac{T}{2T_1}+\frac{2T_d}{T}\right)$$

$$q_2 = K_p\frac{T_d}{T}$$

In the Z domain (Isermann, 1991), discrete PID controller has the following transfer function.

$$\frac{Q(z^{-1})}{P(z^{-1})} = \frac{q_0+q_1 z^{-1}+q_2 z^{-2}}{1-z^{-1}} \tag{7}$$

As well as for continuous-time PID controller, there have been introduced several methods for q_0, q_1, q_2 tuning (Isermann, 1991). Most of them require mathematical model of controlled system (either first principle or experimental one) and if the system is nonlinear, the model has to be linearized around one or several operating points.

In next paragraph, the way how to tune discrete PID controller using Pole Assignment technique is described.

3. Discrete PID controller tuning using Pole Assignment technique

Suppose conventional feedback control loop with discrete PID controller (7) and controlled system described by nominator $B(z^{-1})$ and denominator $A(z^{-1})$ - see Fig. 4.

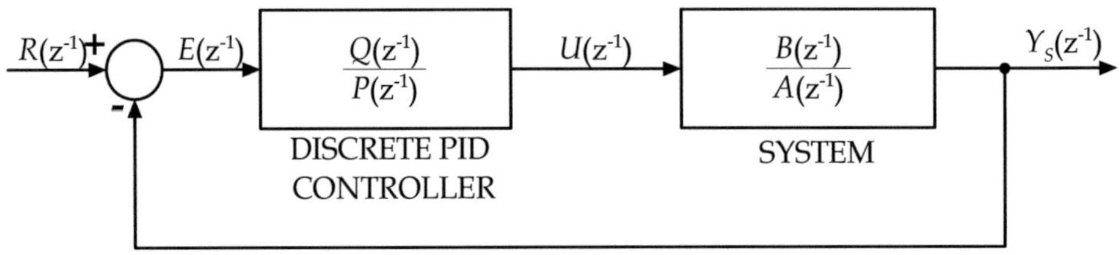

Fig. 4. Feedback control loop with discrete PID controller

Then, Z - transfer function of closed control loop is

$$\frac{Y(z^{-1})}{R(z^{-1})} = \frac{B(z^{-1})Q(z^{-1})}{A(z^{-1})P(z^{-1}) + B(z^{-1})Q(z^{-1})} \tag{8}$$

Denominator of Z - transfer function (8) is the characteristic polynomial

$$D(z^{-1}) = A(z^{-1})P(z^{-1}) + B(z^{-1})Q(z^{-1}) \tag{9}$$

It is well known that dynamics of the closed loop behaviour is defined by the characteristic polynomial (9). It has three tuneable variables which are PID controller parameters q_0, q_1, q_2. The roots of the polynomial (9) are responsible for control dynamics and one can assign those roots (so called poles) (see Fig. 5) by suitable tuning of the parameters q_0, q_1, q_2.
Thus, discrete PID controller tuning using Pole Assignment means choosing desired control dynamics (desired definition of characteristic polynomial) and subsequent computing of discrete PID controller parameters.
Let us show an example: suppose we need control dynamics defined by characteristic polynomial (10), where d_1, d_2, ... are integers (there are many ways how to choose those parameters, one of them is introduced in the case study at the end of this contribution).

$$D(z^{-1}) = 1 + d_1 z^{-1} + d_2 z^{-2} + \cdots \tag{10}$$

So we have to solve Diophantine equation (11) to obtain all controller parameters.

$$1 + d_1 z^{-1} + d_2 z^{-2} + \cdots = A(z^{-1})P(z^{-1}) + B(z^{-1})Q(z^{-1}) \tag{11}$$

If any solution exists, it provides us expected set of controller parameters.
Comprehensive foundation to pole assignment technique is described in (Hunt, 1993).

4. Continuous linearization using artificial neural network

The tuning technique described in section 3 requires linear model of controlled system in form of Z - transfer function. If controlled system is highly nonlinear process, linear model has to be updated continuously with operating point shifting. Except some classical techniques of continuous linearization (Gain Scheduling, Recurrent Least Squares Method, ...), there has been introduced new technique (Doležel et al., 2011), recently. It is presented in next paragraphs.

4.1 Artificial neural network for approximation

According to Kolmogorov's superposition theorem, any real continuous multidimensional function can be evaluated by sum of real continuous one-dimensional functions (Hecht-Nielsen, 1987). If the theorem is applied to artificial neural network (ANN), it can be said that any real continuous multidimensional function can be approximated by certain three-layered ANN with arbitrary precision. Topology of that ANN is depictured in Fig. 6. Input layer brings external inputs $x_1, x_2, \ldots, x_P$ into ANN. Hidden layer contains S neurons, which process sums of weighted inputs using continuous, bounded and monotonic activation function. Output layer contains one neuron, which processes sum of weighted outputs from hidden neurons. Its activation function has to be continuous and monotonic.

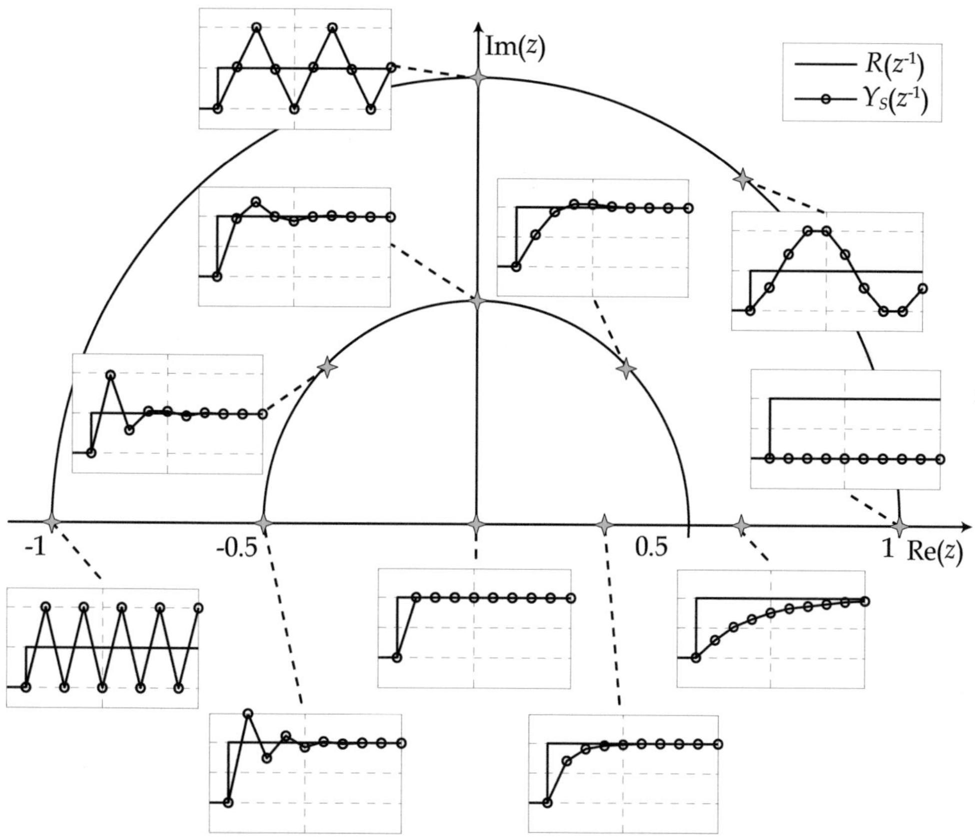

Fig. 5. The effect of characteristic polynomial poles to the control dynamics

So ANN in Fig. 6 takes P inputs, those inputs are processed by S neurons in hidden layer and then by one output neuron. Dataflow between input i and hidden neuron j is gained by weight $w^1_{j,i}$. Dataflow between hidden neuron k and output neuron is gained by weight $w^2_{1,k}$. Output of the network can be expressed by following equations.

$$y_a{}^1{}_j = \sum_{i=1}^{P} w^1{}_{j,i} \cdot x_i + w^1{}_j \tag{12}$$

$$y^1{}_j = \varphi^1\left(y_a{}^1{}_j\right) \tag{13}$$

$$y_a{}^2{}_1 = \sum_{i=1}^{S} w^2{}_{1,i} \cdot y^1{}_i + w^2{}_1 \tag{14}$$

$$y = \varphi^2\left(y_a{}^2{}_1\right) \tag{15}$$

In equations above, $\varphi^1(.)$ means activation functions of hidden neurons and $\varphi^2(.)$ means output neuron activation function.

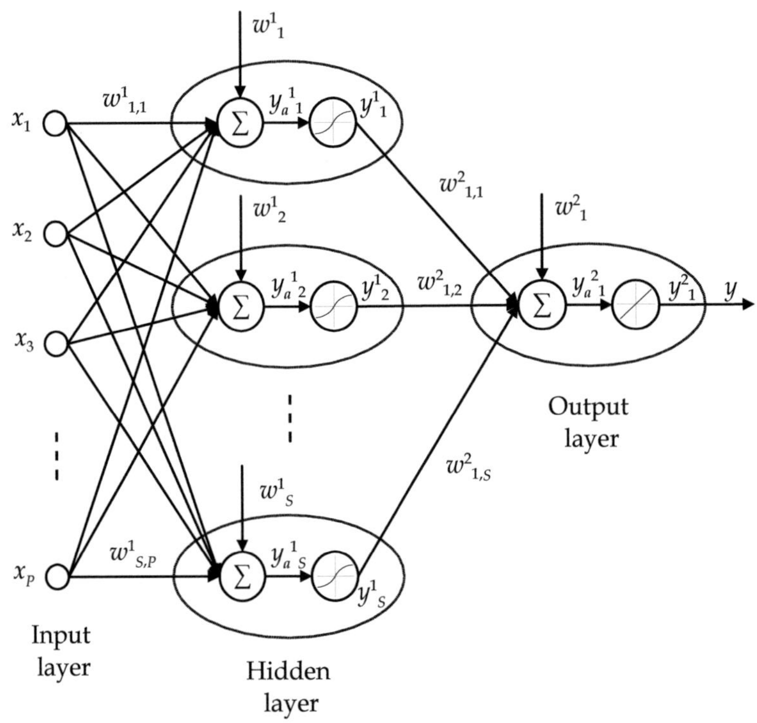

Fig. 6. Three-layered ANN

As it has been mentioned, there are some conditions applicable for activation functions. To satisfy those conditions, there is used mostly hyperbolic tangent activation function (Eq. 16) for neurons in hidden layer and identical activation function (Eq. 17) for output neuron.

$$y^1{}_j = \tanh\left(y_a{}^1{}_j\right) \tag{16}$$

$$y = y_{a\,1}^{2} \tag{17}$$

Mentioned theorem does not define how to set number of hidden neurons or how to tune weights. However, there have been published many papers which are focused especially on gradient training methods (Back-Propagation Gradient Descend Alg.) or derived methods (Levenberg-Marquardt Alg.) - see (Haykin, 1994).

4.2 System identification by artificial neural network

System identification means especially a procedure which leads to dynamic model of the system. ANN is used widely in system identification because of its outstanding approximation qualities. There are several ways to use ANN for system identification. One of them assumes that the system to be identified (with input u and output y_S) is determined by the following nonlinear discrete-time difference equation.

$$y_S(k) = \psi[y_S(k-1),\ldots,y_S(k-n),u(k-1),\ldots,u(k-m)],\ m \le n \tag{18}$$

In equation (18), $\psi(.)$ is nonlinear function, k is discrete time (formally better would be $k \cdot T$) and n is difference equation order.
The aim of the identification is to design ANN which approximates nonlinear function $\psi(.)$. Then, neural model can be expressed by (eq. 19).

$$y_M(k) = \hat{\psi}[y_M(k-1),\ldots,y_M(k-n),u(k-1),\ldots,u(k-m)],\ m \le n \tag{19}$$

In (Eq. 19), $\hat{\psi}$ represents well trained ANN and y_M is its output. Formal scheme of neural model is shown in Fig. 7. It is obvious that ANN in Fig. 7 has to be trained to provide y_M as close to y_S as possible. Existence of such a neural network is guaranteed by Kolmogorov's superposition theorem and whole process of neural model design is described in detail in (Haykin, 1994) or (Nguyen et al., 2003).

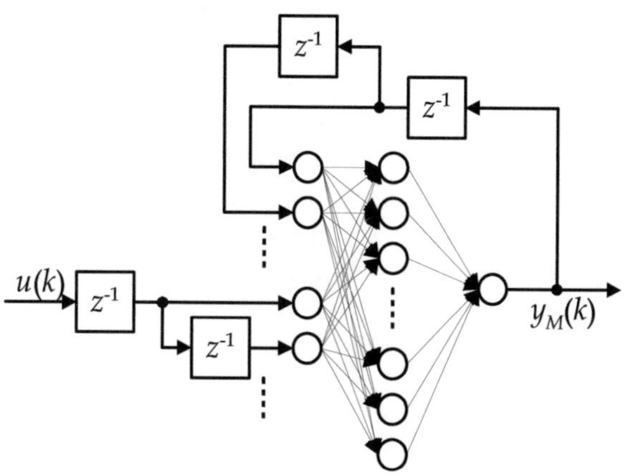

Fig. 7. Formal scheme of neural model

4.3 Piecewise-linear neural model for discrete PID controller tuning

As mentioned in section 4.1, there is recommended to use hyperbolic tangent activation function for neurons in hidden layer and identical activation function for output neuron in

ANN used in neural model. However, if linear saturated activation function (Eq. 20) is used instead, ANN features stay similar because of resembling courses of both activation functions (see Fig. 8).

$$y^1_j = \begin{cases} 1 & \text{for} \quad {y_a}^1_j > 1 \\ {y_a}^1_j & \text{for} \quad -1 \le {y_a}^1_j \le 1 \\ -1 & \text{for} \quad {y_a}^1_j < -1 \end{cases} \tag{20}$$

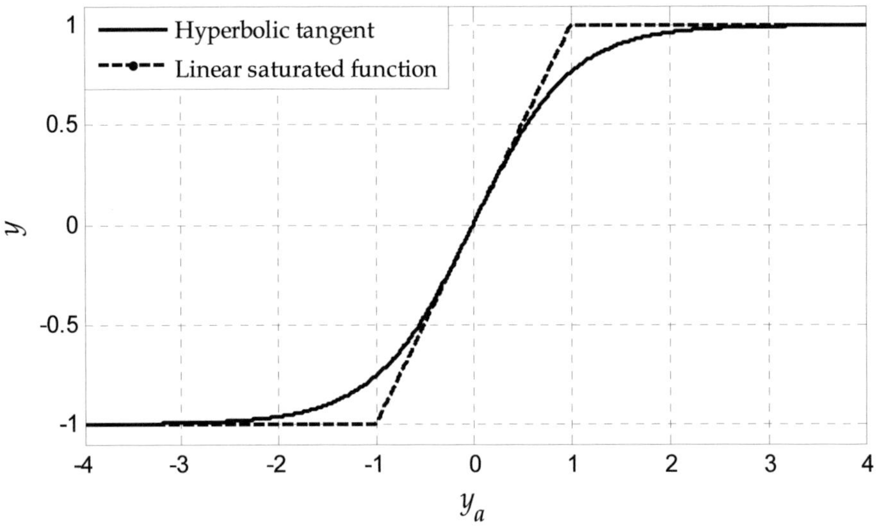

Fig. 8. Activation functions comparison

The output of linear saturated activation function is either constant or equal to input so neural model which uses ANN with linear saturated activation functions in hidden neurons acts as piecewise-linear model. One linear submodel turns to another when any hidden neuron becomes saturated or becomes not saturated.

Let us presume an existence of a dynamical neural model which uses ANN with linear saturated activation functions in hidden neurons and identic activation function in output neuron - see Fig. 9. Let us also presume $m = n = 2$ for making process plainer. ANN output can be computed using Eqs. (12), (13), (14), (15). However, another way for ANN output computing is useful. Let us define saturation vector $\mathbf{z}$ of S elements. This vector indicates saturation states of hidden neurons - see (Eq. 21).

$$z_i = \begin{cases} 1 & \text{for} \quad y^1_i > 1 \\ 0 & \text{for} \quad -1 \le y^1_i \le 1 \\ -1 & \text{for} \quad y^1_i < -1 \end{cases} \tag{21}$$

Then, ANN output can be expressed by (Eq. 22).

$$y_M(k) = -a_1 \cdot y_M(k-1) - a_2 \cdot y_M(k-2) + b_1 \cdot u(k-1) + b_2 \cdot u(k-2) + c \tag{22}$$

where
$$a_1 = -\sum_{i=1}^{S} w^2_{1,i} \cdot \left(1 - |z_i|\right) \cdot w^1_{i,1}$$

$$a_2 = -\sum_{i=1}^{S} w^2_{1,i} \cdot \left(1 - |z_i|\right) \cdot w^1_{i,2}$$

$$b_1 = \sum_{i=1}^{S} w^2_{1,i} \cdot \left(1 - |z_i|\right) \cdot w^1_{i,3}$$

$$b_2 = \sum_{i=1}^{S} w^2_{1,i} \cdot \left(1 - |z_i|\right) \cdot w^1_{i,4}$$

$$c = w^2_1 + \sum_{i=1}^{S} \left(w^2_{1,i} \cdot z_i + \left(1 - |z_i|\right) \cdot w^2_{1,i} \cdot w^1_i \right)$$

Thus, difference equation (22) defines ANN output and it is linear in some neighbourhood of actual state (in that neighbourhood, where saturation vector **z** stays constant). Difference equation (22) can be clearly extended into any order.
In other words, if the neural model of any nonlinear system in form of Fig. 9 is designed, then it is simple to determine parameters of linear difference equation which approximates

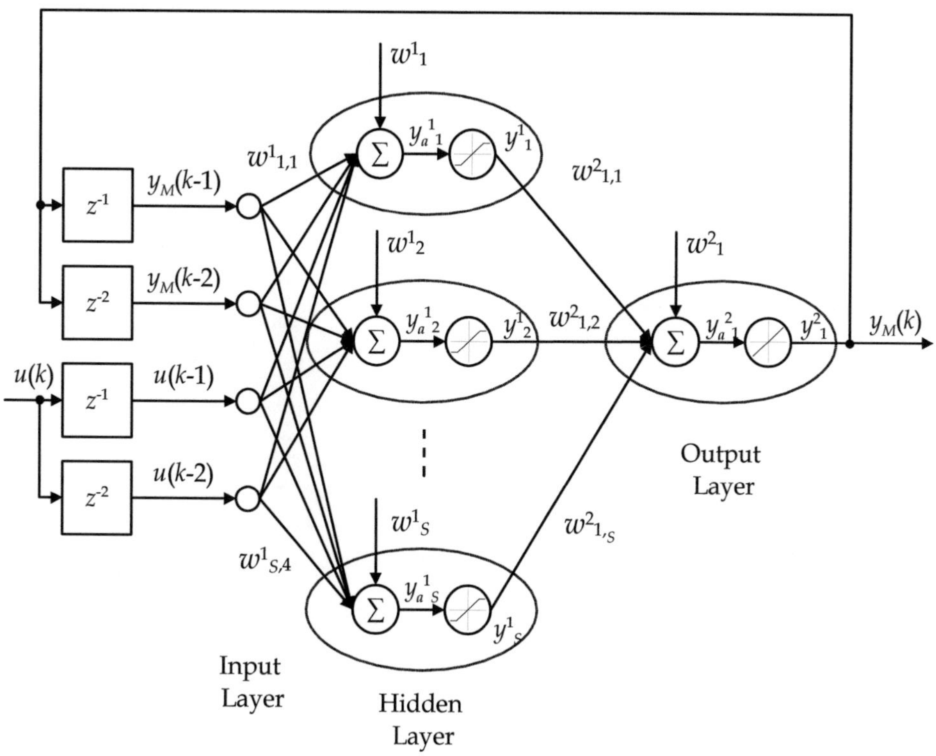

Fig. 9. Piecewise-linear neural model

system behaviour in some neighbourhood of actual state. This difference equation can be used then to the actual control action setting due to many of classical or modern control techniques.

In following examples, discrete PID controller with parameters tuned according to algorithm introduced in paragraph 3 is studied. As it is mentioned above, controlled system discrete model in form of Z - transfer function is required. So first, difference equation (22) should be transformed in following way. Let us define

$$\tilde{u}(k) = u(k) - u_0 \tag{23}$$

where u_0 is constant. Then, (Eq. 22) turns into

$$y_M(k) = -a_1 \cdot y_M(k-1) - a_2 \cdot y_M(k-2) + b_1 \cdot \tilde{u}(k-1) + b_2 \cdot \tilde{u}(k-2) + c + (b_1 + b_2) \cdot u_0 \tag{24}$$

Equation (24) becomes constant term free, if (Eq. 25) is satisfied.

$$u_0 = -\frac{c}{b_1 + b_2} \tag{25}$$

In Z domain, model (24) witch respect to (Eq. 25) is defined by Z - transfer function (26).

$$\frac{Y_M(z^{-1})}{\tilde{U}(z^{-1})} = \frac{b_1 z^{-1} + b_2 z^{-2}}{1 + a_1 z^{-1} + a_2 z^{-2}} \tag{26}$$

5. Algorithm of discrete PID controller tuning using piecewise-linear neural network

Whole algorithm of piecewise-linear neural model usage in PID controller parameters tuning is summarized in following terms (see Fig. 10, too).

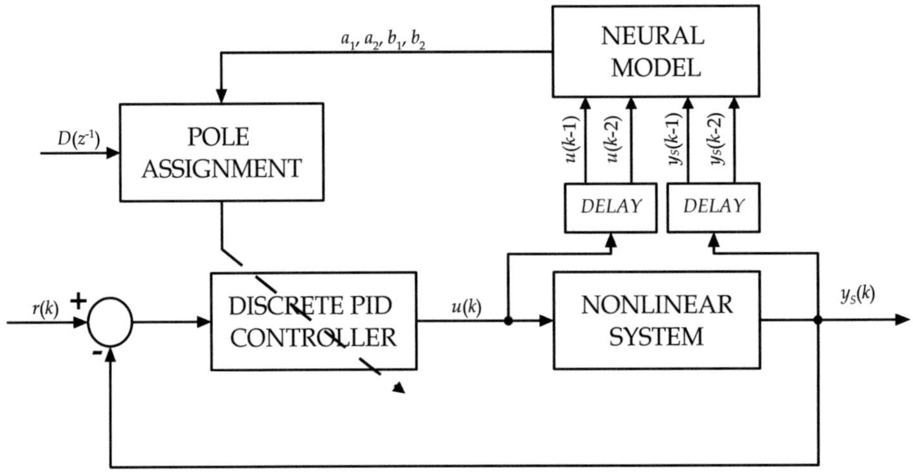

Fig. 10. Control algorithm scheme for second order nonlinear system

1. Create neural model of controlled plant in form of Fig. 9.
2. Determine polynomial $D(z^{-1})$ of (10).
3. Set $k = 0$.
4. Measure system output $y_S(k)$.
5. Determine the parameters a_i, b_i and c of difference equation (22).
6. Transform (Eq. 22) into Z - transfer function (26).
7. Determine discrete PID controller parameters by solving of (Eq. 11) where $A(z^{-1})$ and $B(z^{-1})$ are denominator and nominator of Z - transfer function (26), respectively.
8. Determine $\tilde{u}(k)$ using discrete PID controller tuned in previous step.
9. Transform $\tilde{u}(k)$ into $u(k)$ using (Eq. 23) and perform control action.
10. $k = k + 1$, go to 4.

Introduced algorithm is suitable to control of highly nonlinear systems, especially.

6. Case study

Discrete PID controller tuned continuously by technique introduced above is applied now to control of two nonlinear systems. Both of them are compiled by a combination of nonlinear static part and linear dynamical system - see Fig. 11.

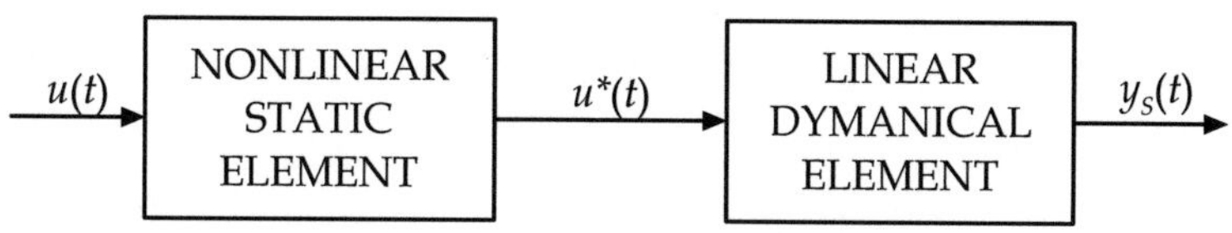

Fig. 11. System to control

6.1 First order nonlinear system

The static element of the first demo system is defined by (Eq. 27) and dynamical system is defined by differential equation (28).

$$u^*(t) = \left(\frac{2}{1+e^{-2u(t)}} - 1\right)^3 \tag{27}$$

$$y(t) + 10\frac{dy(t)}{dt} = u^*(t) \tag{28}$$

Graphic characteristics of the system are shown in Fig. 12.

Control loop is designed as shown in paragraph 5. At first, dynamical piecewise-linear neural model in shape of Fig. 9 is created. This procedure involves training and testing set acquisition, neural network training and pruning and neural model validating. As this sequence of processes is illustrated closely in many other publications (Haykin, 1994), (Nguyen, 2003) it is not referred here in detail. Briefly, training set is gained by controlled system excitation by set of step functions with various amplitudes while both u and y_S are measured (sampling interval $T = 1$ s) - see Fig. 13. Then, order of the neural model is set: $n = 1$ (Eq. 19) because the controlled system is first order one, too. After that, artificial neural network is trained by Backpropagation Gradient Descent Algorithm repeatedly (see Fig. 14) while pruning is applied - optimal neural network topology is determined as two inputs, four neurons in hidden layer and one output neuron. Finally, the neural model is validated (Fig. 15).

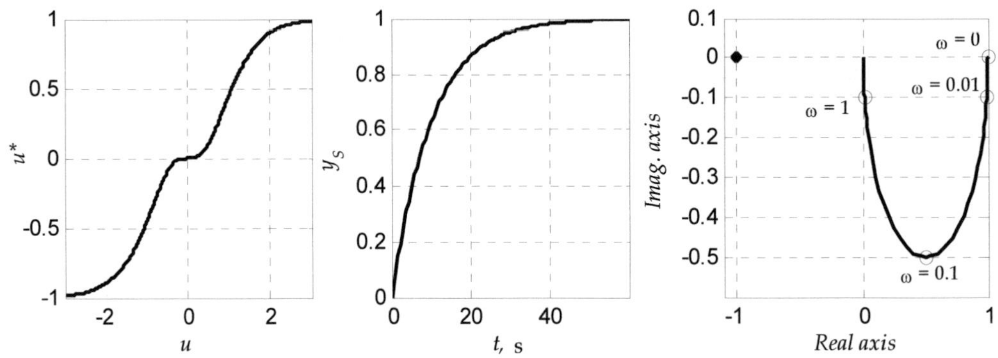

Characteristics of the static element (Eq. 27)

Step response of linear dynamical element (Eq. 28)

Nyquist plot of linear dynamical element (Eq. 28)

Fig. 12. Graphic characteristics of the first order nonlinear system

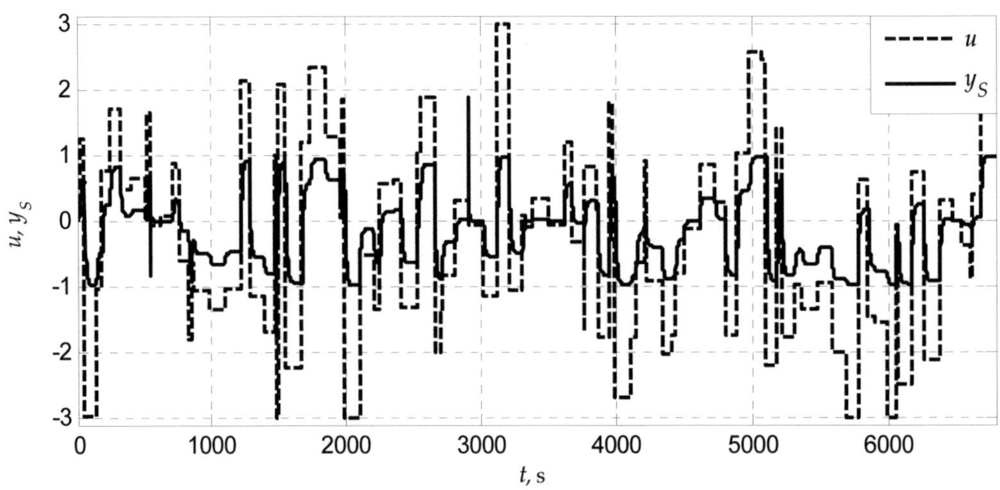

Fig. 13. Training set for the neural model

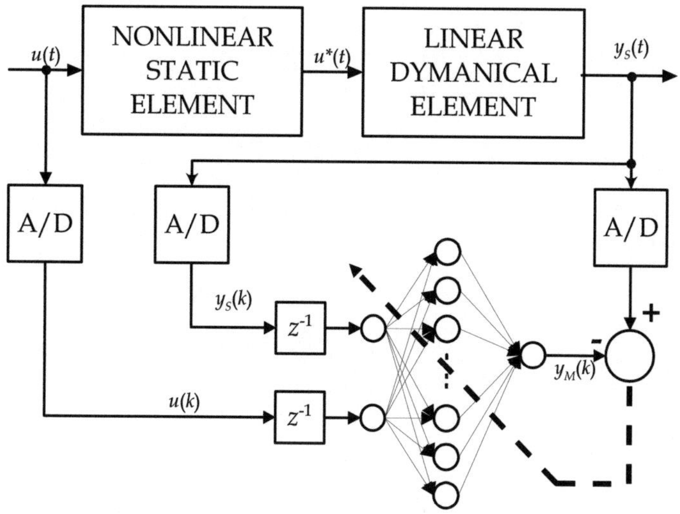

Fig. 14. Neural network training

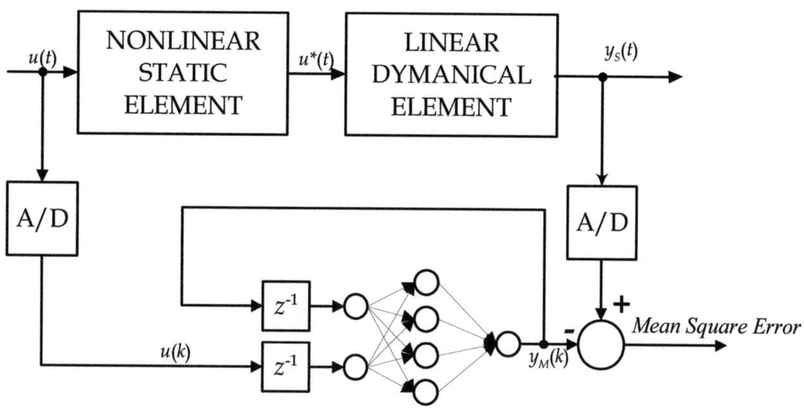

Fig. 15. Neural model validating

Next step is to determine polynomial $D(z^{-1})$. Common ways of $D(z^{-1})$ determination are mentioned below (Hunt, 1993).

- Dead beat is achieved
- Quadratic criterion is satisfied
- Control dynamics of closed loop equals to dynamics of defined second order system
- Special dynamics of closed control loop (defined by customer) is achieved

Let us use the c) possibility and define the standard for control dynamics as second order system with Z – transfer function (29).

$$F(z^{-1}) = \frac{0.2642z^{-1} + 0.1353z^{-2}}{1 - 0.7358z^{-1} + 0.1353z^{-2}} \tag{29}$$

Thus,

$$D(z^{-1}) = 1 + d_1 z^{-1} + d_2\, z^{-2} = 1 - 0.7358z^{-1} + 0.1353z^{-2} \tag{30}$$

Polynomial $D(z^{-1})$ is stable with double pole equal to 0.3679.
Essential part of next three steps of the control algorithm is to solve Diophantine equation (11). In this particular example, (Eq. 31) is to be solved.

$$1+d_1 z^{-1}+d_2 z^{-2}=\left(1+a_1 z^{-1}\right)\left(1-z^{-1}\right)+b_1 z^{-1}\left(q_0+q_1 z^{-1}+q_2 z^{-2}\right) \tag{31}$$

Method of undetermined coefficients is one possibility how to solve this equation. The initial matrix equation is

$$\begin{bmatrix} b_1 & 0 & 0 \\ 0 & b_1 & 0 \\ 0 & 0 & b_1 \end{bmatrix}\begin{bmatrix} q_0 \\ q_1 \\ q_2 \end{bmatrix}=\begin{bmatrix} d_1-a_1+1 \\ d_2+a_2 \\ a_2 \end{bmatrix} \tag{32}$$

And the solution is

$$\begin{aligned} q_0 &= \frac{d_1-a_1+1}{b_1} \\ q_1 &= \frac{a_1+d_2}{b_1} \\ q_2 &= 0 \end{aligned} \tag{33}$$

Now it is possible to perform control simulation. For defined reference variable course (combination of step functions and linearly descending and ascending functions), the control performance is shown in Fig. 16. Comparison of system output to standard (Eq. 29) is shown then in Fig. 17.

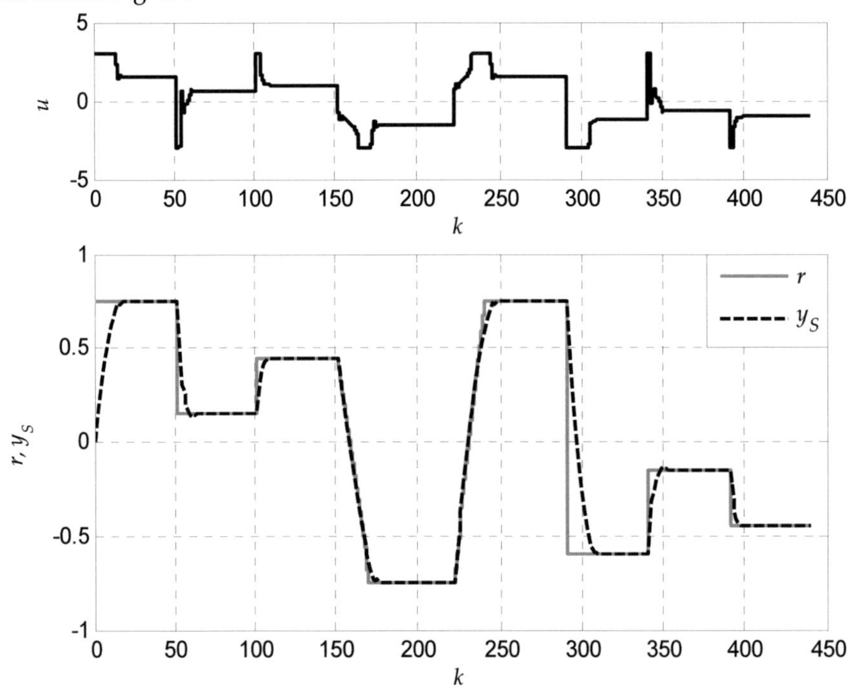

Fig. 16. Control performance – first order nonlinear system

As shown in Figs. 16 and 17, control performance is stable and desired dynamics of the closed loop is close to defined standard.

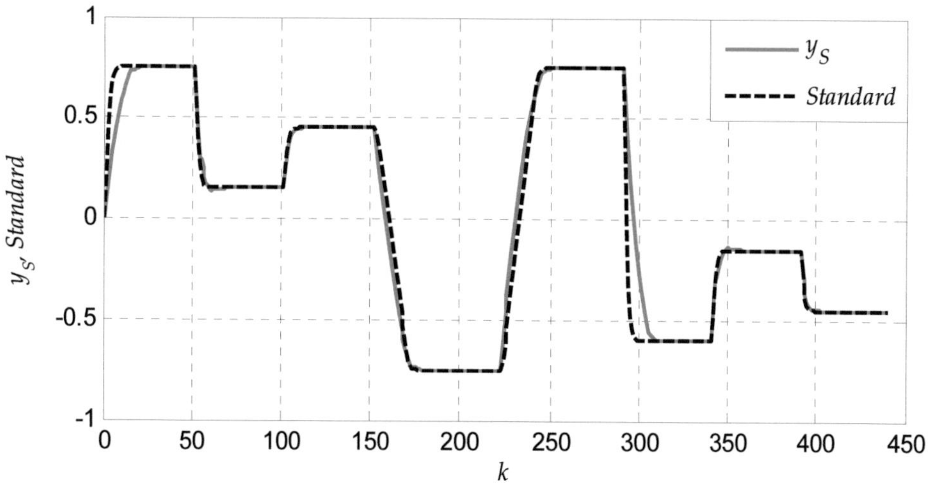

Fig. 17. Comparison to standard – first order nonlinear system

6.2 Second order nonlinear oscillative system

Second demo system is structurally identical as the previous one (Fig. 11). Even the static element is the same. However, the dynamic system is defined now by differential equation (34).

$$y(t)+5\frac{dy(t)}{dt}+50\frac{d^2y(t)}{dt^2}=u^*(t) \tag{34}$$

Graphic characteristics of the system are shown in Fig. 18.

The system is controlled on equal terms as previous one. However, the neural model now has four inputs as original system is second order one. Thus, Diophantine equation (35) should be solved.

$$1+d_1z^{-1}+d_2z^{-2}=\left(1+a_1z^{-1}+a_2z^{-2}\right)\left(1-z^{-1}\right)+\left(b_1z^{-1}+b_2z^{-2}\right)\left(q_0+q_1z^{-1}+q_2z^{-2}\right) \tag{35}$$

However, equation (35) is unsolvable. Thus, algorithm of discrete PID controller has to be extended into Z – transfer function (36) which is kind a filtered discrete PID controller.

$$\frac{Q(z^{-1})}{P(z^{-1})}=\frac{q_0+q_1z^{-1}+q_2z^{-2}}{(1-z^{-1})(1+\gamma z^{-1})} \tag{36}$$

Now, Diophantine equation (11) turns to (Eq. 37).

$$1+d_1z^{-1}+d_2z^{-2}=\left(1+a_1z^{-1}+a_2z^{-2}\right)\left(1-z^{-1}\right)\left(1+\gamma z^{-1}\right)+\left(b_1z^{-1}+b_2z^{-2}\right)\left(q_0+q_1z^{-1}+q_2z^{-2}\right) \tag{37}$$

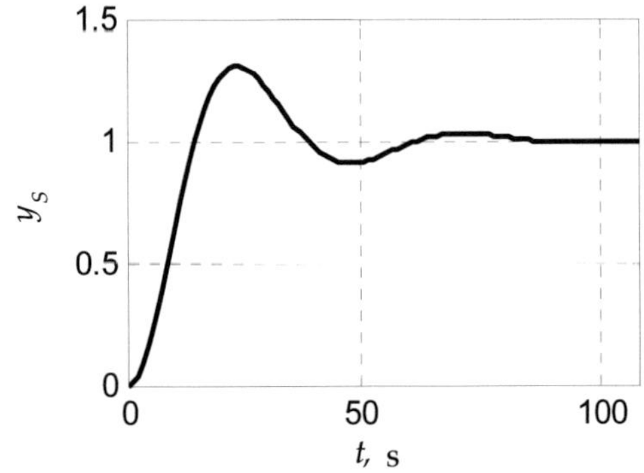

Step response of linear dynamical element (Eq. 34)

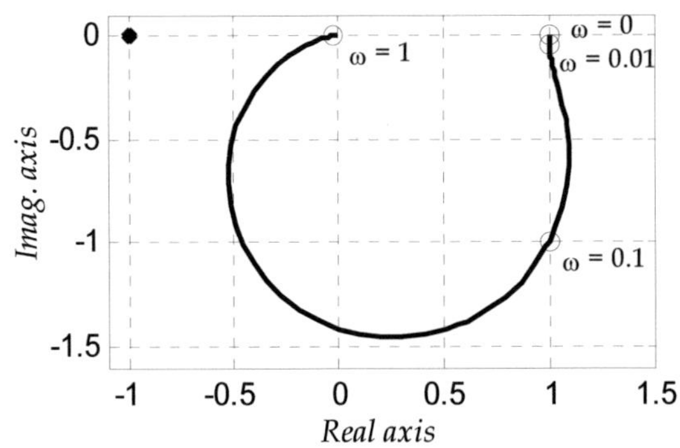

Nyquist plot of linear dynamical element (Eq. 34)

Fig. 18. Graphic characteristics of the second order nonlinear oscillative system

After applying of method of undetermined coefficients, solution can be obtained by solving of following matrix equation.

$$\begin{bmatrix} b_1 & 0 & 0 & 1 \\ b_2 & b_1 & 0 & a_1-1 \\ 0 & b_2 & b_1 & a_2-a_1 \\ 0 & 0 & b_2 & -a_2 \end{bmatrix} \begin{bmatrix} q_0 \\ q_1 \\ q_2 \\ \gamma \end{bmatrix} = \begin{bmatrix} d_1-a_1+1 \\ d_2+a_1-a_2 \\ a_2 \\ 0 \end{bmatrix} \tag{38}$$

And the solution is

$$\begin{bmatrix} q_0 \\ q_1 \\ q_2 \\ \gamma \end{bmatrix} = \begin{bmatrix} b_1 & 0 & 0 & 1 \\ b_2 & b_1 & 0 & a_1-1 \\ 0 & b_2 & b_1 & a_2-a_1 \\ 0 & 0 & b_2 & -a_2 \end{bmatrix}^{-1} \begin{bmatrix} d_1-a_1+1 \\ d_2+a_1-a_2 \\ a_2 \\ 0 \end{bmatrix} \tag{39}$$

Now it is possible to perform control simulation. For defined reference variable course, the control performance is shown in Fig. 19. Comparison of system output to standard (Eq. 29) is shown then in Fig. 20.

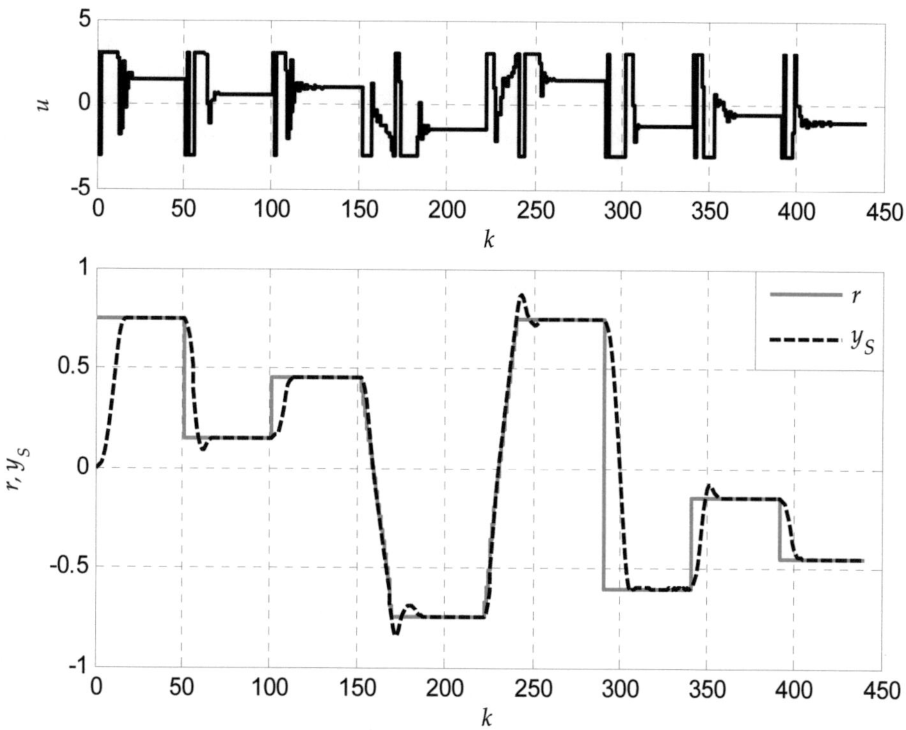

Fig. 19. Control performance - Second order nonlinear oscillative system

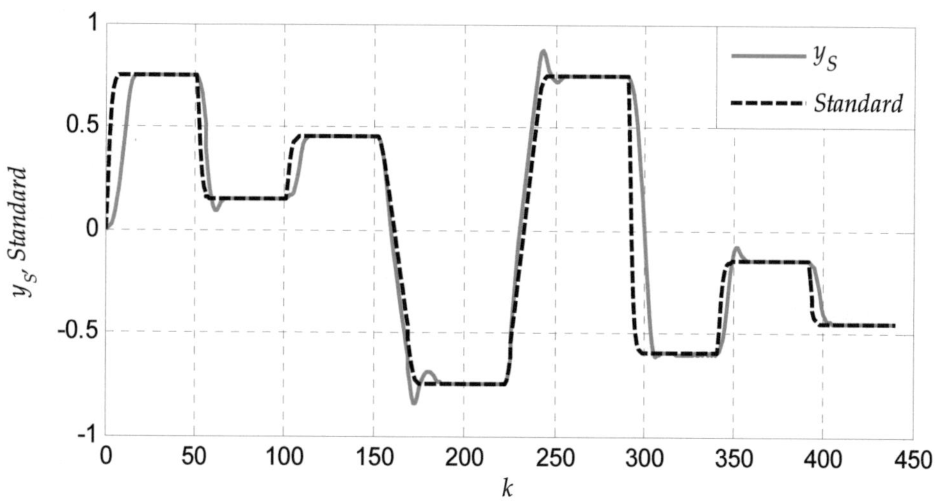

Fig. 20. Comparison to standard - Second order nonlinear oscillative system

As shown in Figs. 19 and 20, control performance is stable and satisfying. On the other hand, oscillative nature of the controlled system is not fully stifled.

7. Conclusion

There is introduced the technique above, which performs continuous adaptation of PID controller via neural model of controlled system. Neural model is used for controlled system continuous linearization and that linearized model is used for discrete PID controller tuning using pole assignment. The technique is suitable for highly nonlinear systems control, while it brings no advantages to control of the systems which are close to linear ones.

8. Acknowledgement

The work has been supported by the funds of the framework research project MSM 0021627505, MSM 6046137306 and by the funds of the project of University of Pardubice SGFEI06/2011 „Artificial Intelligence Control Toolbox for MATLAB". This support is very gratefully acknowledged.

9. References

Astrom, K. J. & Hagglund, T. (1995). *PID controllers: theory, design and tuning,* International Society for Measurement and Control, ISBN 1-55617-516-7, Durham, North Carolina, USA

Bennett, S. (1993). *A history of control engineering, 1930-1955,* IET, ISBN 0-86341-280-8, Stevenage, UK

Doyle, J., Francis, B. & Tannenbaum, A. (1990). *Feedback control theory,* Macmillan Publishing, ISBN 0-02330-011-0, New York, USA

Dwarapudi, S.; Gupta, P. K. & Rao, S. M. (2007). Prediction of iron ore pellet strength using artificial neural network model, ISIJ International, Vol. 47, No 1., pp. 67-72, ISSN 0915-1559

Doležel, P.; Taufer, I. & Mareš, J. (2011). Piecewise-Linear Neural Models for Process Control, *Proceedings of the 18th International Conference on Process Control '11*, pp. 296-300, ISBN 978-80-227-3517-9, Tatranská Lomnica, Slovakia, June 14-17, 2011

Haykin, S. (1994). *Neural Networks: A Comprehensive Foundation,* Prentice Hall, ISBN 0-02352-761-7, New Jersey, USA

Hecht-Nielsen, R. (1987). Kolmogorov's mapping neural network existence theorem, *Proc 1987 IEEE International Conference on Neural Networks,* Vol. 3, pp. 11-13, IEEE Press

Hunt, K. J., Ed. (1993). *Polynomial methods in optimal control and filtering.,* IET, ISBN 0-86341-295-5, Stevenage, UK

Isermann, R. (1991). *Digital Control Systems,* Springer-Verlag, ISBN 3-54010-728-2, Heidelberg, Germany

Montague, G. & Morris, J. (1994). Neural network contributions in biotechnology, *Trends in biotechnology,* Vol. 12, No 8., pp. 312-324, ISSN 0167-7799

Nguyen, H.; Prasad, N.; Walker, C. (2003). *A First Course in Fuzzy and Neural Control,* Chapman & Hall/CRC, ISBN 1-58488-244-1, Boca Raton, USA

Part 6

Fractional Order PID Controllers

9

PID Control Theory

Kambiz Arab Tehrani[1] and Augustin Mpanda[2,3]
[1]*University of Nancy, Teaching and Research at the University of Picardie, INSSET, Saint-Quentin, Director of Power Electronic Society IPDRP,*
[2]*Tshwane University of Technology/FSATI*
[3]*ESIEE-Amiens*
[1,3]*France*
[2]*South Africa*

1. Introduction

Feedback control is a control mechanism that uses information from measurements. In a feedback control system, the output is sensed. There are two main types of feedback control systems: 1) positive feedback 2) negative feedback. The positive feedback is used to increase the size of the input but in a negative feedback, the feedback is used to decrease the size of the input. The negative systems are usually stable. A PID is widely used in feedback control of industrial processes on the market in 1939 and has remained the most widely used controller in process control until today. Thus, the PID controller can be understood as a controller that takes the present, the past, and the future of the error into consideration. After digital implementation was introduced, a certain change of the structure of the control system was proposed and has been adopted in many applications. But that change does not influence the essential part of the analysis and design of PID controllers. A proportional-integral-derivative controller (PID controller) is a method of the control loop feedback. This method is composing of three controllers [1]:

1. Proportional controller (PC)
2. Integral controller (IC)
3. Derivative controller (DC)

1.1 Role of a Proportional Controller (PC)

The role of a proportional depends on the present error, I on the accumulation of past error and D on prediction of future error. The weighted sum of these three actions is used to adjust Proportional control is a simple and widely used method of control for many kinds of systems. In a proportional controller, steady state error tends to depend inversely upon the proportional gain (ie: if the gain is made larger the error goes down). The proportional response can be adjusted by multiplying the error by a constant K_p, called the proportional gain. The proportional term is given by:

$$P = K_P.error(t) \tag{1}$$

A high proportional gain results in a large change in the output for a given change in the error. If the proportional gain is very high, the system can become unstable. In contrast, a

small gain results in a small output response to a large input error. If the proportional gain is very low, the control action may be too small when responding to system disturbances. Consequently, a proportional controller (Kp) will have the effect of reducing the rise time and will reduce, but never eliminate, the steady-state error.
In practice the proportional band (PB) is expressed as a percentage so:

$$PB\% = \frac{100}{K_P} \tag{2}$$

Thus a PB of 10% ⇔ Kp=10

1.2 Role of an Integral Controller (IC)

An Integral controller (IC) is proportional to both the magnitude of the error and the duration of the error. The integral in in a PID controller is the sum of the instantaneous error over time and gives the accumulated offset that should have been corrected previously. Consequently, an integral control (Ki) will have the effect of eliminating the steady-state error, but it may make the transient response worse.
The integral term is given by:

$$I = K_I \int_0^t error(t)dt \tag{3}$$

1.3 Role of a Derivative Controller (DC)

The derivative of the process error is calculated by determining the slope of the error over time and multiplying this rate of change by the derivative gain Kd. The derivative term slows the rate of change of the controller output.A derivative control (Kd) will have the effect of increasing the stability of the system, reducing the overshoot, and improving the transient response. The derivative term is given by:

$$D = K_D . \frac{derror(t)}{dt} \tag{4}$$

Effects of each of controllers Kp, Kd, and Ki on a closed-loop system are summarized in the table shown below in tableau 1.

2. PID controller (PIDC)

A typical structure of a PID control system is shown in Fig.1. Fig.2 shows a structure of a PID control system. The error signal e(t) is used to generate the proportional, integral, and

Parameter	Rise time	Overshoot	Settling time	Steady-state error
K_p	Decrease	Increase	Small change	Decrease
K_i	Decrease	Increase	Increase	Decrease significantly
K_d	Minor decrease	Minor decrease	Minor decrease	No effect in theory

Table 1. A PID controller in a closed-loop system

derivative actions, with the resulting signals weighted and summed to form the control signal u(t) applied to the plant model.

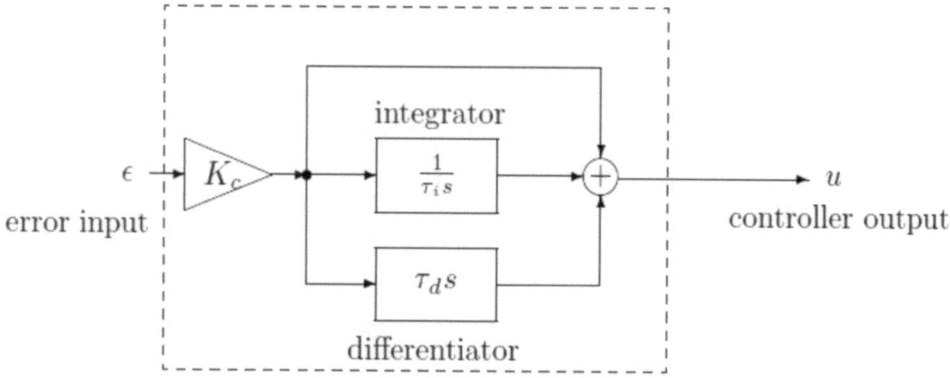

Fig. 1. A PID control system

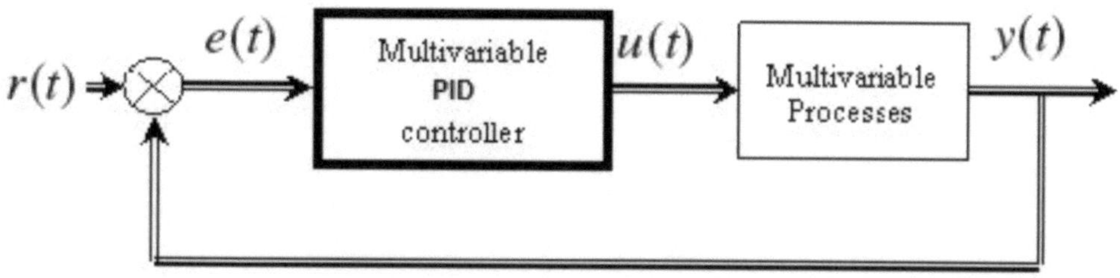

Fig. 2. A structure of a PID control system

where u(t) is the input signal to the multivariable processes, the error signal e(t) is defined as e(t) =r(t) − y(t), and r(t) is the reference input signal.
A standard PID controller structure is also known as the "three-term" controller. This principle mode of action of the PID controller can be explained by the parallel connection of the P, I and D elements shown in Figure 3.

Block diagram of the PID controller

$$G(s) = K_P(1+\frac{1+T_I.T_D.S^2}{T_I.S}) = K_P(1+\frac{1}{T_i s}+T_D s) \tag{5}$$

where K_P is the proportional gain, T_I is the integral time constant, T_D is the derivative time constant, $K_I = K_P / T_I$ is the integral gain and $K_D = K_P T_D$ is the derivative gain. The "three-term" functionalities are highlighted below. The terms K_P, T_I and T_D definitions are:

- The proportional term: providing an overall control action proportional to the error signal through the all pass gain factor.
- The integral term: reducing steady state errors through low frequency compensation by an integrator.
- The derivative term: improving transient response through high frequency compensation by a differentiator.

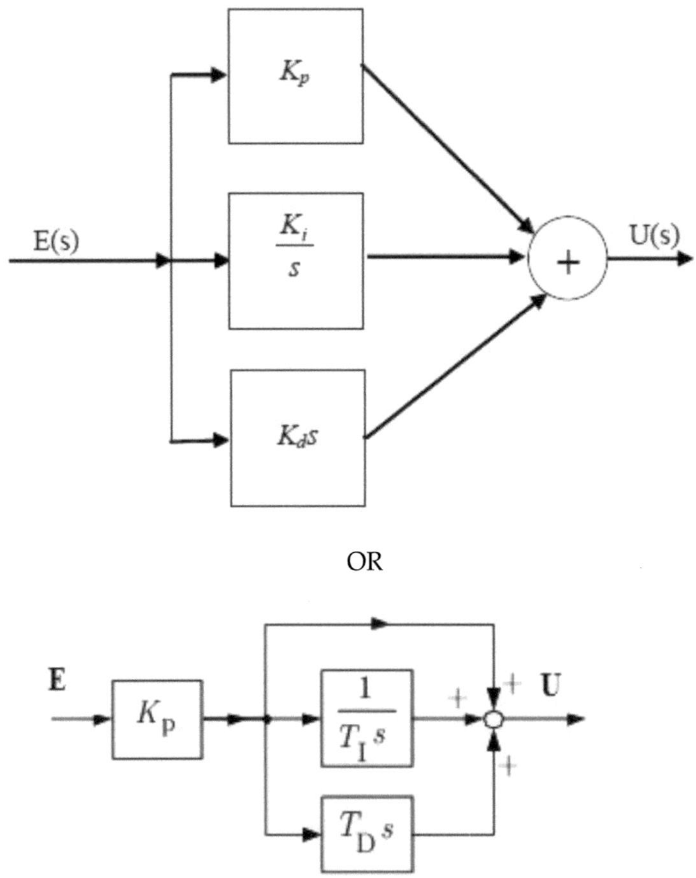

Fig. 3. Parallel Form of the PID Compensator

These three variables K_P, T_I and T_D are usually tuned within given ranges. Therefore, they are often called the *tuning parameters* of the controller. By proper choice of these tuning parameters a controller can be adapted for a specific plant to obtain a good behaviour of the controlled system.
The time response of the controller output is

$$U(t) = K_P(e(t) + \frac{\int_0^t e(t)dt}{T_i} + T_d \frac{de(t)}{dt}) \qquad (6)$$

Using this relationship for a step input of $e(t)$, i.e. $e(t) = \delta(t)$, the step response r(t) of the PID controller can be easily determined. The result is shown in below. One has to observe that the length of the arrow $K_P T_D$ of the D action is only a measure of the weight of the δ impulse.

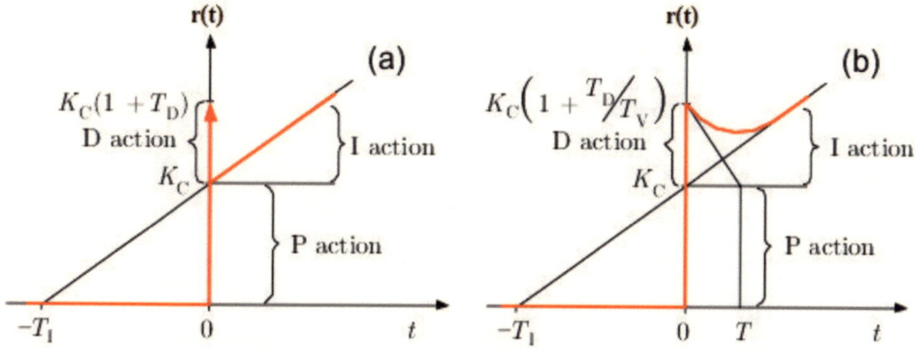

Fig. 4. a) Step response of PID ideal formb) Step response of PID real form

2.1 The transfer function of the PID controller

The transfer function of the PID controller is

$$G(s)=\frac{U(s)}{E(s)} \tag{7}$$

$$G(s)=K_P+\frac{K_I}{S}+K_DS \quad =\frac{K_DS^2+K_PS+K_I}{S} \tag{8}$$

2.2 PID pole zero cancellation

The PID equation can be written in this form:

$$G(s)=\frac{K_d(s^2+\frac{K_p}{K_d}s+\frac{K_i}{K_d})}{s} \tag{9}$$

When this form is used it is easy to determine the closed loop transfer function.

$$H(s)=\frac{1}{s^2+2\xi\omega_0 s+\omega^2{}_0} \tag{10}$$

If

$$\frac{K_i}{K_d}=\omega^2{}_0 \tag{11}$$

$$\frac{K_p}{K_d}=2\xi\omega_0 \tag{12}$$

Then

$$G(s)H(s)=\frac{K_d}{s} \tag{13}$$

This can be very useful to remove unstable poles.
There are several prescriptive rules used in PID tuning. The most effective methods generally involve the development of some form of process model, and then choosing P, I, and D based on the dynamic model parameters.

2.3 Tuning methods

We present here four tuning methods for a PID controller [2,3].

Method	**Advantages**	**Disadvantages**
Manual	Online method No math expression	Requires experienced personnel
Ziegler-Nichols	Online method Proven method	Some trial and error, process upset and very aggressive tuning
Cohen-Coon	Good process models	Offline method Some math Good only for first order processes
Software tools	Online or offline method, consistent tuning, Support Non-Steady State tuning	Some cost and training involved
Algorithmic	Online or offline method, Consistent tuning, Support Non-Steady State tuning, Very precise	Very slow

2.3.1 The Ziegler–Nichols tuning method

The Ziegler–Nichols tuning method is a heuristic method of tuning a PID controller. It was proposed by John G. Ziegler and Nichols in the 1940's. It is performed by setting I (integral) and D (derivative) gains to zero. The P (proportional) gain, Kp is then increased (from zero) until it reaches the ultimate gain Ku, at which the output of the control loop oscillates with a constant amplitude. Ku and the oscillation period Tu are used to set the P, I, and D gains depending on the type of controller used [3,4]:

Ziegler–Nichols method			
Control Type	K_p	K_i	K_d
P	$K_u / 2$	-	-
PI	$K_u / 2.2$	$1.2K_p / T_u$	-
PID	$0.60K_u$	$2K_p / T_u$	$K_pT_u / 8$
Some overshoot	$0.33K_u$	$2K_p / T_u$	$K_pT_u / 3$
No overshoot	$0.2K_u$	$2K_p / T_u$	$K_pT_u / 3$

We can realise a PID controller by two methods:
First, an analog PID controller
Second, a digital PID controller

1. Circuit diagram below (figure.5) shows an analog PID controller. In this figure, we present an analog PID controller with three simple op amp amplifier, integrator and differentiator circuits.

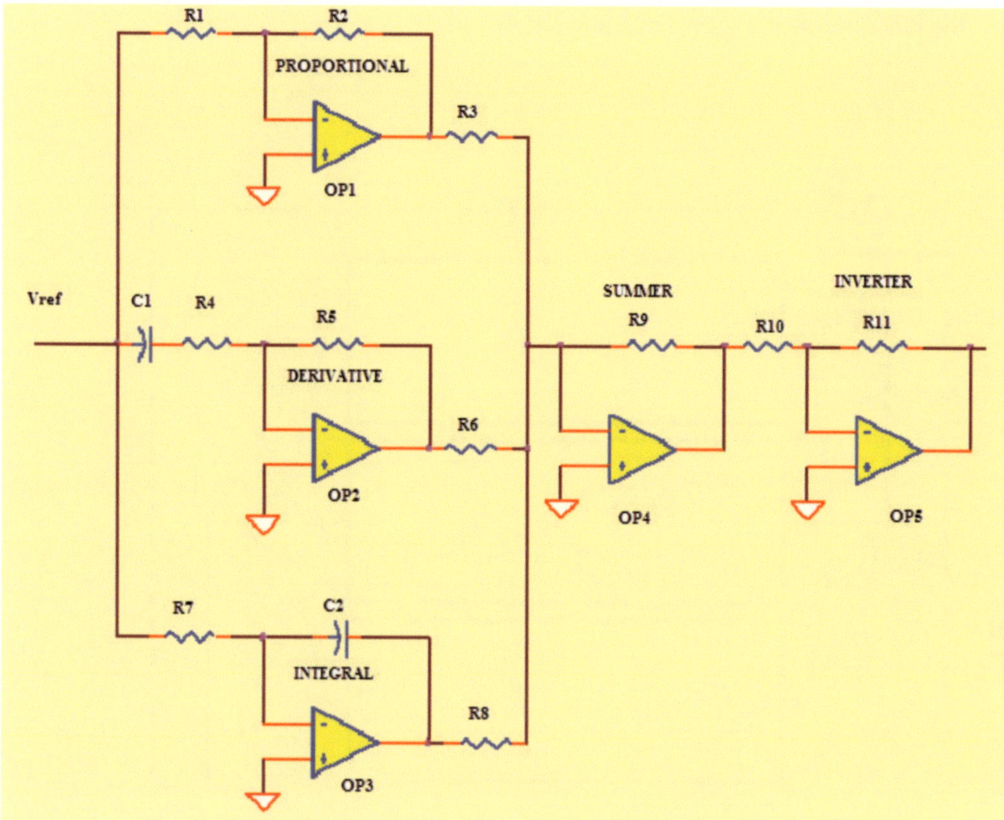

Fig. 5. Electronic circuit implementation of an analog PID controller

TERM	DEFINITION	Op Amp Circuit Function
P	Proportional/ Amplifier	Vo=(R2/R1).Vref
D	Differentiator	Vo=R5.C1.dVref/dt
I	Integrator	Vo=1/(R7.C2). ∫ Vref dt

Finally, we need to add the three PID terms together. Again the summing amplifier OP4 serves us well. Because the error amp, PID and summing circuits are inverting types, we need to add a final op amp inverter OP5 to make the final output positive.

2. Today, digital controllers are being used in many large and small-scale control systems, replacing the analog controllers. It is now a common practice to implement PID controllers in its digital version, which means that they operate in discrete time domain and deal with analog signals quantized in a limited number of levels. Moreover, in such controller we do not need much space and they are not expensive. A digital version of the PID controller is shown in figure 6 [5,6].

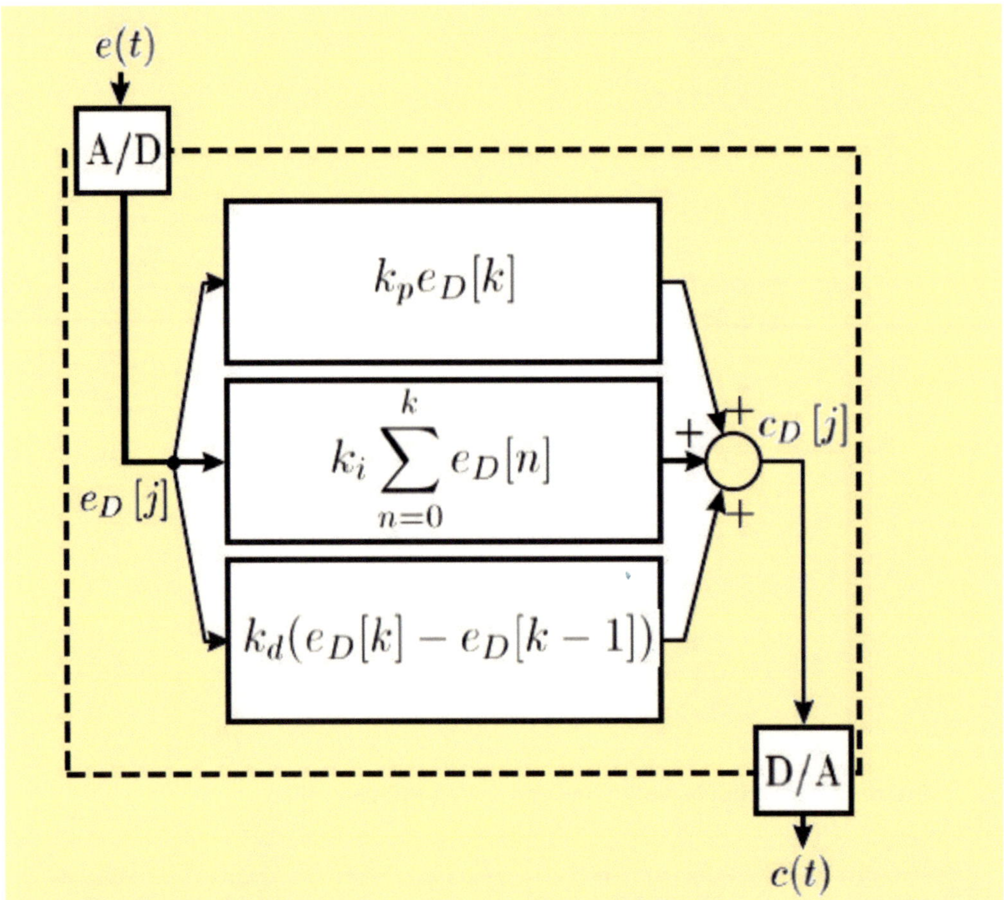

Fig. 6. Digital PID Controller

In its digital version, the integral becomes a sum and the deferential a difference. The continuous time signal e(t) is sampled in fixed time intervals equals a determined sample period, here called Tc (in figure 6 Tc = 1). An A/D (analog to digital) converter interfaces the input and a D/A (digital to analog) converter interfaces the output. This sampled and digitalized input, called eD[j], exists only in time instants $t = kT_C$ for all $k \geq 0 \in Z$. A lower bound for the sample period is the computing time of a whole cycle of the digital PID (which includes the A/D and D/A conversion).

While PID controllers are applicable to many control problems, and often perform satisfactorily without any improvements or even tuning, they can perform poorly in some applications, and do not in general provide optimal control.

3. Fractional systems

Fractional order systems are characterized by fractional-order differential equations. Fractional calculus considers any real number for derivatives and integrals. The FOPID controller is the expansion of the conventional integer-order PID controller based on fractional calculus [7,8].

3.1 Fractional-order PID (FOPID) controller

The PIDs are linear and in particular symmetric and they have difficulties in the presence of non-linearities. We can solve this problem by using a fractional-order PID (FOPID) controller. A FOPID controller is presented below [7-9]:

$$G(s) = K_P + \frac{K_I}{S^{\alpha}} + K_D S^{\beta} = \frac{K_P S^{\alpha} + K_I + K_D S^{\alpha} S^{\beta}}{S^{\alpha}} \tag{14}$$

Figure.7 describes the possibilities a FOPID for the different controllers.

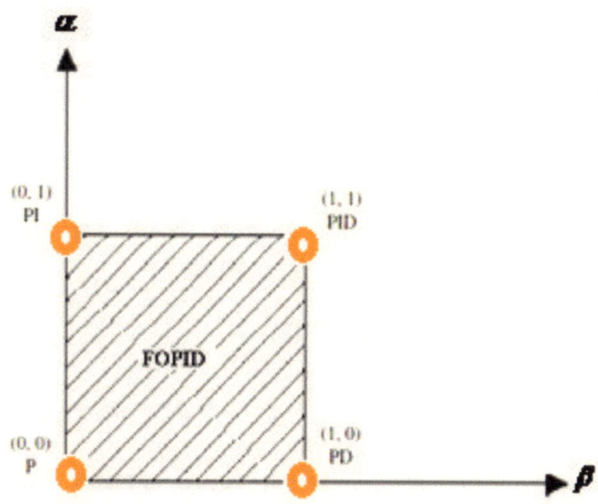

Fig. 7. Generalization of the FOPID controller: from point to plane.

There are several methods to calculate the fractional order derivative and integrator of a fractional order PID controller. For this purpose we present a real order calculus according to the Riemann-Liouville definition.

3.2 Fractional calculus

Fractional calculus is a branch of mathematics dealing with real number powers of differential or integral operators. It generalizes the common concepts of derivative and integral. Among all the different definitions, the definition which has been proposed by Riemann and Liouville is the most usual one [9,10]. The definition is as follows:

$$ {}_cD_x^{-n} f(x) = \int_c^x \frac{(x-t)^{n-1}}{(n-1)!} f(t)dt, \qquad n \in \Re \tag{15}$$

The general definition of D is given by (2):

$$ {}_cD_x^{v} f(x) = \begin{cases} \int_c^x \frac{(x-t)^{-v-1}}{\Gamma(-v)} f(t)dt, & if v < 0 \\ f(x) & if v = 0 \\ D^n[{}_cD_x^{v-n} f(x)] & if v > 0 \end{cases} \tag{16}$$

$$ n = \min\{K \in \Re, K > v\} $$

Where Γ(·) is the well-known Euler's gamma function.

Function $$F(s) = s^{\nu} \tag{17}$$

Function (17) is not only the simplest fractional order transfer function hat may appear but is also very important for applications, as shall be seen subsequently. For that reason, we analyse its time and frequency responses.

Time responses of (17)

The derivatives of the exponential function are given by

$$ {}_0D^{\nu}{}_t e^{at} = E_t(-\nu, a), t > 0 \tag{18}$$

For negative orders, from definition (16) we have:

$$ {}_0D_t^{-\nu} e^{at} = \frac{1}{\Gamma(\nu)} \int_0^t (t-\xi)^{\nu-1} e^{a\xi} d\xi, \nu \in \Re^+ \tag{19}$$

By means of the substitution x = t −ξ , in the first place, and of the substitution ax = y, in the second place, we obtain

$$ \begin{aligned} {}_0D_t^{-\nu} e^{at} &= -\frac{1}{\Gamma(\nu)} \int_t^0 x^{\nu-1} e^{a(t-x)} dx = \frac{e^{at}}{\Gamma(\nu)} \int_0^t x^{\nu-1} e^{-ax} dx = \\ \frac{e^{at}}{\Gamma(\nu)} \int_0^{at} \left(\frac{y}{a}\right)^{\nu-1} e^{-y} \frac{dy}{a} &= \frac{e^{at}}{\Gamma(\nu) a^{\nu}} \int_0^{at} y^{\nu-1} e^{-y} dy = E_t(\nu, a) \end{aligned} \tag{20}$$

For positive orders, the same definition gives

$$ {}_0D_t^{\nu}e^{at} = D^n {}_0D_t^{\nu-n}e^{at} = \frac{d^n}{dt^n}E_t(n-\nu,a) = E_t(-\nu,a), \nu \in \Re^+ \wedge n = \min\{k \in N : k > \nu\} $$

If $v = 0$, we have:

$$ E_t(0,a) = \sum_{k=0}^{+\infty} \frac{(at)^k}{\Gamma(k+1)} \tag{21} $$

which is the series development of e^{at}.
Finally, the Laplace transform of Et is:

$$ \ell[E_t(\nu,a)] = \frac{1}{s^{\nu}(s-a)} \tag{22} $$

The Convolution theorem:

$$ \ell[\int_0^t f(t-\tau)g(\tau)d\tau] = \ell[f(t)]\ell[g(t)] \tag{23} $$

For negative orders, applying the convolution theorem (23) and (19) we obtain

$$ \ell[E_t(\nu,a] = \frac{1}{\Gamma(\nu)}\ell[t^{\nu-1}]\ell[e^{at}] = \frac{1}{s^{\nu}(s-a)} \tag{24} $$

For positive orders, applying the Laplace transform and we have:

$$ \ell[E_t(-\nu,a)] = \ell[\frac{d^n}{dt^n}E_t(n-\nu,a] = \frac{s^n}{s^{n-\nu}(s-a)} = \frac{1}{s^{-\nu}(s-a)} \tag{25} $$

And when $\nu = 0$, we find:

$$ \ell[E_t(0,a)] = \ell[e^{at}] = \frac{1}{s-a} \tag{26} $$

3.3 Approximation of fractional order

Approximation of Fractional Order Derivative and Integral There are many different ways of finding such approximations but unfortunately it is not possible to say that one of them is the best, because even though some are better than others in regard to certain characteristics, the relative merits of each approximation depend on the differentiation order, on whether one is more interested in an accurate frequency behaviour or in accurate time responses, on how large admissible transfer functions may be, and other factors such like these. For that reason this section shall present several alternatives and conclude with a comparison of them.

Approximations are available both in the s-domain and in the z-domain. The former shall henceforth be called continuous approximations or approximations in the frequency domain; the latter, discrete approximations, or approximations in the time domain.
There are 32 approximation methods for fractional order derivative and integral, we present here Crone approximation method [10, 11].

3.3.1 Crone approximation method

The Crone methodology provides a continuous approximation, based on a recursive distribution of zeros and poles. Such a distribution, alternating zeros and poles at well-chosen intervals, allows building a transfer function with a gain nearly linear on the logarithm of the frequency and a phase nearly constant being possible for the values of the slope of the gain and of the phase for any value of ν [12-14].
The functions we are dealing with in this section provide integer-order frequency-domain approximation of transfer functions involving fractional powers of s.
For the frequency-domain transfer function C(s) which is given by:

$$C(s) = Ks^{v} \qquad \nu \in \Re \tag{27}$$

One of the well-known continuous approximation approaches is called Crone. Crone is a French acronym which means 'robust fractional order control'. This approximation implements a recursive distribution of N zeros and N poles leading to a transfer function as (28).

$$C(s) = K' \prod_{n=1}^{N} \frac{1+\dfrac{s}{\omega_{zn}}}{1+\dfrac{s}{\omega_{pn}}} \tag{28}$$

Where K' is an adjusted gain so that both (26) and (27) have unit gain at 1 rad/s. Zeros and poles have to be found over a frequency domain [ω_l, ω_h] where the approximation is valid, they are given for a positive v, by (29), (30) and (31).

$$\omega_{z1} = \omega_l \sqrt{\eta} \tag{29}$$

$$\omega_{pn} = \omega_{z,n-1}\alpha \qquad n = 1 ... N \tag{30}$$

$$\omega_{zn} = \omega_{p,n-1}\eta \qquad n = 2 ... N \tag{31}$$

Where α and η can be calculated thanks to (32) and (33).

$$\alpha = \left(\frac{\omega_h}{\omega_l}\right)^{\frac{v}{N}} \tag{32}$$

$$\eta = \left(\frac{\omega_h}{\omega_l}\right)^{\frac{1-v}{N}} \tag{33}$$

For negative values of ν, the role of the zeros and the poles is swapped. The number of poles and zeros is selected at first and the desired performance of this approximation depends on the order N. Simple approximation can be provided with lower order N, but it can cause ripples in both gain and phase characteristics. When $|\nu|>1$, the approximation is not satisfactory. The fractional order ν usually is separated as (34) and only the first term s^{ρ} needs to be approximated.

$$s^{v} = s^{\rho} s^{n}, \quad v = n + \rho, \quad n \in \Re, \quad \rho \in [0,1] \tag{34}$$

3.4 Bode and Nichols plots of s^{ν} for real orders

The frequency response of s^{ν} is:

$$\begin{aligned} &F(j\omega) = (j\omega)^{\nu} \\ &|F(j\omega)| = \left|j^{\nu}\omega^{\nu}\right| = \left|\omega^{\nu}\right| = \omega^{\nu} \\ &\arg[F(j\omega)] = \arg(j^{\nu}\omega^{\nu}) = \arg(j^{\nu}) \end{aligned} \tag{35}$$

Now there are several complex numbers z with different arguments such that $z = j^{\nu}$; by choosing the one with a lower argument in interval $[0; 2\pi[$, we will obtain:

$$\arg[F(j\omega)] = \nu\pi / 2 \tag{36}$$

The gain in decibel shall be

$$|F(j\omega)| = 20\log\omega^{\nu} = 20\nu\log\omega \quad (dB) \tag{37}$$

Thus the Bode and Nichols plots of $F(s) = s^{\nu}$ are those shown in Figure 8 and Figure 9:

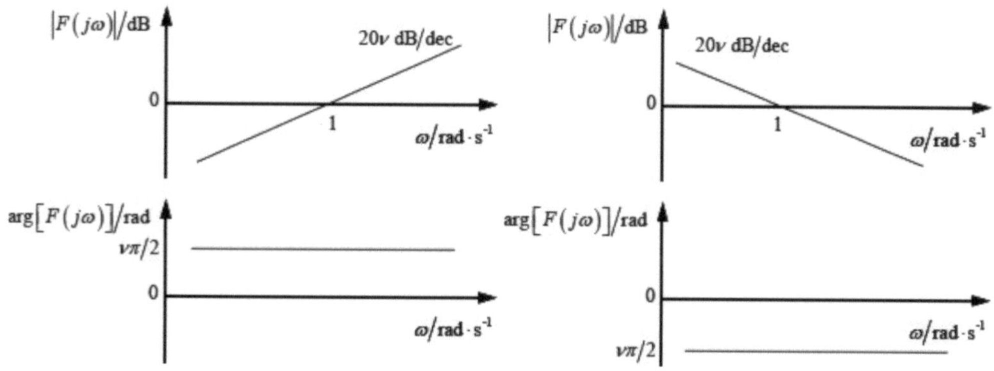

Fig. 8. Bode diagrams

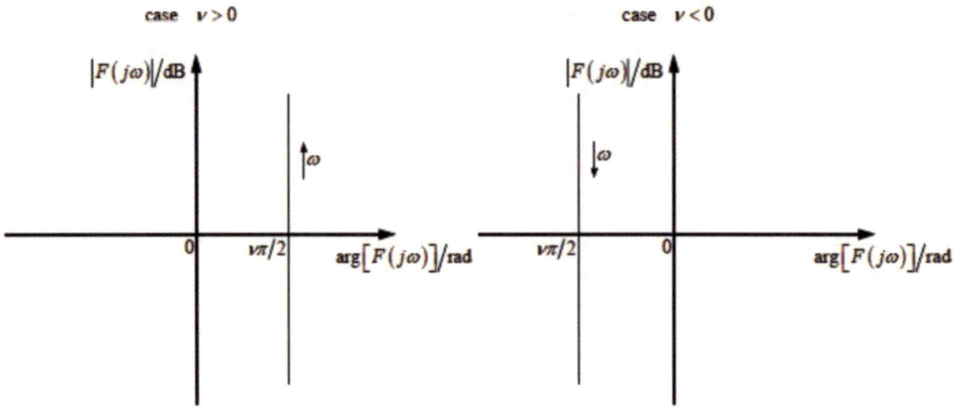

Fig. 9. Nichols diagrams

4. Conclusions

Manny industrial processes are nonlinear and thus complicate to describe mathematically. However, it is known that a good many nonlinear processes can satisfactory controlled using PID controllers providing that controller parameters are tuned well. PID controller and its different types such as P, PI and PD controllers are today basic building blocks in control of various processes. In spite their simplicity; they can be used to solve even a very complex control problems, especially when combined with different functional blocks, filters (compensators or correction blocks), selectors etc. One of the ways to improve the traditional PID controllers is to use fractional order controllers with non integer derivation and integration parts. However, the difficulties of designing Fractional Order PID (FOPID) is relatively higher because these controllers include derivative order and integer order in comparison with traditional PID controllers. As for a linear system, if the dynamic characteristics are basically the same using either integer-order PID controller or FOPID controller, the result of using FOPID controller may provide a better robustness. We get the optimal control with a FOPID than a conventional PID controller.

5. Annex

Controller type	K_p	K_i	K_d	$C(s)$
P (Proportional)	≠0	zero	zero	K_p
I (Integral)	zero	≠0	zero	$\frac{K_i}{s}$
PI (Proportional plus Integral)	≠0	≠0	zero	$\frac{K_p s + K_i}{s} = \frac{K_p\left(s + \frac{K_i}{K_p}\right)}{s}$
PD (Proportional plus Derivative)	≠0	zero	≠0	$K_d s + K_p = K_d(s + \frac{K_p}{K_d})$
PID (Proportional + Integral + Derivative)	≠0	≠0	≠0	$\frac{K_d s^2 + K_p s + K_i}{s} = \frac{K_d\left(s^2 + \frac{K_p}{K_d}s + \frac{K_i}{K_d}\right)}{s}$

V	Approximated transfer function for an FOPID
0.1	$\frac{1584.8932(s+0.1668)(s+27.83)}{(s+0.1)(s+16.68)(s+2783)}$
0.2	$\frac{79.4328(s+0.05623)(s+1)(s+17.78)}{(s+0.03162)(s+0.5623)(s+10)(s+177.8)}$
0.3	$\frac{39.8107(s+0.0416)(s+0.3728)(s+3.34)(s+29.94)}{(s+0.02154)(s+0.1931)(s+1.73)(s+15.51)(s+138.9)}$
0.4	$\frac{35.4813(s+0.03831)(s+0.261)(s+1.778)(s+12.12)(s+82.54)}{(s+0.01778)(s+0.1212)(s+0.8254)(s+5.623)(s+38.31)(s+261)}$
0.5	$\frac{15.8489(s+0.03981)(s+0.2512)(s+1.585)(s+10)(s+63.1)}{(s+0.01585)(s+0.1)(s+0.631)(s+3.981)(s+3.981)(s+25.12)(s+158.5)}$
0.6	$\frac{10.7978(s+0.04642)(s+0.3162)(s+2.154)(s+14.68)(s+100)}{(s+0.01468)(s+0.1)(s+0.631)(s+4.642)(s+31.62)(s+215.4)}$
0.7	$\frac{9.3633(s+0.06449)(s+0.578)(s+5.179)(s+46.42)(s+416)}{(s+0.01389)(s+0.1245)(s+1.116)(s+10)(s+89.62)(s+803.1)}$
0.8	$\frac{5.3088(s+0.1334)(s+2.371)(s+42.17)(s+749.9)}{(s+0.01334)(s+0.2371)(s+4.217)(s+74.99)(s+1334)}$
0.9	$\frac{2.2675(s+1.292)(s+215.4)}{(s+0.01292)(s+2.154)(s+359.4)}$

Table 2. Approximation of $1/S^v$ for different v values

6. References

[1] K. Astrom, K. and T.Hagglund, ''PID Controllers: Theory, Design, and Tuning'', Instrument Society of America, ISBN 1-55617-516-7, 1995

[2] Barbosa, Ramiro S.; Machado, J. A. Tenreiro; FERREIRA, Isabel M, A fractional calculus perspective of PID tuning. In Proceedings of ASME 2003 design engineering technical conferences and Computers and information in engineering conference. Chicago: ASME, 2003.

[3] Barbosa, Ramiro S.; Machado, J. A. Tenreiro; Ferreira, Isabel M, Tuning of PID controllers based on Bode's ideal transfer function. Nonlinear dynamics. 38 (2004a) 305-321.

[4] D. Maiti, A. Acharya M. Chakraborty, A. Konar, R. Janarthanan, "Tuning PID and Fractional PID Controllers using the Integral Time Absolute Error Criterion" 4th International Conference on Information and Automation for Sustainability, 2008 , pp. 457-462

[5] Z. Yongpeng , Sh. Leang-San , M. A.Cajetan , and A. Warsame, '' Digital PID controller design for delayed multivariable systems'', Asian Journal of Control, Vol. 6, No. 4, Dec 2004, pp. 483-495.

[6] Pierre, D.A. and J.W. Pierre, "Digital Controller Design-Alternative Emulation Approaches," ISA Trans.Vol. 34, No. 3, (1995), pp. 219-228.

[7] D. Xue, C. N. Zhao and Y. Q. Chen, "Fractional order PID control of a DC-motor with an elastic shaft: a case study," Proceedings of American Control Conference, pp. 3182-3187, June 2006.

[8] L. Debnath, "A brief historical introduction to fractional calculus," Int. J. Math. Educ. Sci. Technol., vol. 35, no. 4,pp. 487-501, 2004.

[9] Y. Q. Chen, D. Xue and H. Dou, "Fractional Calculus and Biomimetic Control," Proc. of the First IEEE Int. Conf. on Robotics and Biomimet- ics (RoBio04), pp. robio2004-347, August 2004.

[10] D. Valrio, "Fractional Robust System Control," PhD thesis, Instituto Superior Tcnico, Universidade Tcnica de Lisboa, 2005.

[11] TSENG, Chien-Cheng ,"Design of fractional order digital FIR differentiators". IEEE signal processing letters. 8:3 (2001) 77-79.

[12] Concepcion Alicia Monje Micharet. "Design Methods of Fractional Order Controllers for Industrial Applications". Ph.D. thesis, University of Extremadura, Spain, 2006.

[13] Youxin Luo, Jianying Li, "The Controlling Parameters Tuning and its Application of Fractional Order PID Bacterial Foraging-based Oriented by Particle Swarm Optimization" IEEE International Conference on Intelligent Computing and Intelligent Systems, 2009, pp. 4-7.

[14] Tehrani, K.A. Amirahmadi, A. Rafiei, S.M.R. Griva, G. Barrandon, L. Hamzaoui, M. Rasoanarivo, I. Sargos, F.M.," Design of fractional order PID controller for boost converter based on Multi-Objective optimization", Power Electronics and Motion Control Conference (EPE/PEMC), 2010 14th International, Ohrid 6-8 Sept. 2010,pp: 179-185.

Part 7

Extended Applications of PID

10

An Innovative Systematic Approach to Financial Portfolio Management via PID Control

Gino Gandolfi[1], Antonella Sabatini[1,2] and Monica Rossolini[3]
[1]*University of Parma*
[2]*M.I.T.*
[3]*"Banking and Finance" Tor Vergata University, SDA Bocconi*
[1,3]*Italy*
[2]*USA*

1. Introduction

Portfolio management is the art and science of modifying the asset allocation of a financial portfolio in response to and/or in anticipation of market conditions and dynamics of financial markets. The modification of the asset allocation is obtained by rebalancing and varying the relative weights of the assets comprising the portfolio on a periodic basis. The asset manager considers two distinct portfolios: the financial portfolio subject to his management technique (referred to here as the experimental portfolio, or Portfolio "A"), and a benchmark (or comparison) portfolio called Portfolio "B". The asset manager composes his experimental portfolio, also referred to as the benchmark-based portfolio, following, generally, two different types of strategies: active and passive (indexed) strategy. In this work, we analyze a fundamental aspect of portfolio management: the active asset allocation. The objective of this writing is to illustrate a new asset allocation technique to compose an experimental portfolio, which uses the Proportional, Integral, Derivative (PID) controller aiming to overcome a benchmarked portfolio. Therefore, the two portfolios taken into consideration are the experimental portfolio subject to the PID controlling methodology and a buy-and-hold diversified portfolio as the benchmark portfolio. The technique consists in managing portfolio asset-allocation revisions through PID control, a tool that is highly utilized and implemented in the engineering, industrial processing units and in production plants. The goal is to achieve a good portfolio performance trying to control volatility; in other words, the goal is to obtain good performance of risk adjusted returns. Thus, in finance, financial market assets forming a portfolio or a market benchmark represent the process plant controlled by the PID controller.

A brief literature review covering the comparison between strategic and tactical asset allocation introduces the topic, followed by some examples of tactical asset allocation techniques. Subsequently, this article illustrates how the PID controller functions. Then, it exemplifies the new asset allocation technique, functioning, and methodology. This work shows how a portfolio managed by this new technique attains fine results of risk adjusted returns compared with a benchmark.

2. Strategic and tactical asset allocation

Asset allocation can be defined as the action of allocating the various components of a financial portfolio in different asset classes according to the investor risk/return profile level. The portfolio construction is an articulated process based on the identification of the optimal asset mix, given a desired time horizon (holding period) and given the investor's risk aversion level.
The activity of asset allocation is a 3-phase procedure: analysis of investors' needs, consideration of investor's choices and inclinations, and investor's portfolio performance monitoring. At first, it is necessary to analyze investor's needs in order to understand his/her risk aversion level. The investment subsequent choices depend on the latter analysis, which is not so straightforward and easy to perform. The second phase, illustrated in more detail in the following sections, consists in the actual choice of the asset classes in which to invest, the determination of the relative weights assigned to each asset class and the choice of the securities to be bought and included in the portfolio management process. The third phase consists in the monitoring of the portfolio performance through the utilization of specific indicators enabling the observation of the return and the risk of the managing activity. In this phase, the risk-adjusted return indices (Sharpe Ratio, Sortino Ratio, etc) become important; they specify the return of the portfolio adjusted by the implicit and inherent risk underlying that specific asset management strategy.
As specified herein, the central activity of asset allocation is strictly bound to the investment choices. The portfolio manager first defines the macro asset classes to be considered. The macro asset classes are a set of financial activities or real activities with adequate future potential growth. Upon the definition of such macro asset classes, relative weights shall be determined strategically in order to obtain a diversified portfolio consistent and in line with the return/risk profile of the investor. This asset allocation can be achieved by using quantitative strategies, such as the implementation and utilization of Markowitz's efficient frontier technique (Markowitz, 1952), or qualitative approaches and methodologies based on the individual managers' expectations, experience, and estimates on future market conditions. This primary activity of asset allocation is called strategic asset allocation.
The definition of strategic asset allocation is a component of asset allocation, implemented by the identification of the optimal long-term mix, in compliance with the investor risk/return profile.
A second component of asset allocation is defined as the tactical asset allocation. This is an activity that aims to take, periodically, the most interesting investment opportunities by temporarily and partially deviating from the main strategic portfolio structure.
If in the long term, the adherence to investors' risk profile levels must be maintained; in the short term, the tactical asset allocation manager may deviate from the strategic asset allocation technique aiming to take further advantage from certain market conditions. For example, the tactical asset allocation manager may slightly vary the weights of the various asset classes or the individual securities contained in them, targeting to further increase portfolio returns.
Relative to strategic asset allocation, a fundamental choice to make is the adoption of a particular style of management relative to a benchmark. In defining the strategic asset allocation, the manager must decide which style of management to use relative to a benchmark. In fact, managers differentiate between active and passive strategies by analyzing the portfolio management strategy compared to a benchmark. Passive strategies aim to obtain benchmark returns, structuring a portfolio analogous to the benchmark composition. The asset manager chooses the same asset classes and the same relative or absolute weights as the benchmark. In this case, the risk/return profile level is consistent with the benchmark

risk/return level. On the contrary, an active strategy aims to reach an active return compared to the benchmark. The active manager can select different asset classes relative to the benchmark, or different weights. In this case, it is the manager's responsibility to construct the portfolio based on his expectations. In literature, a vivid debate about the superiority of passive vs. active strategies and vice versa, comes forwards. The issue starts with the Efficient Market Hypothesis (Fama, 1965, 1970). This theory assumes that under strong efficient information conditions, it is not possible to have mispriced securities; all prices in the market are fair and balanced; therefore, it is impossible to outperform the market by using active strategies (Samuelson, 1974). Another important factor to consider is the transaction costs (Sharpe, 1991). In fact, even if active and passive strategies are able to achieve the same returns (market returns), the first strategy has unavoidably a diminished total performance, since transaction costs and research costs worsen the outcome. Normally, many active managers manage portfolios formed by index asset classes and liquidity; hence, outperformance compared to the benchmark results. When the market makes a severe downtrend, active portfolios achieve a better performance than the market thanks to the liquidity portion of the portfolios. Not all authors concur in the use and benefits of active strategies. Some authors (Gruber, 1996; Carhart, 1997) state that the active strategies' outperformance has no persistence and exhibits random behavior. Other authors confirm that active strategies produce an effective investment methodology (Gold, 2004).

In order to implement an active strategy, asset managers can apply different tactical asset allocation methods. Each of these active strategies aims to take opportunities when markets are non-aligned (Anson, 2004). Tactical asset allocation can be defined as "active strategies which seek to enhance performance by opportunistically shifting the asset mix of a portfolio in response to changing patterns of reward available in capital market" (Arnott & Fabozzi, 1988). Tactical asset allocation establishes the variations in the asset weights in a portfolio. The rebalancing is performed at different time intervals: on a monthly basis, quarterly or annually. Tactical asset allocation methodologies can be divided into two macro categories: dynamic asset allocation and pure tactical asset allocation (Sampagnaro, 2006). Dynamic asset allocation consists in a series of modifications following a set of precise rules (algorithms). The manager implements such rules such that the portfolio weight rebalancing allows the manager to achieve a predetermined target: to regain alignment to the strategic asset allocation weights, or to apply portfolio protection strategies (portfolio insurance).

Pure strategies of tactical asset allocation, on the other hand, include all those methodologies in which the manager aims to maximize the absolute return of the portfolio or the relative return of the portfolio compared to a benchmark. The manager could change the portfolio composition by removing securities and adding others, selecting those securities that present the best expected future returns. The manager could also modify the weights of the current securities producing a distance from the original strategic allocation weight determination. In literature, an extensive variation of methodologies to take advantage of financial markets is available. Some authors (MacBeth & Emanuel, 1993) suggest to use dividend yield price/earning ratio and price/book ratio to estimate market overvaluation or undervaluation. Others use the spreads between the earning/price ratio of the S&P 500 index and interest rates (Shen, 2003), or present the use of Beta drivers to decide the exposure to the financial market and Alpha drivers to underweight or overweight relative to the benchmark (Anson, 2004). As a final point, a research paper (Gandolfi et al., 2007) pioneers an innovative tactical asset allocation technique. The novelty embedded in this model consists in the application of the well-known PID feedback controlling mechanism, used in industrial plant production and engineering, to tactical financial portfolio asset allocation. The goal of their model was to attain long-term performance steadiness over time by controlling the risk adjusted return variable of portfolios. The main attribute to perceive was the achieved constancy and consistency of the Sharpe Ratio of the experimental portfolio (i.e. the portfolio managed by the PID methodology) in comparison to the benchmark. In the present work, the authors build up a new application based on this novel strategy. The target here is to seek a portfolio (Portfolio "A") capable of enhanced long-term risk adjusted performance and risk stability than the Buy-and-Hold portfolio (Portfolio "B").

3. The PID controller acting on the experimental portfolio

The most important attributes of the PID controller are illustrated in this section. It is vital to understand the functioning of this engineering feedback system since it underlies and stands at the basis of the new asset allocation technique presented herein. The PID (Proportional-Integral-Derivative) controller is broadly used and implemented in several industrial production plants; "it is been successfully used for over 50 years and it is used by more than 95% of the plants processes. It is a robust and easily understood algorithm, which can provide excellent control performance in spite of the diverse dynamic characteristics of the process plant" (Gandolfi et al., 2007). In industrial environments such as chemical plants, power plants, and engineering industries, numerous processes need to be accurately controlled to conform to the required specifications of the resulting products. PID control is straightforward, easily implementable method, still currently preferred by engineers and scientists to more complex systems (Skogestad, 2010). In finance, financial market assets comprising a portfolio or a market benchmark represent the process plant, controlled by the PID controller.

The PID controller is a feedback system. It has an input and returns an output. An iterative process forms it. The inputs of the system are the set-point, or desired value, and the controlled variable that is subject to the effect of the PID controller. The PID controller, working on the input variable, returns as output the same variable operated on by the PID operators. The output variable, in turns, is fed back as an input during the following iteration. The simplest and most basic PID control is formed by the linear combination of three components: the Proportional (P), Integral (I), and Derivative (D) components. During each iteration, the current output is compared to the set-point yielding an error. The goal of the PID control is to diminish this error to the minimum (Gandolfi et al., 2007). The continuous time expression of the PID controller is given by:

$$u(t) = k_p\left(e(t) + k_i \int e(\tau)d\tau + k_d \frac{de(t)}{dt} \right)$$

where:

$u(t) = \text{output}$ (1)

$k_p = \text{Proportional Constant}$

$k_i = \text{Integral Constant}$

$k_d = \text{Derivative Constant}$

$e(t) = \text{error}$

In this present work, the following recurrence relation, obtained by discrete time formulation and simple-lag implementation of the integral part (Gandolfi et al., 2007) yields:

$$u_n = \left(k_p e_n + k_i(e_n - u_{n-1}) + u_{n-1} + k_d(e_n - e_{n-1})\right) \tag{2}$$

where:

u_n=output at time n

u_{n-1}=output at time n-1

k_p=Proportional Constant

k_i=Integral Constant

k_d=Derivative Constant

e_n=error at time n

e_{n-1}=error at time n-1

A block diagram of the PID controller follow:

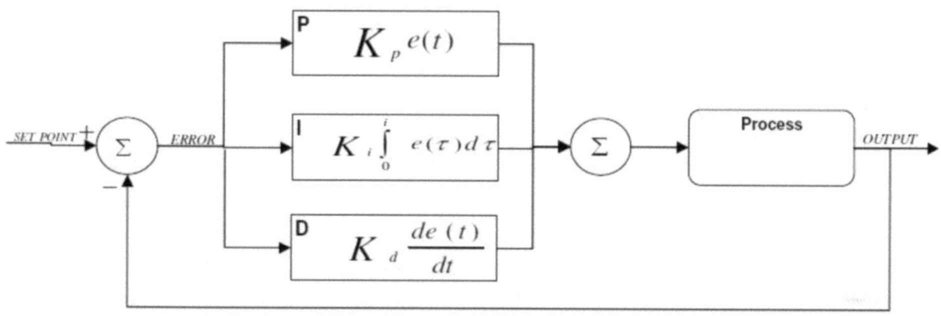

Fig. 1. PID control block diagram - This figure presents dynamics and processing of the error, Set-Point and controlled variable while subjected to the PID control action.
Set-point = Desired value. Error = (Output – Set-Point).

4. Mechanisms of action of the new asset allocation technique

This section presents an original method and system for allocating numerous assets in portfolios, via tactical asset allocation in order to achieve better return and long-term target stability (volatility control) over a desired time horizon. In particular, the present work illustrates a method and system for asset allocation of the 20 securities having each one, its own level of risk and return. The methodology consists in stabilizing the portfolio return . hence the decreasing of portfolio volatility based on the PID feedback control. By applying our strategy to a financial portfolio, financial market assets represent the process plant, controlled by PID controlling action. The assets mix of the portfolio determines the total portfolio return. The action of rebalancing the portfolio alters its return. In various aspects, this work offers methods and systems as an innovative approach to active strategy portfolio management. It is worth noting that the rebalancing of the experimental portfolio (Portfolio "A") is not dictated by a forecast analysis of the various prices of the assets belonging to the portfolio. There is no use of a vector of expected returns and there is no need of determining a variance-covariance matrix. The rebalancing is rather driven by an asset selection technique consisting in the stabilization of return by means of the PID feedback control modeling procedure. The new model simply tends to follow and not predict the financial market oscillations and market variability, adjusting to such variations and oscillations. It takes into consideration past and current portfolio dynamics. It tunes to financial market fluctuations by performing smoothing and anticipatory actions in the attempt to hold as close as possible to the target, hence minimizing the error generated by the difference between the set-value and current portfolio return. The controlled process plant, namely the return variable, does not need to be modeled or defined by a mathematical closed form equation; assumptions, linearization, and simplification procedures on the dynamics of the plant are not required. The PID control modifies the portfolio asset weights, according to the PID algorithm. The methodology starts by presenting two initially identical portfolios: the benchmark, namely Portfolio "B", and the experimental portfolio, or Portfolio "A". The procedure uses a 12-year monthly frequency time-series per each of the securities of the Portfolio "B", covering the period February 1999 - February 2011. Portfolio "A" assets are rebalanced at the end of each month, according to the PID procedure. At the end of the observation period, namely in February 2011, the two portfolios, the Benchmarked Portfolio "B" and the experimental portfolio, Portfolio "A", are observed and compared, targeting to verify the efficiency of the new model compared to benchmarking. In this work, the comparison is carried out without taking into consideration tax and transaction costs. Portfolio "B" , namely the benchmark is composed by 20 assets chosen in such a way to form a well diversified portfolio. In particular, the following assets have been considered: a monetary index, 4 fixed-income (or bonds) indices, 7 stocks (equity) indices, 6 commodities indices, gold and a risk-free asset class denominated "cash". The inclusion and use of a risk-free asset in the experimental portfolio is been indicated by the consideration that the new model permits partial disinvestment of the risky portfolio by partially reallocating risky assets in risk-free assets (Qian, 2003). The following table illustrates how the strategic asset allocation of the well-diversified portfolio has been defined. The right-end-side column indicates the respective weights of each asset class:

Asset class	Weight
Monetary	6%
Bonds	40%
Equity	35%
Commodities	12%
Gold	5%
Cash	2%

Table 1. Strategic composition for macro-asset class of Portfolio "B". The table illustrates Portfolio "B" composition, namely the benchmark composition. It specifies the various macro-asset classes and their relative assigned weights.

After having presented which the strategic macro-asset classes are, for the benchmark portfolio, the following table is presented. It exhibits for each asset class, which are the selected indices in order to form the well diversified portfolio with its respective assigned weights.
Firstly, Portfolio "A" has the identical composition as that of Portfolio "B". Next, Portfolio "A" asset weights are varied following the PID signals. The rebalancing occurs on a monthly

basis. The constraints for rebalancing are the following: every asset can take on a minimum or a maximum weight within the portfolio. The minimal weight has been defined to be equal to 1% and the maximal weight has been defined to be equal to 20% under the effect of the PID control action.

The set-point value of this procedure, in order for the new model to achieve its target, is set to be equal to 0.5% monthly target portfolio return. The mechanism of action of this model is similar to a dynamic Exchanged Traded Fund (ETF), replicating an index in terms of underlying assets. On the opposite, it is different in terms of relative weights and, therefore, the model is a dynamic strategy.

The algorithm and implementation of the new model is the outlined in the following steps, using the expression:

Asset class	Index	Weight
Monetary	Deutsche Borse EUROGOV Germany Money Market (TR)	6%
Bonds	iBoxx Euro Index World Wide Performance Overall	10%
	Market iBoxx € Financials Total Return Index	10%
	Market iBoxx € Non Financials Total Return Index	10%
	Market iBoxx € Euro Sovereign Overall Total Return Index	10%
Equity	MSCI Daily TR Gross Europe Local Currency	5%
	STOXX 600 Total Return Index EUR	5%
	STOXX Style Index TMI Growth Return Index EUR	5%
	STOXX Europe Total Market Value (Net Return) EUR	5%
	MSCI Daily TR Gross Total Return World USD	5%
	MSCI Emerging Markets Daily Gross Total Return USD	5%
	MSCI Daily TR Gross North America Total Return USD	5%
Commodities	S&P GSCI Tot Return Indx	2%
	S&P GSCI Energy Tot Ret	2%
	S&P GSCI Industrial Metals Index Total Return	2%
	S&P GSCI Agricultural Index Total Return CME	2%
	S&P GSCI Livestock Index Total Return.	2%
	S&P GSCI Crude Oil Total Return CME	2%
Gold	S&P GSCI Gold Index Total Return	5%
Cash	Out of the market	2%

Table 2. Strategic Portfolio "B" composition: index specification. The table presents, for any macro-asset class, the specification of which particular selected indices form each macro-asset class. Furthermore, the relative weights are indicated.

$$return_n = \left(k_p e_n + k_i(e_n - return_{n-1}) + return_{n-1} + k_d(e_n - e_{n-1})\right)$$

where:

$return_n$=output= return at time n

u_{n-1}=output= return at time n-1

k_p=Proportional Constant = 0,5 (3)

k_i=Integral Constant = 0,6

k_d=Derivative Constant = 0,5

e_n=($return_n$ - 0,005) at time n

e_{n-1}=($return_{n-1}$ - 0,005) at time n-1

SetPoint = desired return = 0,005

- Define set-point = Desired Return = 0,005.
- Calculate portfolio return (controlled variable), $return_0$, for the initial portfolio, given current market conditions.
- At each iteration n, the PID controller designates a controlled value for the portfolio return, called $return_n$ given by equation [3]. The making of such rebalancing is necessary to minimize the error between current return (determined by current market conditions) and $return_n$ and set-point. Since the objective is to reduce the error, e_n defined by the difference between current return, $return_n$, and the set-point or desired return defined as 0,005, each iteration contributes in reducing e_n. The error decrease is generally counteracted by the dynamics of the markets. Given ideal market conditions, e_n approaches zero after the transient system response has died out.
- New market data acquisition and corresponding portfolio return, $return_n$, is calculated at end of each period (monthly).
- The previous items are iteratively re-executed until the end of the observation period.
- The PID parameters, chosen to be constant for all market conditions, are set to be:
  K_p = 0,5
  K_i = 0,6
  K_d = 0,5
- In this work, the parameters values were set according to an empirical criterion: under risk-free market conditions (portfolio with zero exposure to financial markets), the selection of a transient time domain response with a slight oscillatory response, exhibiting reasonable overshoot, and approaching set-point value within a small number of iterations was adopted.
- The objective of each iterations is to make returns as stable and consistent as possible given the contributions and interactions of the controller and the market dynamics influence. The change in asset mix is dictated by the controller indications and the market behavior of the underlying securities.

The main results of this methodology are illustrated in the following paragraph.

5. Portfolio "A" vs. Portfolio "B"

This section recapitulates the main results of the new model comparing the returns of Portfolio "A" to the returns of Portfolio "B". The comparison is performed in terms of return and volatility for the observation period.

Table 3 illustrates information about return and volatility. Portfolio "A" has an annualized return of 7,25% compared to 5,14% of Portfolio "B". The cumulative return in the observation period (1999-2011) is 86,96% for Portfolio "A" and 61,66% for the benchmark. In terms of portfolio risk, the experimental portfolio realizes an annualized volatility of 7,93%, indicatively in line and consistent with 7,01% recorded by Portfolio "B". Portfolio "A", with only a slightly higher volatility, is able to obtain more satisfying results both in annualized and in cumulative data analysis.

	Portfolio "A"	Portfolio "B"
Annualized Return	7,25%	5,14%
Cumulative Return	86,96%	61,66%
Annualized Volatility	7,93%	7,01%

Table 3. Return and Volatility data. This table presents the comparison of annualized return, cumulative return and annualized volatility of Portfolio "A" and Portfolio "B". Period of observation: February 1999-February 2011.

After having analyzed the data in the observation period, it is considered interesting to analyze the data on a monthly basis.

Table 4 demonstrates monthly data; scrupulously, it is evident that the mean monthly return of the Portfolio "A"(0,60%) is superior to the Portfolio "B" mean monthly return (0,43%). The set-point or target value for the model was 0,5% monthly; thus, the experimental portfolio reaches the ideal target. The mean monthly volatility for Portfolio "A" is 2,29%, whereas the benchmark (Portfolio "B") exhibits a volatility of 2,02%.

	Portfolio "A"	Portfolio "B"
Mean Monthly Return	0,60%	0,43%
Mean Monthly Volatility	2,29%	2,02%

Table 4. Monthly Return and Volatility information. This table presents the comparison of average monthly returns and average monthly standard deviations of Portfolio "A" and Portfolio "B". Period of observation: February 1999-February 2011.

Table 5 shows, in the first and second column respectively, Portfolio "A" returns and Portfolio "B" returns for each year of the observation period. It is important to specify that each year is considered by counting from February (t-1) to February (t). This allows the yearly periods to be defined by 12 periods of 12 month each one, considering that the given time series starts in February. This table demonstrates that the new model performance is, in most cases, equivalent or better than the benchmark portfolio performance for each analyzed year, except for three years 2004-2005, 2006-2007 and 2009-2010, where Portfolio "A" underperforms Portfolio "B". The third and fourth columns of Table 5 display the annual volatility for the two portfolios. We can see that in many years, the new model presents higher volatility than Portfolio "B", but it is necessary to remember what mentioned herein, that performances are also superior.

	Portfolio "A" Return	Portfolio "B" Return	Portfolio "A" Annual Volatility	Portfolio "B" Annual Volatility
1999-2000	16,05%	19,25%	7,62%	7,49%
2000-2001	6,07%	2,08%	8,67%	6,38%
2001-2002	2,97%	-1,25%	6,11%	6,90%
2002-2003	4,54%	-8,71%	7,90%	7,63%
2003-2004	8,27%	13,21%	11,94%	5,82%
2004-2005	-1,32%	8,06%	4,70%	2,57%
2005-2006	18,28%	14,74%	8,13%	4,99%
2006-2007	2,19%	3,55%	5,20%	2,84%
2007-2008	9,98%	1,99%	7,91%	4,83%
2008-2009	-16,27%	-24,22%	6,21%	9,94%
2009-2010	19,85%	22,83%	6,77%	5,62%
2010-2011	16,34%	10,13%	5,35%	4,16%

Table 5. Portfolio "A" and Portfolio "B" annual returns and volatilities. This table presents annual returns and annual standard deviations of Portfolio "A" and Portfolio "B" for each observed year. Period of observation: from February 1999 - February 2011.

The following chart illustrates, graphically, the dynamics of volatility of Portfolios "A" and "B". It can be noticed that the continuous line representing the volatility of Portfolio "A" is often higher than that of the benchmark. However, it is interesting to underline the stabilization effect starting from 2004 and becoming evident under the PID control action. As it is well known, this instrument needs a history before it can enable its efficient control action and make it functional.

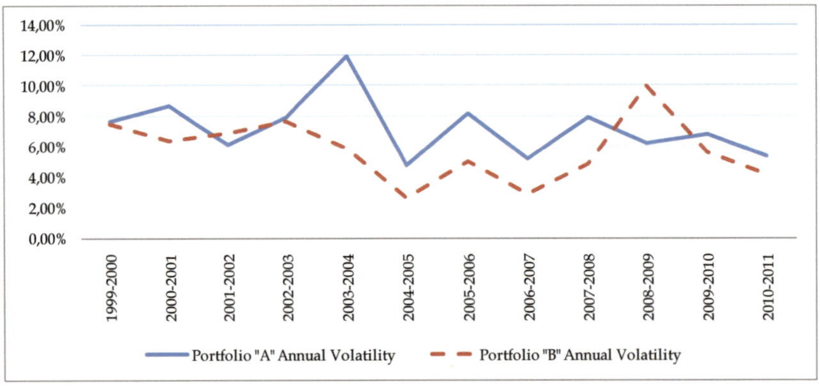

Fig. 2. Annual Volatility of Portfolio "A" and Portfolio "B". This chart presents the annual volatility dynamics of the two portfolios in the observation period February 1999-February 2011. The continuous line represents Portfolio "A"; the dotted line represents Portfolio "B".

After having calculated the return and risk of the two portfolios, a comparison of the two portfolios is performed by using a risk adjusted return indicator, the Sharpe Ratio. This indicator is defined as the ratio of the difference between return and risk free return at the numerator, divided by the standard deviation of the portfolio returns. In order to define the risk free rate the average of the Libor values in the 12 years (1999-2011) of observation period are calculated. This calculation has yielded a value equal to 2,80%. The results are depicted in the table below:

	Portfolio "A" Sharpe Ratio	Portfolio "B" Sharpe Ratio
1999-2000	1,74	2,20
2000-2001	**0,38**	Negative
2001-2002	**0,03**	Negative
2002-2003	**0,22**	Negative
2003-2004	0,46	1,79
2004-2005	Negative	2,05
2005-2006	1,90	2,39
2006-2007	Negative	0,26
2007-2008	**0,91**	Negative
2008-2009	Negative	Negative
2009-2010	2,52	3,56
2010-2011	**2,53**	1,76

Table 6. Sharpe Ratio of Portfolio "A" and Portfolio "B". This table presents the results of a risk adjusted return indicator, namely the Sharpe Ratio applied to the two portfolios for every year in the observation period. In bold are illustrated the cases in which Portfolio "A" has outperformed Portfolio "B".

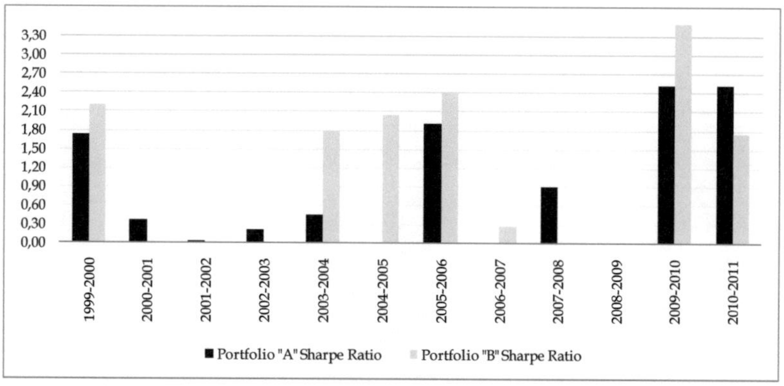

Fig. 3. Sharpe Ratio of Portfolio "A" and Portfolio "B". The chart presents, for each observation period, the Sharpe Ratio values of the two portfolios. In particular, in black the values belonging to Portfolio "A" are represented. In grey, the corresponding values for Portfolio "B" are illustrated. The absence of a column shows that the indicator value is negative, hence non-interpretable.

When in table 6, the word "Negative" is present, it means that for that specific year, it was not possible to record the indicator due to its negative value. The Sharpe Ratio is not defined for negative values. Hence, the Sharpe Ratio becomes meaningless since a return net of the risk free is negative. The case of a negative numerator in the Sharpe Ratio formulation can occur in two situations: when the portfolio return for that period is negative, or when the portfolio return for that period is positive but inferior to the risk free rate of return.

The analysis of table 6 allows the reader to notice that Portfolio "A" is able to obtain better results than Portfolio "B" in 5 instances out of 11 (the observation for year 2008-2009 is eliminated since both portfolios have negative Sharpe Ratios). The consistent returns of Portfolio "A" in many cases, allow the overcoming of the risk free return when Portfolio "B" is not able to do so; hence, Portfolio "B" presents negative Sharpe Ratios (examples in the range 2000-2003).

Figure 3 represents the trend of Sharpe Ratios of the 2 portfolios.

When a column of one of the two portfolios is not visible, it means that one of the two values is negative.

It was considered interesting to investigate another risk adjuster return indicator: Sortino. This indicator of risk adjusted return, is defined as the ratio of the difference between the return and the risk free return, and, at the denominator, a risk measure defined as the Down Side Risk (DSR). The Down Side Risk is a measure of risk that considers only the volatility of the returns inferior to the risk free return. By calculating the Down Side Risk, we investigated the type of reduced risk, up or downside risk. We have analyzed if the new model acts more successfully in decreasing positive risk or downside risk.

	Portfolio "A" Annual DSR	Portfolio "B" Annual DSR
1999-2000	8,84%	8,19%
2000-2001	11,44%	11,06%
2001-2002	**10,74%**	12,20%
2002-2003	**11,43%**	14,40%
2003-2004	13,81%	8,27%
2004-2005	11,12%	7,81%
2005-2006	8,64%	7,38%
2006-2007	10,45%	9,13%
2007-2008	**10,18%**	10,33%
2008-2009	**15,68%**	19,42%
2009-2010	7,21%	5,76%
2010-2011	**7,09%**	7,87%

Table 7. The Down Side Risk of Portfolios "A" and "B". The table represents for every year in the observation period the comparison between the Down Side Risk of the two portfolios. The DSR is calculated considering the volatility of returns inferior to the risk free rate relative to the risk free rate itself.

The Down Side Risk (DSR) of Portfolio "A" and of Portfolio "B" was calculated and analyzed for this purpose. The main results of this study on downside risk are depicted in Table 7. As illustrated in this table, the new model exhibits a DSR lower than the benchmark in 5 cases out of 12.

This situation is interesting and it is visible in figure 4. It illustrates the stabilization effect of Portfolio "A" on Down Side Risk. The continuous line (new model) tends visibly to smooth out the extreme values better than the movement of the benchmark.

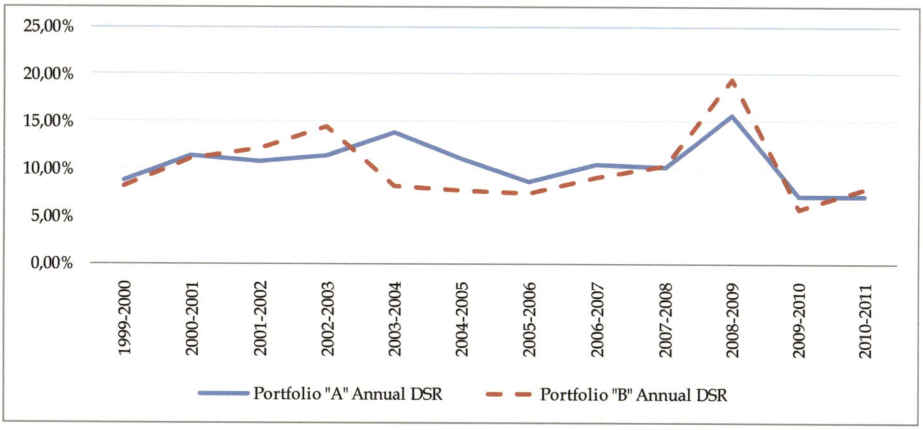

Fig. 4. Comparison between the Down Side Risk of Portfolio "A" and Portfolio "B". In the figure, the continuous line illustrates the DSR Portfolio "A". The dotted line serves for Portfolio "B".

After having calculated the value of the Down Side Risk, it is possible to calculate the risk adjusted return indicator defined above, Sortino ratio. Differently from Sharpe, this indicator has at the denominator, not the standard deviation, hence the volatility of the portfolio, but rather uses the DSR, hence the volatility defined for the returns below the risk free rate. As it can be observed from table 8, Portfolio "A" obtains better results than Portfolio "B" in 6 years out of 11 (the year 2008-2009 is not considered since both portfolios

	Portfolio "A" Sortino	Portfolio "B" Sortino
1999-2000	1,50	2,01
2000-2001	**0,29**	Negative
2001-2002	**0,02**	Negative
2002-2003	**0,15**	Negative
2003-2004	0,40	1,26
2004-2005	Negative	0,67
2005-2006	**1,79**	1,62
2006-2007	Negative	0,08
2007-2008	**0,71**	Negative
2008-2009	Negative	Negative
2009-2010	2,36	3,48
2010-2011	**1,91**	0,93

Table 8. Sortino ratio for portfolios "A" and "B". This table presents the results of the risk-adjusted return Sortino, applied to the two portfolios, for the whole observation period. In bold, the cases when Portfolio "A" over performs Portfolio "B" are highlighted. The indication "Negative" shows the fact that for a negative numerator, the indicator is not defined.

exhibited negative values). This indicates that, selecting the criterion of the most negative of the risk factors, the DSR, (that is the returns inferior to the risk free rate) the new model is bale to guarantee a better performance in comparison to the benchmark.
The following chart allows the visualization of the comparison of the two portfolios. It is to be remembered that when a column is missing, it indicates that its corresponding value is negative. In year 2001-2002, the column of portfolio A since its value is negligible. However, Sortino's value in that year is relevant.

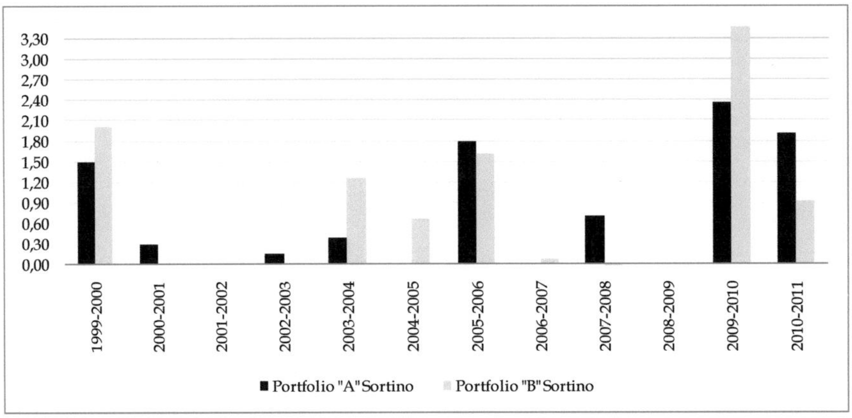

Fig. 5. Sortino ratio of portfolios "A" and "B". The chart represents per each year of observation, Sortino values for the two portfolios. In particular, in black the results for Portfolio "A" are represented. In grey, the results of Portfolio "B" are illustrated. The absence of a column indicates that the indicator value is negative, hence non-interpretable.

If Sortino and Sharpe Ratio results are compared it is evident the ability of Portfolio "A" to better perform in comparison of Portfolio "B". Since the difference between Sortino and Sharpe resides in the definition of the denominator portion of the formula, it is apparent that Portfolio "A" acts more efficiently on the DSR than on the total volatility. Hence, this selectivity capability of the model is a good feature. The PID control action on financial portfolios seems to function as a stabilizer of returns. Above all, it diminishes the worst component of the returns, namely the ones inferior to the risk free rate.

6. Conclusion

This work illustrates a portfolio management model with the aim to obtain good returns and decrease portfolio risk through stabilization of returns, by means of the PID control applied to pure returns. As demonstrated in the previous sections, the new model is able to obtain returns that are satisfactory in the observation period. In addition, it is able, in about half of the analyzed cases, to diminish the volatility relative to the benchmark. In particular, the best results are exhibited when the Down Side Risk is considered instead of the whole volatility. The results illustrated herein relative to the Down Side Risk are of a good quality. The new model, through asset rebalancing, in the observation period, successfully reduces the negative volatility factor in 5 cases out of 11 more than the negative volatility of the benchmark. This research work furthers the analysis of two indicators of risk adjusted returns: Sharpe and Sortino. Confirming and reiterating what just said, Sortino, which uses the DSR in its denominator, obtained the best performances.

Portfolio "A" presents, in 6 years out of 11, a risk adjusted return value for the Down Side Risk better than the benchmark. These initial results confirm that the PID based asset allocation technique seems to be a good instrument, adapt for adverse market conditions. It effectively controls and bounds negative volatility. At the light of the current results herein achieved, the authors desire to further and develop the model in the attempt to seek and understand relations, functions and interacting factors among the managed portfolio characteristics and intrinsic and endogenous parameters of the model, such as the set-point, aiming to maximize returns' stabilization effects.
The authors will further the model verifying and testing its applicability on various financial market indices and diversified portfolios, including the impact of transaction costs. The goal is to confirm broad-spectrum negative volatility controllability, steadiness and performance stabilization for financial portfolio managers.

7. References

Amenc, N., Malaise P. & Martellini, L. (2004). Revisiting Core-Satellite Investing. A dynamic model of relative risk management. *The Journal of Portfolio Management,* Vol 31, No. 1, (Fall, 2004), pp- 64-75, ISSN 00954918.

Amman, M., Kessler, S. & Tobler, J. (2006). Analyzing Active Investment Strategies. Using tracking error variance decomposition. *The Journal of Portfolio Management,* Vol. 33, No.1, (Fall, 2006), pp. 56-67, ISSN 00954918.

Anson, M. (2004). Strategic versus Tactical Asset Allocation. Beta versus alpha drivers. *The Journal of Portfolio Management*. Vol.30, No. 2, (Winter, 2004), pp. 8-22, ISSN 00954918.

Arnott, D. R. & Fabozzi, F.J. (eds) (1988). Asset *allocation: A Handbook of Portfolio Policies, Strategies and Tacties,* Probus Professional Publishers, ISBN 1557380139, USA.

Arshanapalli, B., Switzer, N. L. & Hung, T. S. L. (2004). Active versus Passive Strategies for EAFE and the S&P500. *The Journal of Portfolio Management*. Vol. 30, No. 4, (Summer, 2004), pp. 51-60, ISSN 00954918.

Carhart M. (1997). On persistence in mutual funds performance. *The Journal of Finance,* Vol.52, No. 1, (March, 1997), pp. 57-82, ISSN 0022-1082.

Da Silva, S., A., Lee, W. & Pornrojnangkool, B. (2009). The Black-Litterman Model for Active Portfolio Management. *The Journal of Portfolio Management,* Vol.35, No. 2, (Winter, 2009), pp. 61-70, ISSN 00954918.

Don, P. & Lee, J. (1989) Current issue: Tactical Asset Allocation. *Financial Analyst Journal* Vol. 45, No. 2, (March-April, 1989), pp. 14-16, ISSN 0015-198X.

Faff, R., Gallagher, R. D. & Wu, E. (2005). Tactical Asset Allocation: Australian Evidence. *Australian Journal of Management,* Vol. 30, No. 2, (December, 2005), pp. 261-282, ISSN 1320-5161.

Fama, F. E. (1965). The Behaviour of Stock Market Prices. *The Journal of Business,* Vol. 38, No.1, (January, 1965), pp. 34-105, ISSN 1573-0697.

Fama, F. E. (1970). Efficient Capital Markets: review of theory and empirical work. *The Journal of Finance,* Vol. 25, No. 2, (May, 1970), pp. 383-417, ISSN 0022-1082.

Gandolfi, G., Sabatini, A. & Rossolini, M. (2007) PID feedback controller used as a tactical asset allocation technique: The G.A.M. model. *Physica A,* Vol. 383, No. 1, (September, 2007), pp. 71-78, ISSN 0378-4371.

Gold, L. M. (2004). Investing in pseudo-science: the active versus passive debate. *Journal of the Securities Institute of Australia,* Vol.3, No. 3, (Summer 2004), pp. 2-6, ISSN 0313-5934.

Gruber, J. M. (1996). Another puzzle: the growth in actively managed funds. *The Journal of Finance,* Vol. 51, No. 3, (July, 1996), pp.783-810, ISSN 0022-1082.

MacBeth, J. & Emanuel, C. D. (1993) Tactical Asset Allocation: Pros and Cons. *Financial Analysts Journal,* Vol. 49, No. 6, (November-December, 1993), pp. 30-43, ISSN 0015-198X.

Markowitz, H. (1952). Portfolio Selection. *The Journal of Finance,* Vol. 7, No. 1, (March, 1952), pp. 77-91, ISSN 0022-1082.

Qian, E. (2003). Tactical Asset Allocation with Pairwise Strategies. Using pairwise information to influence weights. *The Journal of Portfolio Management,* Vol.30, No.1, (Fall, 2003), pp. 39-48, ISSN 00954918.

Sampagnaro, G. (Ed.). (2006). Asset Management: tecniche e stile di gestione di portafoglio, Franco Angeli, ISBN 8846472829, Milan.

Samuelson, A. P. (1974). Challenge to Judgment. *The Journal of Portfolio Management,* Vol. 1, No. 1, (Fall, 1974), pp.17-19, ISSN 00954918.

Samuelson, A. P. (2004). The Backward Art of Investing Money. *The Journal of Portfolio Management,* Vol. 30, No. 5, (30th anniversary, 2004), pp- 30-33, ISSN 00954918.

Sharpe, F. W. (1991). The arithmetic of active management. *Financial Analyst Journal,* Vol. 47, No. 1, (January-February, 1991), pp. 7-9, ISSN 0015-198X.

Shen, P. (2003). Market Timing Strategies That Worked. Based on the E/P Ratio of the S&P500 and interest rates. *The Journal of Portfolio Management,* Vol. 29, No. 2, (Winter, 2003), pp. 57-68, ISSN 00954918.

Skogestad, S. (2010) Feedback: still the simplest and best solution. Paper presented at International Conference Cybernetics and Informatics, 10 February, Bratislava, Slovac Republic.

Part 8

Practical Applications

11

Relay Methods and Process Reaction Curves: Practical Applications

Manuela Souza Leite and Paulo Jardel P. Araújo
Tiradentes University (UNIT), Aracaju, Brazil

1. Introduction

Proportional-integral-derivative (PID) controllers are the most adopted controllers in industrial settings because of the advantageous cost/benefit ratio they are able to provide (Astrom & Hanglund, 2006). Its function is very to explain and in most cases it is the easiest controller to adjust. Tuning controllers can significantly improve control performance.
PID controller is to be applied in practical cases. It is seen that many PID variants have been developed in order to improve transient performance, such as biotechnological processes and chemical processes.
Automation and process control can significantly influence the yield and final quality of products. However, there are few studies on the application of automatic controllers in the experimental plants. Most works focus on results obtained from computational simulations, that indeed do not represent these processes in all their complexity. The transient behavior and nonlinearities of these processes make the design of classical control dependent on trial-and-error methodology.
In this context, this topic concerns in show some practical applications of use PID Controller. The development of a design and tuning method for use with PID controllers in experimental processes for temperature control.

2. Tuning methods for pid controller

The primary function of a close-loop system is to make the controlled variable a desired value established by the set-point. Whenever the controlled variable becomes different then the set-point, the objective of the closed-loop system is to make then the same as quickly as possible. The controlled variable becomes different than the set-point under tree conditions:

- Set-point change;
- Disturbance;
- Load demand change.

One of the traditional ways to design a PID controller was to use empirical tuning rules based on measurements made on the real plant. Today is preferable for the PID designer to employ model based techniques. There is a large number of tuning methods, but in this chapter we describes for calculating proper values of the PID parameters (kc, ti, td) two methods: Relay Methods and Process Reaction Curve.

3. Relay methods

To understand the relay method is necessary first to explain the ultimate gain method (Oscillation method) proposed by Ziegler Nichols (Z-N). This procedure is only valid for open loop stable plants and it is carried out through the following steps:

a. Set the true plant under proportional control, with a very small gain;
b. Increase the gain until the loop starts oscillating;
c. Record the controller critical gain $K_p = K_u$ and the oscillation period of the controller output, P_u ;
d. Adjust the controller parameters according to Table 3.1.

	K_p	τ_i	τ_d
P	$0{,}5K_u$		
PI	$0{,}45K_u$	$\frac{1}{1{,}2}P_u$	
PID	$0{,}60K_u$	$\frac{1}{2}P_u$	$\frac{1}{8}P_u$

Table 3.1. Ziegler Nichols tuning using the ultimate gain method

Note that linear oscillation is required and that it should be detected at the controller output. In fact the Ziegler - Nichols tuning scheme, where the controller gain is experimentally determined to just bring the plant to the brink of instability is a form of model identification. This is known as the ultimate gain K_u. Relay-based auto tuning is a simple way to tune PID controller that minimizes the possibility of operating the plant close to the stability limit.

As it turns out, under relay feedback, most plants oscillate with a modest amplitude fortuitously at the critical frequency. The procedure is now the following:

a. Substitute a relay with amplitude d for the PID controller as shown in Figure 3.1;
b. Kick into action, and record the plant output amplitude a and period P (Fig. 3.2).
c. The ultimate period is the observed period, $P_u = P$, while the ultimate gain is inversely proportional to the observed amplitude,

$$K_u = \frac{4d}{\pi a} \tag{3.1}$$

Having established the ultimate gain and period with a single succinct experiment, we can use the Ziegler - Nichols tuning rules (or equivalent) to establish the PID tuning constants.

The Figure 3.1 shows a plant with the PID regulator temporarily disabled and the Figure 3.2 shows a plant oscillating under relay feedback.

The settings in Table 3.1 obtained by Ziegler and Nichols, can be used to make the model response of a PID controller:

$$u_{PID}(t) = K_p e(t) + \frac{K_p}{\tau_i}\int_{to}^{t} e(t)dt + K_p \tau_d \frac{de(t)}{dt} \tag{3.2}$$

Many plants, particularly the ones arising in the process industries, can be satisfactorily described by the model in Equation 3.3.

$$G_0(s) = \frac{K_0 e^{-\tau s}}{\gamma_0 s+1};\ \gamma_0 > 0 \tag{3.3}$$

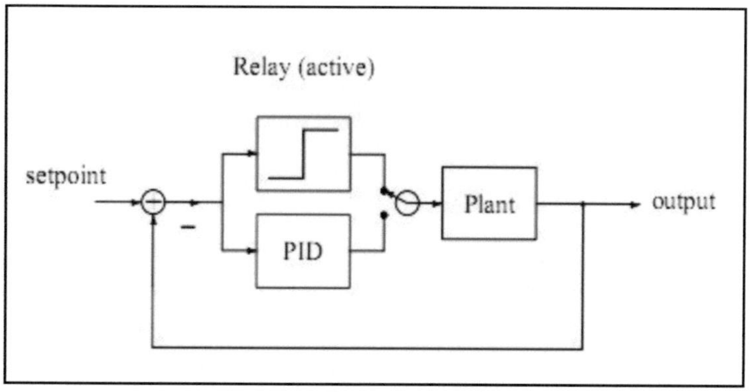

Fig. 3.1. Plant with the PID regulator temporarily disabled

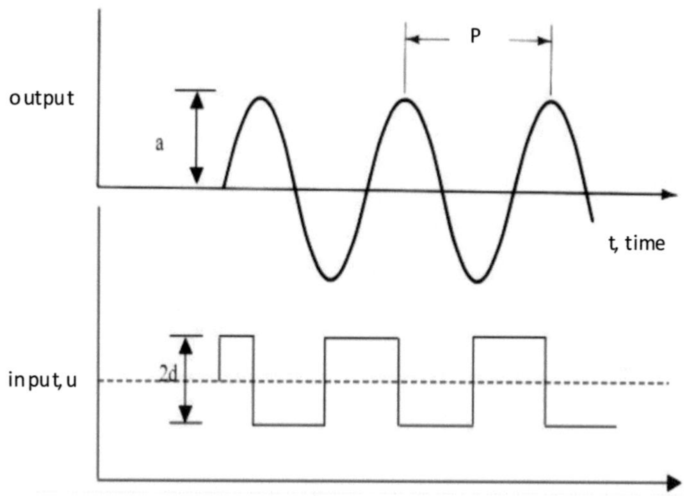

Fig. 3.2. Plant oscillating under relay feedback

The one can obtain the PID settings via Ziegler-Nichols tuning for different values of τ and γ_0. These parameters can be calculated using:

$$\tau_1 = \frac{T_u}{2\pi}\sqrt{(K_u K_p)^2 - 1}\ ;\ \tau_1 = \tau \tag{3.4}$$

$$\theta_1 = \frac{T_u}{2\pi}\left(\pi - \arctan\frac{2\pi}{T_u}\tau_1\right); \tag{3.5}$$

Ku and Tu parameters are obtained from the experiment using the relay method.

3.1 Case study

The use of polymers has been growing gradually in many industrial products, such as: automobile, electronic devices, food packaging, and building and medicine materials. Among these products stands the polystyrene, usually produced in batch or semi-batch reactors.

Temperature variation in polymerization reactor systems greatly affects the kinetics of polymerization and consequently changes the physical properties and quality characteristics of the produced polymer (Ghasem et al., 2007; Lepore et al., 2007). In order to ensure the maintenance of the final product quality is crucial to keep suitable operating conditions during the polymerization reaction process.

3.2 PID controller design

The PID controller is designed for temperature control of an experimental process of polymerization (Leite et al., 2010a; Leite et al., 2011). The developed models will can be online implemented to a pilot plant. A pilot plant was built specifically to evaluate the polymerization reaction performance. It consists essentially of a stirred batch reactor, an oil storage tank, a positive displacement pump and temperature sensors. Thermal oil was used as heat transfer medium in the jacket. The polymerization reaction is exothermic.
Using a PCL (Programable controller logic), a thermal fluid variable speed pump will be driven by the controller, to maintain the temperature constant into the reactor. The flow of thermal fluid (manipulated variable) was step of 30 and 100%. The maximum pump flow rate equivalent to approximately 900 L/H. Disturbances in the manipulated variable were performed in a short time interval (P=300 s).
The Figure 3.3 shows response of the experiment using the relay method.

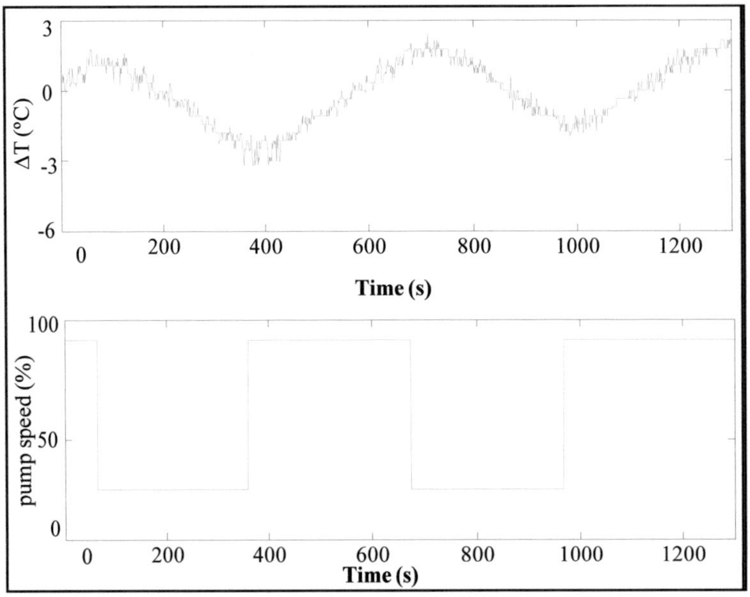

Fig. 3.3. Response of the experiment using the relay method

According to the tuning method used, we found the initial control parameters as shown in Table 3.2.

Parameter obtained from Relay Method		
a = 3	2d = 70	P=300
Controller	PI	PID
K_c	6,68 %/°C	8,91 %/°C
τ_i	0,004 s	0,007 s
τ_d	0 s	37,5 s

Table 3.2. Initial parameters PID controller (Relay method).

From these results it is possible to implement an on-line PID controller in the experimental polymerization process.

4. Process reaction curve

The closed-loop system will respond in a desirable way only if its controller is properly tuned. This means that its proportional, integral and derivative (PID) settings are properly made. A popular procedure for tuning a controller is the Ziegler-Nichols Reaction Curve Tuning Method.
This procedure requires a step change of the controllers output alters the controlled variable. The Figure 4.1 shows the resultant closed loop step.
The method used to make the step change and measure the controlled variable is called the Process Identification Procedure. This controller setting puts the system into an open-loop condition. Based on the shape and magnitude of the controlled variable's reaction curve in reference to the step change, value are obtained and used in mathematical formulas. These values are then used to determine the PID settings.

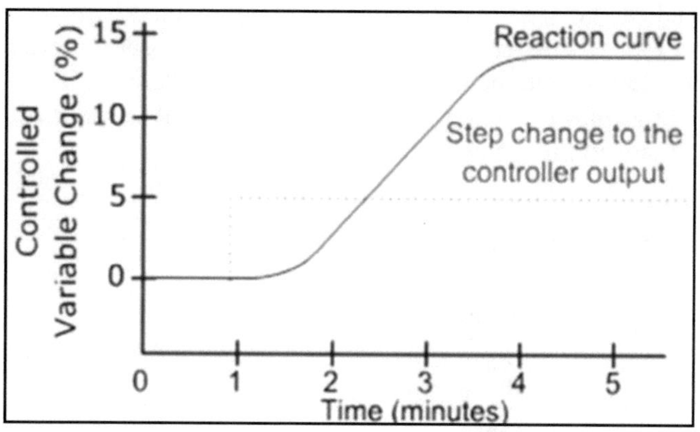

Fig. 4.1. Resultant closed loop step

Loop responses for a unit step reference are shown in Figure 2 (similar to Figure 1). A linearized quantitative version of the model in Equation 3.3 can be obtained with an open loop experiment, using the following procedure:

a. With the plant in open loop, take the plant manually to a normal operating point. Say that the plant output settles at $y(t) = y_o$ for a constant plant input u (t) = u_o.
b. At an initial time, t_0, apply a step change to the plant input, from u_0 to u_∞ .
c. Record the plant output until it settles to the new operating point. Assume you obtain the curve shown in Figure 2. This curve is known as the process reaction curve.

d. Compute the parameter model as follows:

$$K_0 = \frac{y_\infty - y_0}{u_\infty - u_0} \tag{4.1}$$

$$\tau_0 = t_1 - t_0 \tag{4.2}$$

$$\gamma_0 = t_2 - t_1 \tag{4.3}$$

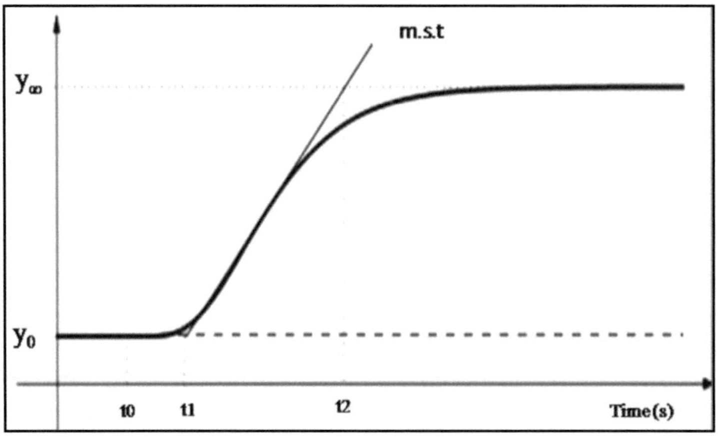

m.s.t stands for maximum slope tangent

Fig. 4.2. Reaction curve: Process Identification Procedure

The model obtained can be used to derive various tuning methods for PID controllers. This method was proposed by Ziegler and Nichols. In their proposal the design objective is to achieve a particular damping in the loop response to a step reference.
The parameter setting rules proposed in Table 4.1 are applied to the model (Eq.3.3), where we have again normalized time in delay units.

	K_p	τ_i	τ_d
P	$\frac{\gamma_0}{K_0\tau_0}$		
PI	$\frac{0{,}9\gamma_0}{K_0\tau_0}$	$3\tau_o$	
PID	$\frac{1{,}2\gamma_0}{K_0\tau_0}$	$2\tau_o$	$0{,}5\tau_o$

Table 4.1. Ziegler-Nichols tuning using the reaction curve.

4.1 Case study

Bromelain is widely used in the chemical and pharmaceutical industries. It is employed not only for its pharmacological effects, but also in food industry activities such as brewing and meat processing (Kelly, 1996). Currently there were no experimental studies about automation and process control in the production of bromelain, despite the growing number of scientific papers related to this enzyme. Temperature control during the recovery process of the bromelain from pineapple fruits is an extremely important practice, because the

temperature directly affects the final activity of the enzyme precipitated. The use of controllers to maintain the temperature of this process prevents the denaturation of the enzyme, improving the quality of the product. It is also important to emphasize that the design of the developed controllers can be easily extended to similar processes in which some transient and nonlinear behavior are found.
The robust PID controller is designed for temperature control of an experimental process of enzyme recovery from pineapple rind. To assess the performance of the controllers the following parameters were used: ITAE (integral of Time multiplied by Absolute Error), response time, saturation of the final element of control, enzymatic activity of the product and electric power consumption of the cooling system.

4.2 PID controller design

Conventional controller was implemented in experimentally tested in a pilot plant of the precipitation process (Leite et al., 2010b; Leite et al., 2010c; Silva et al., 2010).
The proteolytic enzyme bromelain (EC 3.4.22.4[*]) is precipitated with alcohol at low temperature in a fed-batch jacketed tank. Temperature control is crucial to avoid irreversible protein denaturation. Using a Fieldbus network architecture, a coolant variable speed pump was driven by the controller, to maintain the temperature constant into the tank.
Tuning the controllers proved to be a difficult task in this fed-batch nonlinear process. To tune the controller, by Ziegler and Nichols, a new methodology for the experimental procedure was designed and implemented (Leite et al., 2010c).
In order to evaluate the influence of the variation of the tank volume on the precipitation process, and to obtain the process reaction curve samples containing extract and ethanol in different proportions (from 1:1 to 1:3 v/v) were used in the pseudo-steady state operation.
Positive and negative disturbances were then applied (± 30%) to the initial conditions of the speed of the coolant pump (manipulated variable). The data obtained from the reaction curve (Figure 4.3) for this process allowed to find initial values of the process parameters K_p (static gain), τ_p (time constant) and τ_d (time delay).

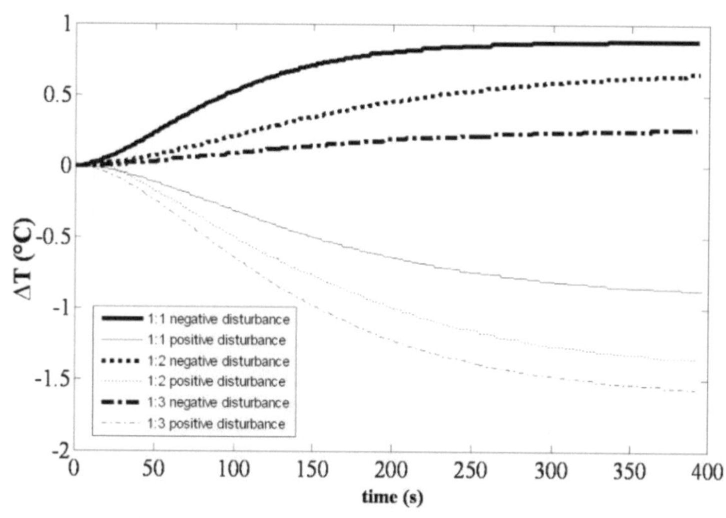

[*]*The Enzyme Commission number (EC number) is a numerical classification scheme for enzymes, based on the chemical reactions they catalyze.*

Fig. 4.3. Reaction curves obtained from disturbances in the manipulated variable.

Fine tuning was then conducted to adjust these parameters by trial-and-error procedure.
In these closed loop experiments, the following indices of performance were considered: ITAE, response time and saturation of the final element of control.
The best parameters found after this fine tuning were: K_c=35%/°C, τ_i = 28s and τ_d = 7s (PID_2). Figure 4.4 shows the behavior of the tank temperature under well-tuned conventional PID.

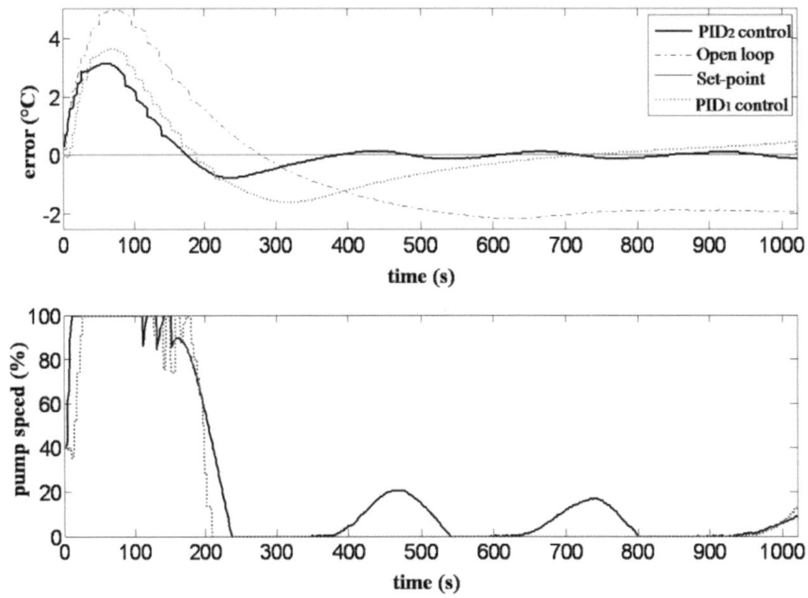

Fig. 4.4. Behavior of the controlled and manipulated variables under PID1 control (Kc=8%/°C, τ_i = 28s e τ_d = 1,5s) and PID2 (Kc=35%/°C, τ_i = 28s e τ_d = 7s).

Table 4.2 presents quantitative and qualitative analyses of the performance of the implemented controllers.

Performance parameters	Controller		
	Open-loop	PID_1	PID_2
Overshoot (°C)	5.0	3.9	3.1
Rise time (s)	281	200	171
Response time (s)	-	710	400
Pump saturation time (s)	-	141	130
ITAE (x10^3)	950.5	187	80.3
Specific enzymatic activity (U/g)	0.32	0.96	1.03
Eletric energy comsuption (kWh)	42.00	5.75	9.11

Table 4.2. Performance parameters of the PID controllers.

From these results, it is clear that PID controllers performed satisfactorily in controlling the temperature of the precipitation process. However, the PID2 controller kept the variation closer to the set-point, which is important for enzyme activity recovery, since the enzyme is highly sensitive to temperature changes. The early stage of ethanol addition is critical. In order to keep the overshoot to a minimum, intense controller response is required, causing pump saturation.

Despite the PID_1 controller have lower power consumption, the PID controller showed better global performance criteria: small overshoot, small rise time, small ITAE, short response time and pump saturation time and higher enzyme activity in the product.

The adaptative PID tuning procedure, based on the analysis of the process reaction curves, can be an attractive strategy to provide a suitable non-linear controller design for transient processes. The further development of the adaptive PID controller can contributed to improving the performance of the conventional PID controller.

5. Conclusions

PID control tuning are popular and offer many benefits such ease of use, new development help to implement other PID controller variants, and control for common industry applications.

In this chapter, two techniques from PID tuning were applied for the temperature control of the practical applications: 1-polymerization system and 2-bromelain precipitation. The main feature of these process is its complex nonlinear behavior, wich poses a challenging control system design for the batch reactor.

In the first case a PID controller experiment was designed to be implemented later in the pilot plant. The controller was developed from the relay method proposed by Astrom and Haglund.

In the second case the controller was designed based on reaction curve method of Ziegler and Nichols, by disturbances in a real experimental system bromelain precipitation. The authors carried out fine-tuning of this controller, which was subsequently implemented efficiently in maintaining the process temperature.

The methods performed well for estimation of the PID controller, easy to apply and prove to be an effective option in practical cases will help achieve the proposed objectives. There is a large number of tuning methods, but related methods cover most practical cases and common industry applications.

6. References

Åström, K. J. & Hägglund, T. (2004). Revisiting the Ziegler-Nichols step response method for PID control. *Journal of Process Control*, Vol. 14, pp. 635-650.

Ghasem, N. M., Sata, S. A. & Hussain, M. A. (2007). Temperature control of a bech-scale batch polymerization reactor for polystyrene production. *Chemical Engineering Technology*, Vol. 3, No. 9, pp. 1193-1202.

Kelly, G. S. (1996). Bromelain: A literature review and discussion of its therapeutic applications. *Alternative Medicine Review*, Vol. 1, No. 4, pp. 243-257.

Leite, M. S. ; Fileti, A. M. F. & Silva, F. V. (2010c). Development and experimental application of fuzzy and conventional controllers to a bioprocess. *Revista Controle & Automação*, Vol. 21, No. 2, March and April 2010, pp. 147-158, ISSN 0103-1759.

Leite, M. S.; Fileti, A. M. F. & Silva, F. V. (2010a). Design, assembly and instrumentation of an experimental prototype for the application of automation techniques and development of control strategies in a polymerization process. Proceedings of XVIII COBEQ, Brazil, Foz do Iguaçu, 2010, Vol. 1, p. 7539-7548., ISSN 2178-3659.

Leite, M. S.; Santos, B. F.; Lona, L. M. F.; Silva, F. V. & Fileti, A. M. F. (2011). Application of Artificial Intelligence Techniques for Temperature Prediction in a Polymerization Process. *Chemical Engineering Transactions*, Vol. 24, 2011, pp. 385-390, ISSN 1974-9791.

Leite, M. S; Fujiki, T. L.; Silva, F. V. & Fileti, A. M. F. (2010b). Online Intelligent Controllers for an Enzyme Recovery Plant: Design Methodology and Performance. *Enzyme Research*, Vol. 2010, November 2010, pp. 1-13, DOI 10.4061/2010/250843.

Lepore, R., Wouwer, A. V., Remy, M., Findeisen, R., Nagy, Z. & Allgower, F. (2007). Optimization stategies for a MMA polymerization reactor. *Computers and Chemical Engineering*, Vol. 31, pp. 281-291.

Silva, F. V.; Santos, R. L. A.; Leite, M. S.; Fujiki, T. L. & Fileti, A. M. F. Design of automatic control for the precipitation of bromelain from the extract of pineapple wastes. *Ciência e Tecnologia de Alimentos*, Vol. 30, 2010, pp. 1033-1040, ISSN 0101-2061.